Development and Application of Bio-Based Polymers

Development and Application of Bio-Based Polymers

Guest Editors

Masoud Ghaani
Stefano Farris

MDPI

Basel • Beijing • Wuhan • Barcelona • Belgrade • Novi Sad • Cluj • Manchester

Guest Editors
Masoud Ghaani
Department of Civil, Structural & Environmental Engineering
Trinity College Dublin
Dublin
Ireland

Stefano Farris
Department of Food, Environmental and Nutritional Sciences
Università degli Studi di Milano
Milan
Italy

Editorial Office
MDPI AG
Grosspeteranlage 5
4052 Basel, Switzerland

This is a reprint of the Special Issue, published open access by the journal *Polymers* (ISSN 2073-4360), freely accessible at: https://www.mdpi.com/journal/polymers/special_issues/1US150D1UP.

For citation purposes, cite each article independently as indicated on the article page online and as indicated below:

Lastname, A.A.; Lastname, B.B. Article Title. *Journal Name* **Year**, *Volume Number*, Page Range.

ISBN 978-3-7258-2403-8 (Hbk)
ISBN 978-3-7258-2404-5 (PDF)
https://doi.org/10.3390/books978-3-7258-2404-5

Contents

Pratthana Chomchalao, Nuttawut Saelim, Supaporn Lamlertthon, Premnapa Sisopa and Waree Tiyaboonchai
Mucoadhesive Hybrid System of Silk Fibroin Nanoparticles and Thermosensitive In Situ Hydrogel for Amphotericin B Delivery: A Potential Option for Fungal Keratitis Treatment
Reprinted from: *Polymers* **2024**, *16*, 148, https://doi.org/10.3390/polym16010148 . . . . **1**

Aizhamal Usmanova, Yelena Brazhnikova, Anel Omirbekova, Aida Kistaubayeva, Irina Savitskaya and Lyudmila Ignatova
Biopolymers as Seed-Coating Agent to Enhance Microbially Induced Tolerance of Barley to Phytopathogens
Reprinted from: *Polymers* **2024**, *16*, 376, https://doi.org/10.3390/polym16030376 . . . . **18**

Madara Žiganova, Remo Merijs-Meri, Jānis Zicāns, Agnese Ābele, Ivan Bochkov and Tatjana Ivanova
Accelerated Weathering Testing (AWT) and Bacterial Biodegradation Effects on Poly(3-hydroxybutyrate-co-3-
hydroxyvalerate) (PHBV)/Rapeseed Microfiber Biocomposites Properties
Reprinted from: *Polymers* **2024**, *16*, 622, https://doi.org/10.3390/polym16050622 . . . . **35**

Gregory M. Glenn, Gustavo H. D. Tonoli, Luiz E. Silva, Artur P. Klamczynski, Delilah Wood, Bor-Sen Chiou, et al.
Effect of Starch and Paperboard Reinforcing Structures on Insulative Fiber Foam Composites
Reprinted from: *Polymers* **2024**, *16*, 911, https://doi.org/10.3390/polym16070911 . . . . **57**

Kansiri Pakkethati, Prasong Srihanam, Apirada Manphae, Wuttipong Rungseesantivanon, Natcha Prakymoramas, Pham Ngoc Lan and Yodthong Baimark
Improvement in Crystallization, Thermal, and Mechanical Properties of Flexible Poly(L-lactide)-*b*-poly(ethylene glycol)-*b*-poly(L-lactide) Bioplastic with Zinc Phenylphosphate
Reprinted from: *Polymers* **2024**, *16*, 975, https://doi.org/10.3390/polym16070975 . . . . **74**

Oroitz Sánchez-Aguinagalde, Eva Sanchez-Rexach, Yurena Polo, Aitor Larrañaga, Ainhoa Lejardi, Emilio Meaurio and Jose-Ramon Sarasua
Physicochemical Characterization and In Vitro Activity of Poly(ε-Caprolactone)/Mycophenolic Acid Amorphous Solid Dispersions
Reprinted from: *Polymers* **2024**, *16*, 1088, https://doi.org/10.3390/polym16081088 . . . . **91**

Zydrune Gaizauskaite, Renata Zvirdauskiene, Mantas Svazas, Loreta Basinskiene and Daiva Zadeike
Optimised Degradation of Lignocelluloses by Edible Filamentous Fungi for the Efficient Biorefinery of Sugar Beet Pulp
Reprinted from: *Polymers* **2024**, *16*, 1178, https://doi.org/10.3390/polym16091178 . . . . **107**

Masoud Ghaani, Maral Soltanzadeh, Daniele Carullo and Stefano Farris
Development of a Biopolymer-Based Anti-Fog Coating with Sealing Properties for Applications in the Food Packaging Sector
Reprinted from: *Polymers* **2024**, *16*, 1745, https://doi.org/10.3390/polym16121745 . . . . **121**

Annalisa Apicella, Konstantin V. Malafeev, Paola Scarfato and Loredana Incarnato
Generation of Microplastics from Biodegradable Packaging Films Based on PLA, PBS and Their Blend in Freshwater and Seawater
Reprinted from: *Polymers* **2024**, *16*, 2268, https://doi.org/10.3390/polym16162268 . . . . **133**

polymers MDPI

Article

Mucoadhesive Hybrid System of Silk Fibroin Nanoparticles and Thermosensitive In Situ Hydrogel for Amphotericin B Delivery: A Potential Option for Fungal Keratitis Treatment

Pratthana Chomchalao [1,2], Nuttawut Saelim [3], Supaporn Lamlertthon [4], Premnapa Sisopa [5] and Waree Tiyaboonchai [1,*]

1 Department of Pharmaceutical Technology, Faculty of Pharmaceutical Sciences, Naresuan University, Phitsanulok 65000, Thailand; pratthana.c@ubu.ac.th
2 College of Medicine and Public Health, Ubon Ratchathani University, Ubon Ratchathani 34190, Thailand
3 Department of Pharmacy Practice, Faculty of Pharmaceutical Sciences, Naresuan University, Phitsanulok 65000, Thailand
4 Centre of Excellence in Fungal Research, Department of Microbiology and Parasitology, Faculty of Medical Science, Naresuan University, Phitsanulok 65000, Thailand
5 Department of Health and Cosmetic Product Development, Faculty of Food and Agricultural Technology, Pibulsongkram Rajabhat University, Phitsanulok 65000, Thailand
* Correspondence: wareet@nu.ac.th

Abstract: The purpose of this work was to investigate the feasibility of a novel ophthalmic formulation of amphotericin B-encapsulated silk fibroin nanoparticles incorporated in situ hydrogel (AmB-FNPs ISG) for fungal keratitis (FK) treatment. AmB-FNPs ISG composites were successfully developed and have shown optimized physicochemical properties for ocular drug delivery. Antifungal effects against *Candida albicans* and in vitro ocular irritation using corneal epithelial cells were performed to evaluate the efficacy and safety of the composite formulations. The combined system of AmB-FNPs-ISG exhibited effective antifungal activity and showed significantly less toxicity to HCE cells than commercial AmB. In vitro and ex vivo mucoadhesive tests demonstrated that the combination of silk fibroin nanoparticles with in situ hydrogels could enhance the adhesion ability of the particles on the ocular surface for more than 6 h, which would increase the ocular retention time of AmB and reduce the frequency of administration during the treatment. In addition, AmB-FNP-PEG ISG showed good physical and chemical stability under storage condition for 90 days. These findings indicate that AmB-FNP-PEG ISG has a great potential and be used in mucoadhesive AmB eye drops for FK treatment.

Keywords: mucoadhesive eye drops; amphotericin B; silk fibroin nanoparticles; composite formulations; ocular drug delivery

Citation: Chomchalao, P.; Saelim, N.; Lamlertthon, S.; Sisopa, P.; Tiyaboonchai, W. Mucoadhesive Hybrid System of Silk Fibroin Nanoparticles and Thermosensitive In Situ Hydrogel for Amphotericin B Delivery: A Potential Option for Fungal Keratitis Treatment. *Polymers* **2024**, *16*, 148. https://doi.org/10.3390/polym16010148

Academic Editors: Stefano Farris and Masoud Ghaani

Received: 22 November 2023
Revised: 24 December 2023
Accepted: 28 December 2023
Published: 3 January 2024

1. Introduction

The incidence of corneal infection by fungi or fungal keratitis (FK) has increased worldwide in recent years and can cause blindness [1]. Several risk factors, such as corneal injury, extended wear of contact lenses, ocular surface diseases, and topical steroid use have been associated with the development of FK [2]. In tropical regions, filamentous fungi, such as *Fusarium* and *Aspergillus* species, are predominant causes of FK, whereas yeasts, such as *Candida* species, are a less frequent cause [3]. However, corneal infections caused by *Candida* spp. are complicated to treat due to their ability to form a biofilm that leads to antifungal resistance [4]. Amphotericin B (AmB) is an effective drug in the treatment of FK because it has broad spectrum activity against most fungi, especially *Candida* spp. with a low incidence of clinical resistance [5]. Unfortunately, the therapeutic use of AmB is limited by its toxicity and poor solubility, which make it difficult to fabricate an optimum ophthalmic formulation. Furthermore, AmB eye drops are not commercially available, and the conventional intravenous dosage form of AmB deoxycholate (Fungizone®, Bristol-Myers Squibb, Montreal, QC, Canada) is mostly off-label and used for FK treatment [6,7]. The main problem of this extemporaneous preparation is eye irritation from the sodium deoxycholate in the formulation, leading to poor patient compliance [8]. Less toxic lipid-based formulations of AmB have been developed, but these forms are very expensive, have low stability in aqueous solution, and are unavailable in developing countries, thus restricting their use [9]. In addition, AmB delivered through conventional eye drops is rapidly eliminated from the ocular surface due to the protective mechanisms of the eye, such as reflex blinking and tear dilution, and entry into the systemic circulation via nasolacrimal drainage [10]. High doses and high-frequency application have been used to achieve therapeutic efficacy of treatment of the fungal infection, which can result in serious systemic side effects [11–13]. Accordingly, many studies have focused on developing AmB ophthalmic formulations with high effectiveness, low toxicity, and good stability using drug delivery systems such as polymeric nanoparticles [14,15], nanostructure lipid carriers [16,17], liposomes [18], microneedles [19], and nanofiber with in situ gelling [20]. Nevertheless, no licensed AmB topical ophthalmic formulations are available.

Consequently, our research focuses on the use of combination systems of nanoparticles and thermosensitive in situ hydrogels to fabricate a promising ophthalmic AmB. Nanoparticles offer various outstanding advantages for ocular drug delivery including improved bioavailability of the drugs and extended drug retention time, enabling targeted delivery and reduced side effects [21]. Several studies reported that small nanosized particles, ranging from 50 to 400 nm, could overcome the ocular barrier, enhancing bioavailability of poorly soluble drugs and prolonging the contact time of the drugs on the eye [22,23]. Furthermore, nanoparticles shield the drug from interacting directly with normal cells, thus minimizing its side effects. As is known, AmB activity and toxicity depend on the aggregation state of AmB, which is influenced by carriers. The incorporation of AmB into nanoparticles having a monomeric form has been shown to enhance antifungal activity and reduce the irritation induced by AmB [24]. However, low viscosity nanodispersion can be quickly cleared from the eye by tears or blinking, resulting in inadequate drug retention on the ocular surface. To overcome this drawback, combining the nanoparticles with the in situ gelling system can be an effective approach to increase viscosity upon application. This system has the capability to undergo a phase transition in response to environmental triggers such as temperature, pH, or ions [25]. The liquid form of in situ hydrogel can be easily administered as eye drops and it can transform into a gel after contact with the ocular surface, thereby increasing the drug's residence time, and decreasing the frequency of administration and dosing [12,26].

Based on our previous studies, we successfully developed a ready-to-use AmB ophthalmic formulation by formulating a combined system of AmB-encapsulated silk fibroin nanoparticles incorporated in thermosensitive in situ hydrogel (AmB-FNPs ISG) [27] to enhance ocular bioavailability and increase precorneal residence time of AmB. To create the thermosensitive in situ gelling system, Pluronic F127-based formulations were utilized in this work. Pluronic F127 is a thermoresponsive polymer widely employed in pharmaceutical formulations. At low temperature, Pluronic exists in a liquid state and can self-assemble to form small micelles due to their amphiphilic structure. With a temperature increase, the micellar structures pack closely together to form a three-dimensional network, leading to the formation of the gel [28]. AmB-FNPs ISG possesses a liquid form at ambient temperature and rapidly converts to a gel at ocular temperature, offering ease of administration while prolonging its retention on the eye. These novel formulations present pale yellowish solutions, high transparency, and optimum pH and osmolality. All AmB-FNPs incorporated into the thermosensitive in situ hydrogel showed high entrapment efficiency with a mean particle size of ~200 nm, which could enhance AmB bioavailability and cause no ocular irritation. Moreover, the results from FTIR, XRD, and molecular aggregation studies revealed that highly hydrophobic AmB was encapsulated in FNPs in an amorphous form, which could reduce the aggregated toxicity of AmB. These findings demonstrated that

AmB-FNPs ISG formulations possess great potential physicochemical properties for topical ocular application. Accordingly, the present study aims to demonstrate the antifungal activity, mucoadhesive properties, reduced ocular irritation, and stability of the AmB-FNPs ISG for the treatment of FK.

2. Materials and Methods

2.1. Materials

Amphotericin B, as the active pharmaceutical ingredient, was bought from Bio Basic Inc. (Toronto, ON, Canada). Intravenous amphotericin B was supplied by Biolab (Samutprakarn, Thailand). Polyethylene glycol 400 (PEG) was purchased from Nam Siang Co. Ltd. (Bangkok, Thailand). Branched polyethylenimine (PEI), hyaluronic acid, mucin from porcine stomach type II, hydrocortisone, insulin from bovine pancreas, and fluorescein isothiocyanate were ordered from Sigma-Aldrich (St. Louis, MO, USA). Poloxamer 407 (Pluronic® F127) was acquired from BASF (Florham Park, NJ, USA). Muller-Hinton agar and Sabouraud Dextrose agar were purchased from HIMEDIA (Mumbai, India). Roswell Park Memorial Institute (RPMI) 1640 medium, keratinocyte serum free medium (K-SFM), bovine pituitary extract (BPE), epithelial growth factor (EGF), and Penicillin/ Streptomycin solution were ordered from Gibco (New York, NY, USA). Thiazolyl blue tetrazolium bromide was acquired from Amresco® (Solon, OH, USA). Crystal violet was supplied by Riedel-de Haën (Munich, Germany). All other reagents were of analytical grade or higher.

2.2. Preparation and Characterization of AmB-FNPs ISG Composites

The soluble silk fibroin (SF) was extracted as reported before by dissolving small fibers of degummed silk yarn (Bodin Thai Silk Khorat Co., Ltd., Nakhon Ratchasima, Thailand) in a mixture of $CaCl_2$:H_2O:$Ca(NO_3)_2$:EtOH (30:45:5:20 weight ratio) solvent at 80 °C for 4 h [29]. Then, pure SF solution was obtained after dialysis in distilled water for 3–5 days using snakeskin dialysis tubing (10,000 Da MWCO). Three different amphotericin B-encapsulated silk fibroin nanoparticles (AmB-FNPs)—uncoated AmB-FNP (AmB-FNP), AmB-FNP crosslinked with PEI (AmB-FNP-PEI), and AmB-FNP coated with PEG 400 (AmB-FNP-PEG) were prepared following our previous study using the desolvation method [27]. To enhance the entrapment efficiency of AmB and promote the interaction between AmB-loaded nanoparticles and mucin at the ocular surface, cationic polymer PEI and a mucoadhesive polymer PEG 400 were utilized for modification of the surface of silk fibroin nanoparticles. Briefly, an aqueous 1% w/v SF solution (2% w/v SF was used instead of 1% w/v SF for AmB-FNP-PEI) was injected dropwise into mild stirred absolute ethanol, ethanol/1% w/v PEI solution (pH 7.0), or ethanol/1% w/v PEG 400 solution with AmB (15 mg per 30 mL). The SF:Ethanol ratio tested was 10: 20 v/v. The spontaneously formed particles were centrifuged at 12,000 rpm for 60 min, washed thrice with DI water, and sonicated at 40% amplitude for 60 s. Finally, all AmB-FNPs were lyophilized and kept in the refrigerator for further experiment.

Mean particle size, polydispersity index (PDI), and zeta potential of all AmB-FNPs were measured at 25 °C using a Zetasizer Ultra (Malvern Panalytical Ltd., Malvern, UK) by diluting the samples in DI water and examining them in triplicate. In addition, entrapment efficiency (EE) and drug loading capacity (DL) were evaluated using an indirect method. After centrifugation, the amount of unentrapped AmB in the supernatant was analyzed using a UV–Visible spectrophotometer (Genesys 10 s, Thermo Scientific, Waltham, MA, USA) at 405 nm to calculate %EE and %DL.

To determine the crystallinity of AmB in the FNPs, X-ray diffractometry (XRD) analysis was performed. Pure AmB powder, SF, freeze-dried powder of AmB-FNPs, blank FNPs, and physical mixes of AmB and blank FNPs were examined with an X-ray diffractometer (D2 Phaser, Bruker AXS Inc., Madison, WI, USA) at 45 kV and 36 mA with a scan speed of 2°/min and scanning from 10–30°.

The prepared AmB-FNPs were incorporated in 2 optimal in situ hydrogel bases, namely, 19% w/v Pluronic® F127 (F127) and 18% w/v Pluronic® F127 blended with 0.2%

w/v hyaluronic acid (F127/HA). Briefly, AmB-FNP dispersion and the in situ hydrogel solution were separately prepared using the cold method. Lyophilized AmB-FNPs were re-dispersed in deionized (DI) water and sonicated at 40% amp for 60 s. The F127 and F127/HA solution were prepared by dispersing F127 powder in cold DI water and hyaluronic acid (HA) aqueous solution, respectively. Then, both polymer dispersions were kept in a refrigerator for at least 24 h for completely dissolution. The AmB-FNP dispersions were mixed into an equal volume of F127 or F127/HA solution at 4 °C by constant stirring until homogenous, and the final dose of AmB was equivalent to 150 μg/mL. All AmB-FNPs ISG formulations were kept in a refrigerator and protected from light for further investigation. The physical properties of AmB-FNPs ISG formulations were characterized in term of clarity, gelling capacity, pH, osmolality, optical transmittance, viscosity, rheological behavior, and sol–gel transition temperature.

Visual examination against black and white backgrounds was used to determine the clarity of the composite formulations.

The gelling capacity was performed by dropping 30 μL of each composite formulation into a test tube containing 2 mL of stimulated tear fluid (STF) at pH 7.4 equilibrated at 35 ± 1 °C. The appearance of gel forming was visually evaluated.

The pH and osmolality of the composite formulation were measured using a pH meter (SK20, Mettler-Toledo, Zurich, Switzerland) and freezing point depression osmometer (Osmomat® 030, Gonotec, Berlin, Germany), respectively. All measurements were made in triplicate and the data are reported as a mean ± SD.

The optical transmittance was performed on a UV–Visible spectrophotometer. Fifty microliters of each composite formulation was smeared on the outside surface of a quartz cuvette and maintained at 35 ± 1 °C. Then, the % transmittance was measured under the visible wavelength from 381 to 780 nm and the empty cuvette was used as a blank. All tests were conducted in triplicate and the data are expressed as mean ± SD.

The viscosity and rheological behavior of AmB-FNPs ISG formulations were investigated using a cone and plate viscometer (DV3T model, Brookfield, MA, USA). The viscosity of each formulation was measured at 25 ± 1 °C (room temperature) or 35 ± 1 °C (ocular surface temperature) with a constant shear rate at 20 s^{-1}. The rheology tests were conducted at 35 ± 1 °C with increasing shear rate from 1 to 80 s^{-1} and the rheograms were plotted between the viscosity and shear rate. The sol–gel transition temperatures of both AmB-FNPs-F127 and AmB-FNPs-F127/HA were measured at a 10 rpm spindle speed using a Brookfield rheometer (RST-CVS-PA, Brookfield, USA) with increasing temperatures from 20 °C to 40 °C controlled by a Peltier plate. The data were plotted as viscosity versus temperature and the gelling temperature was defined as the temperature when the rigid gel state formed.

2.3. *Stability Study*

A stability study was performed to determine the physicochemical stability of the formulations under storage conditions. AmB-FNP-PEG F127 ISG and AmB-FNP-PEG F127/HA ISG were represented as a candidate ophthalmic formulation for the stability study. Both formulations were stored at 4 ± 1 °C (refrigerated temperature) in darkness for a period of 3 months and evaluated at intervals of 7, 14, 21, 30, 60, and 90 days for clarity, pH, gelling capacity, and drug content.

2.4. *In Vitro Antifungal Activity*

2.4.1. Screening of the Antifungal Activity

The antifungal effects of the AmB-FNPs dispersion and AmB-FNPs-ISG were tested against a standard strain of *Candida albicans* (TISTR 5779). The agar well diffusion technique was used to screen the antifungal effects of all formulations [30]. Briefly, a *C. albicans* suspension (1 × 10^6 CFU/mL) was prepared in 0.85% NaCl and swabbed evenly on the surface of Muller–Hinton agar containing 2% glucose and methylene blue. Then, 10 μg/mL of AmB stock solution of standard AmB, AmB deoxycholate, AmB-FNPs, and AmB-FNPs-

ISG was prepared in sterile water. Subsequently, holes having a diameter of 6 mm were made in the inoculated agar plates and filled with 20 μL of stock solution of the samples. In addition, sterile water and plain in situ hydrogels were also added to the wells as a negative control. The samples were then incubated for 24 h at 37 °C. Antifungal effects were determined as the absence of fungal growth in the area surrounding the hole and the diameter of inhibition zone was measured using a Vernier caliper. These analyses were performed in triplicate.

2.4.2. Minimum Inhibitory and Minimum Fungicidal Concentration Test

The minimal inhibitory concentration (MIC) was evaluated using the broth dilution technique according to the EUCAST guidelines [31]. The stock solution of standard AmB powder was dissolved in DMSO while AmB deoxycholate, AmB-FNPs, and AmB-FNPs-ISG were prepared in DI water. All samples were two-fold serial diluted in a 96-well plate with RPMI 1640 medium, 2% glucose, and 25 mM HEPEs, pH 7.0, and with the AmB concentration ranging from 0.0312 to 16 μg/mL. The *C. albicans* suspension was prepared in RPMI 1640 medium and further inoculated in each well of the 96-well plates to obtain a final inoculum concentration of ~1–5 × 10^5 CFU/mL. Consequently, the final AmB concentration ranged from 0.0156 to 8 μg/mL. A *Candida* suspension cultured in the media without the samples served as a positive control, and the mixture of media and the samples without the *Candida* suspension served as a negative control. The assay plates were used to measure the optical density at 530 nm using a microplate reader (Synergy H1 Hybrid Reader, BioTek, Agilent, CA, USA) prior to and after incubation at 37 °C, for 24 h. The MIC_{90}, defined as the lowest test concentration that inhibited 90% of the fungal growth, was calculated for all formulations.

After MIC determination, 10 μL aliquots of media from each well with no fungal growth were dropped on Sabouraud Dextrose Agar (SDA) plates and incubated at 37 °C for 24 h. Finally, the minimal fungicidal concentration (MFC) was defined as the lowest AmB concentration that showed no detectable growth on the SDA surfaces. All experiments were conducted in triplicate.

2.5. *In Vitro Mucoadhesive Study*

All blank FNPs were covalently bound with fluorescent dye of fluorescein isothiocyanate (FITC) for both in vitro and ex vivo mucoadhesive studies. Briefly, 10 mg of lyophilized blank particles was re-dispersed in 1 mL of carbonate buffer pH 9. Then, 50 μL of 1 mg/mL FITC solution was slowly added into the dispersion with gentle and continuous stirring. The mixture was incubated in the dark for 24 h at 4 °C. The FITC conjugated with FNPs (FITC-FNPs) was then separated from unbound FITC by centrifugation at 16,000 rpm for 30 min and washed 2 times with DI water. The FITC-FNPs were freshly made before each experiment.

The flowing liquid test was performed to investigate the mucoadhesive properties of the prepared formulations. This method involves washing the formulation with an appropriate artificial fluid at a constant flow rate while the residence time of the formulation is determined visually or fluorometrically [32]. The in vitro mucoadhesive test was conducted under a condition mimicking the pre-ocular surface. A mucus layer was prepared by soaking a hydrophilic membrane (polycarbonate membrane, 0.2 μm pore size, 6 mm diameter, Isopore™, Merck, Germany) in 0.1% of aqueous mucin solution for 24 h. Then, 10 μL of the FITC solution, FITC-FNPs dispersion, and FITC-FNPs-ISG was applied as a single drop at the center of the membrane and incubated at 35 ± 1 °C for 2 min to induce a gel forming an in situ hydrogel formulation. The membrane was then immediately washed with a continuous flow of STF solution (pH 7.4, 35 ± 1 °C) at a flow rate of 10 μL/min controlled by a peristaltic pump. Then, STF containing the eliminated FITC was collected at 5, 15, 30, 60, 120, 240, and 360 min, and the fluorescence intensity was measured using a microplate reader at excitation (Ex) and emission (Em) wavelengths of 495 nm and 525 nm, respectively. The amount of eliminated FITC in STF was calculated according to the standard curve of

FITC (concentration range 0.25–20 ng/mL) and the percentage of FITC remaining on the mucus membrane was calculated as follows:

$$\%FITC\ remaining = \frac{\text{Initial amount of FITC} - \text{Amount of eliminated FITC at time point}}{\text{Initial amount of FITC}} \times 100 \quad (1)$$

2.6. Ex Vivo Mucoadhesive Study

The ex vivo mucoadhesive test using fresh porcine cornea and an apparatus setup modified from Chiyasan et al. [33] was performed. Porcine eyes were obtained from the local slaughterhouse, Phitsanulok, Thailand (license number PC 06 47001/2536), and kept in ice-cold PBS, pH 7.4, containing 1% *v/v* antibiotic solution until used (less than 8 h after death). Six millimeters of the corneal tissue was excised with a surgical blade and mounted onto a glass slide; 48 porcine corneas were obtained in this experiment. Ten microliters of the FITC solution, FITC-FNP-PEG dispersion, and FITC-FNP-PEG-ISG was dropped on the corneal surface. The samples were incubated at 35 °C for 5 min to ensure the gelation form of in situ hydrogel. Then, the tissue was placed in contact with a continuous stream of STF pH 7.4 at 35 ± 1 °C with a flow rate of 10 μL/min to mimic the eye blinking. At the time points of 30, 120, 240, and 360 min, cryostat sections of the cornea tissue were removed and imaged using fluorescence microscopy to visualize green fluorescence of FITC adhering to the tissue.

2.7. In Vitro Irritation Study

2.7.1. MTT Assay

The human corneal epithelial (HCE) cell line (ATCC CRL-11135) was used for cell viability investigation. HCE cells were cultured in the complete growth media of K-SFM supplemented with 0.05 mg/mL BPE, 5 ng/mL of EGF, 500 ng/mL hydrocortisone, 0.005 mg/mL insulin, and 1% *v/v* Penicillin/Streptomycin at 37 °C with 5% CO_2. The influence of the prepared formulations on cell viability was investigated using an MTT assay, which was modified following the short time exposure (STE) protocol recommended for an alternative ocular irritation assay [34]. Briefly, HCE cells were trypsinized and seeded into a 96-well plate at 1×10^4 cells/well and cultured in the complete growth media at 37 °C with 5%CO_2 for 24 h. Then, the culture media was removed and the HCE cells were exposed to AmB-FNPs and AmB-FNPs-ISG formulations for comparison with the control. After 12 h, the medium containing the samples was discarded carefully, and each well was washed with sterile PBS, pH 7.4. The HCE cells were further incubated with MTT solution (final concentration of 0.5 mg/mL) at 37 °C in the dark for 2 h. Finally, the MTT solution was removed, and the formazan crystals were dissolved with dimethyl sulfoxide (DMSO). The absorbance of each well was read on a microplate reader at 595 nm. The percentage of cell viability was calculated in comparison to the vehicle treated cells.

2.7.2. Crystal Violet Staining

The HCE cells were cultured and treated with the prepared formulations for 12 h, similar to the MTT assay. After the treatment times, the supernatant was discarded, and the cells were washed 3 times with PBS pH 7.4. The treated cells were fixed with 4% paraformaldehyde for 3 h at room temperature. After fixation, the HCE cells were stained with 0.5% *w/v* crystal violet solution and incubated for 30 min, then washed with tap water to remove excess staining. Cell samples were air dried and visualized under a light microscope.

2.8. Statistical Analysis

The mean ± SD (standard deviation) is presented for quantitative experiments. The statistical analysis was conducted using SPSS 17.0 software (Chicago, IL, USA). The significance was evaluated using one-way analysis of variance (1-way ANOVA) along with Tukey's post hoc test, and $p < 0.05$ was considered as statistically significant.

3. Results and Discussion

3.1. AmB-FNPs ISG Composites Characterization

From our previous study, we successfully prepared the novel formulation of AmB-FNPs-ISG. Their physicochemical properties are summarized in Table 1. All AmB-FNPs exhibited a mean particle size of ~200 nm with high entrapment efficiency of up to 63%. The aggregation study of AmB and XRD results from our previous study indicated that AmB was entrapped in the FNPs with an amorphous form [27]. The decrease in the crystallinity of AmB may be attributed to the interaction of AmB and silk fibroin via hydrophobic interaction. All prepared AmB-FNPs-ISG formulations possessed optimized osmolality, pH, and viscosity for ocular application. The rheological evaluation of all composite formulations exhibited pseudoplastic flow behavior after gel formation at 35 ± 1 °C, which allowed good spreadability on the ocular surface and ease of eye blinking.

Table 1. Physicochemical characterization of AmB-FNPs and AmB-FNPs-ISG composites.

Parameters	AmB-FNP-ISG		AmB-FNP-PEI-ISG		AmB-FNP-PEG-ISG	
	F127	F127/HA	F127	F127/HA	F127	F127/HA
Particle size, shape	206.8 ± 5.6 nm, spherical		209.0 ± 14.4 nm, cubic		214.7 ± 15.9 nm, spherical	
Zeta potential	−23.13 ± 2.69 mV		35.87 ± 1.37 mV		−22.04 ± 1.81 mV	
PDI	0.11 ± 0.05		0.18 ± 0.03		0.15 ± 0.01	
EE/DL	63.2%/8.7%		72.6%/5.2%		71.3%/9.7%	
Gelling capacity	++	++	++	++	++	++
pH	7.2 ± 0.2	6.7 ± 0.1	6.9 ± 0.1	6.7 ± 0.1	6.9 ± 0.1	6.9 ± 0.1
Osmolality (mOsm/kg)	338 ± 12	328 ± 9	348 ± 14	347 ± 11	332 ± 8	323 ± 7
%T (381–780 nm)	98 ± 5	97 ± 6	97 ± 5	96 ± 6	98 ± 4	97 ± 4
Viscosity (mPa·s)						
at 25 ± 1 °C	104 ± 2	324 ± 43	125 ± 31	280 ± 16	101 ± 12	315 ± 40
at 35 ± 1 °C	8214 ± 256	2706 ± 1349	7329 ± 1557	4140 ± 916	8233 ± 325	3571 ± 984

Note: PDI = polydispersity index, EE = entrapment efficiency, DL = drug loading capacity, gelling capacity: ++ = sol–gel transition within 30 s, and %T = % transmittance.

In addition, they undergo temperature-dependent sol–gel transition, from a flowing solution at ambient temperature (25 ± 1 °C) to a non-flowing gel at ocular temperature (35 ± 1 °C), making them easy to administer while enhancing the retention time of the drug on the eye surface. According to the viscosity vs. temperature curve, it was observed that both types of in situ hydrogels, AmB-FNPs-F127 and AmB-FNPs-F127/HA ISG, exhibited a constant viscosity at low temperature and their viscosity significantly increased as the temperature increased, indicating the gelation process (Figure 1). Furthermore, AmB-FNPs-F127 ISG showed a lower sol–gel transition temperature (~28 °C) than AmB-FNPs-F127/HA ISG (~31 °C), indicating AmB-FNPs-F127 ISG underwent gelation faster than AmB-FNPs-F127/HA ISG. After completed gel formation, AmB-FNPs-F127 ISG revealed higher viscosity than AmB-FNPs-F127/HA ISG, suggesting higher gel strength. These results may be related to the higher Pluronic concentration in the AmB-FNPs-F127 ISG formulation. Pluronic molecules can promptly self-assemble to form micelles in the aqueous media due to their amphiphilic nature, and the number of micelles increased when Pluronic concentration increased. The increasing temperature results in packing of micelles to form a large micellar crosslinked network, leading to gel formation [35]. Hence, the number and size of Pluronic micelles in the AmB-FNPs-F127 ISG formulation increased, resulting in a higher number of crosslinked micelles, and then leading to faster gelation and greater viscosity.

There is limited literature available on the development of composite systems involving nanoparticles and hydrogels for the ophthalmic delivery of AmB. Göttel et.al. developed AmB-loaded in situ gelling nanofiber to enhance the solubility of AmB for keratomycoses treatment [20]. However, the entrapment efficiency of AmB in PLGA nanoparticles for this study was ~40% and no mucoadhesive study was conducted. In addition, Elhabal et.al.

recently examined a thermosensitive hydrogel of AmB and Lactoferrin combination-loaded PLGA-PEG-PEI nanoparticles for eradication of ocular fungal infections [36]. This study exhibited high entrapment efficiency of AmB > 90% and good stability, but the mucoadhesion of the formulation was not reported in this study.

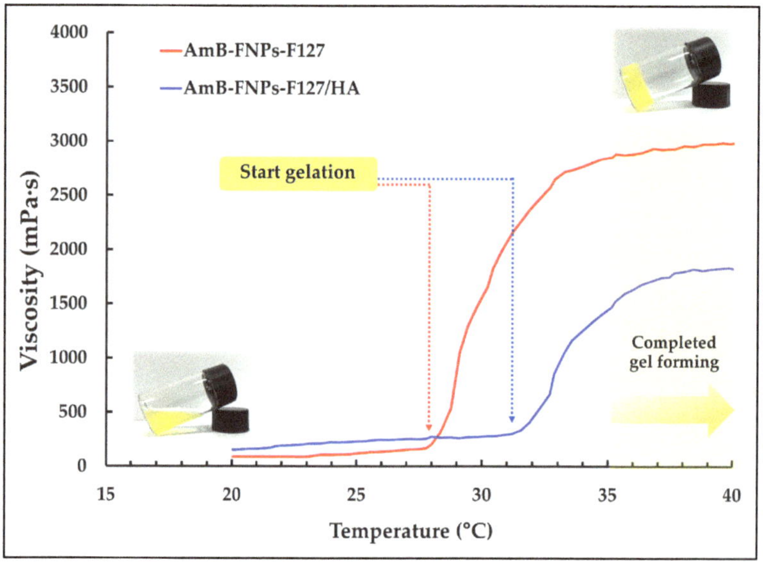

Figure 1. Average viscosity vs. temperature profiles of AmB-FNPs ISG.

3.2. Stability Study

The stability of AmB in topical ophthalmic dosage form is a challenge for pharmaceutical research. As is known, the stability of AmB in aqueous solution under heat condition is very low; hence, AmB products are stored at a low temperature of ~2–8 °C. Curti et al. reported that conventional AmB eye drops are stable for fewer than 15 days under ambient temperature and for 60 days under refrigeration conditions (2–8 °C) [37]. Moreover, Chanell et al. investigated the stability of ready-to-use amphotericin B solubilized in 2-hydroxypropyl-γ-cyclodextrin (AB-HP-γ-CD) formulations. They found that their formulation showed AmB instability after 28 and 56 days at 25 °C and 5 °C, respectively [9].

In this study, AmB-FNP-PEG-ISG was selected as a representative to study the stability of the composite formulations. Samples were kept at 4 °C for 90 days. At all predetermined time points, both AmB-FNP-PEG-ISG samples showed good gelling capacity with pH ~7. However, both formulations exhibited phase separation when the storage time was increased (Figure 2a). This result may be associated with agglomeration of the particles in the colloidal system. Interestingly, AmB-FNP-PEG-F127 ISG showed higher sedimentation of nanoparticles on the bottom of vial than AmB-FNP-PEG-F127/HA ISG. This result could be attribute to electrostatic stabilization because the negative charge of hyaluronic acid enhances the strong repulsion of the AmB-FNP-PEG particles in the F127/HA ISG network [38]. Although AmB-FNP-PEG-ISG showed a phase separation, all AmB-FNPs-ISG could transform to a homogeneous dispersion after mild shaking. Interestingly, Figure 2b shows no significant difference in the drug remaining of AmB-FNP-PEG-ISG between the beginning and end of the storage time, indicating good stability of AmB in the prepared formulation. This result may be attributed to the incorporation of AmB into silk fibroin nanoparticles, which could protect AmB degradation from the hydrolysis mechanism in the aqueous media, leading to increased stability of AmB.

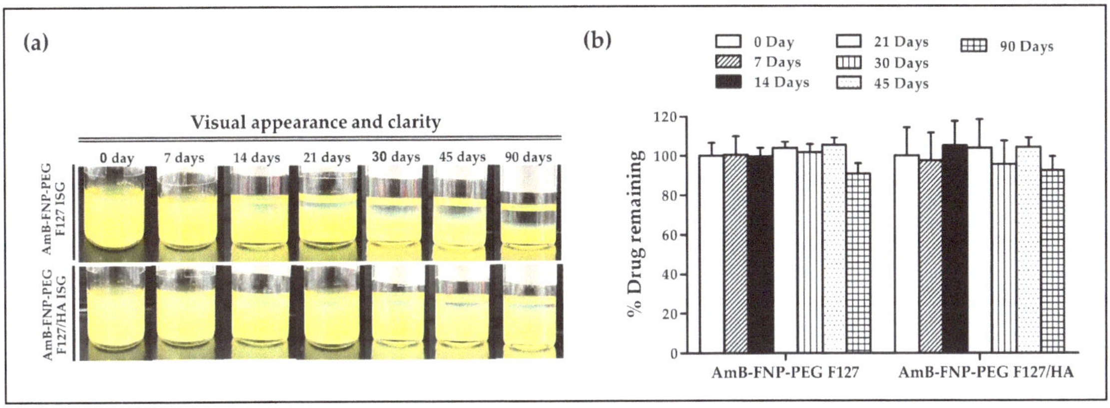

Figure 2. Stability of AmB-FNP-PEG-F127 ISG and AmB-FNP-PEG-F127/HA ISG under storage conditions (4 °C). (**a**) Visual appearance and (**b**) drug remaining of AmB in the formulations at different time points (mean ± SD, n = 3).

3.3. Antifungal Efficacy

The potential of using the novel AmB-FNPs-ISG as topical ophthalmic formulation for fungal keratitis was investigated. The agar well diffusion technique was used to screen the antifungal effect of the prepared formulations compared with AmB deoxycholate, a commercial formulation. This study was conducted by measuring the inhibition zone of the sample against *C. albicans*. As expected, the inhibition zone was observed around the holes of AmB-FNPs, AmB-FNPs-ISG, and the positive controls (standard AmB and AmB deoxycholate), whereas the negative controls (sterile water, blank F127 ISG, and blank F127/HA ISG) showed no inhibition zone, indicating no antifungal activity (Figure 3). AmB deoxycholate showed the highest inhibition zone (~20 mm), which was significantly different from that of standard AmB (~17 mm), AmB-FNP dispersion (~17 mm), AmB-FNP-PEI (~16 mm), AmB-FNP-PEG dispersion (~17 mm), and AmB-FNP-PEI F127/HA (~17 mm). Interestingly, the inhibition zone of all AmB-FNPs-ISG formulations, with the exception of AmB-FNP-PEI F127/HA, was slightly greater than that of their FNP counterparts. These results may be associated with the effect of surfactants in the formulations, deoxycholate in the commercial AmB and Pluronic in the in situ hydrogels, which could enhance the membrane-associated target of AmB and increase the permeability of the fungal membrane [39]. Although the agar well diffusion technique is a widely used for assessing the antimicrobial activity of various drug formulations, the outcomes of this technique are variable due to inherent limitations in the hydrophilicity and viscosity of the formulations, as well as the interaction with the agar component [40,41]. To this end, we combined the agar diffusion test with MIC and MFC tests to confirm the potential of the prepared formulations for treatment of ocular infections.

In addition, MIC_{90} and MFC of all formulations against *C. albicans* were investigated (Table 2). AmB deoxycholate and standard AmB showed similar MIC_{90} and MFC values of 0.0625 μg/mL and 0.5 μg/mL, respectively. However, AmB-FNPs exhibited higher MIC_{90} and MFC values of 0.25 μg/mL and 1 μg/mL, respectively. The lower antifungal activity could be explained by some of the AmB in FNP formulations being restricted in the particles due to strong interaction between AmB and silk fibroin, as observed from dissolution studies, which showed ~50% AmB release within 5 h [27].

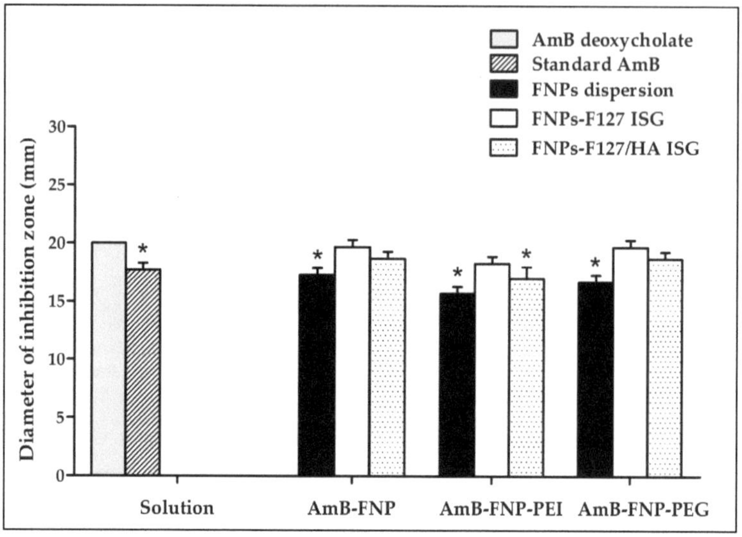

Figure 3. The antifungal activity of AmB deoxycholate, standard AmB, AmB-FNPs, and AmB-FNPs ISG against *C. albicans* according to the agar well diffusion technique. * Significant at $p < 0.05$ compared with AmB deoxycholate.

Table 2. Minimum inhibitory concentration (MIC90) and minimum fungicidal concentration (MFC) of the formulations against *C. albicans* ($n = 3$).

Formulation	MIC_{90} (μg/mL)	MFC (μg/mL)
AmB Deoxycholate	0.0625	0.5
Standard AmB	0.0625	0.5
AmB-FNP	0.250	1
AmB-FNP-PEI	0.250	2
AmB-FNP-PEG	0.250	1
AmB-FNP-F127 ISG	0.125	0.5
AmB-FNP-PEI-F127 ISG	0.125	0.5
AmB-FNP-PEG-F127 ISG	0.125	0.5
AmB-FNP-F127/HA ISG	0.125	0.5
AmB-FNP-PEI-F127/HA ISG	0.125	0.5
AmB-FNP-PEG-F127/HA ISG	0.125	0.5

Interestingly, AmB-FNPs-ISG formulations exhibited higher antifungal activity than AmB-FNPs with the MIC_{90} value of 0.125 μg/mL and MFC value of 0.5 μg/mL, close to those of AmB deoxycholate. This may be due to the support from Pluronic, a surfactant in the hydrogel matrix, which could enhance the diffusion of AmB across the fungal cells, leading to an increase in the fungal cell death. These results indicate that the prepared AmB-FNPs-ISG have potential antifungal activity similar to that of the marketed AmB deoxycholate.

3.4. *In Vitro Mucoadhesive Study*

To track the mucoadhesion of particles on the mucus layer, all FNPs were labeled with FITC, a green fluorescent dye. Generally, FITC is widely used to label proteins via the reaction of isothiocyanate groups of FITC and amine groups in the protein. Therefore, isothiocyanate groups of FITC could attach with residual amine groups of silk fibroin. In this study, FITC-labeled FNP, FNP-PEI, and FNP-PEG were prepared following the method of Pham et al. [42]. To examine the in vitro mucoadhesive properties of FITC-labeled FNPs and FITC-labeled FNP in situ hydrogels, the hydrophilic membrane soaked with mucin solution was used to mimic a mucus membrane. Figure 4 illustrates the percentage of FITC remaining on the mucus membrane after continuous flow of STF. After 5 min of fluid flow, the membrane instilled with FITC solution showed only ~13% remaining, while those instilled with FITC-FNP, FITC-FNP-PEG, and FITC-FNP-PEI showed 38%, 65%, and 88% remaining, respectively (Figure 4a). Moreover, nearly 100% loss occurred after 1 h of fluid flow when instilled with FITC solution, whereas all FITC-FNPs showed % FITC remaining of up to 30%. These results indicated that small particles of FNPs could enhance FITC adherence to the mucus membrane because the large surface area of the particles can increase the adhesion to the mucus membrane [43].

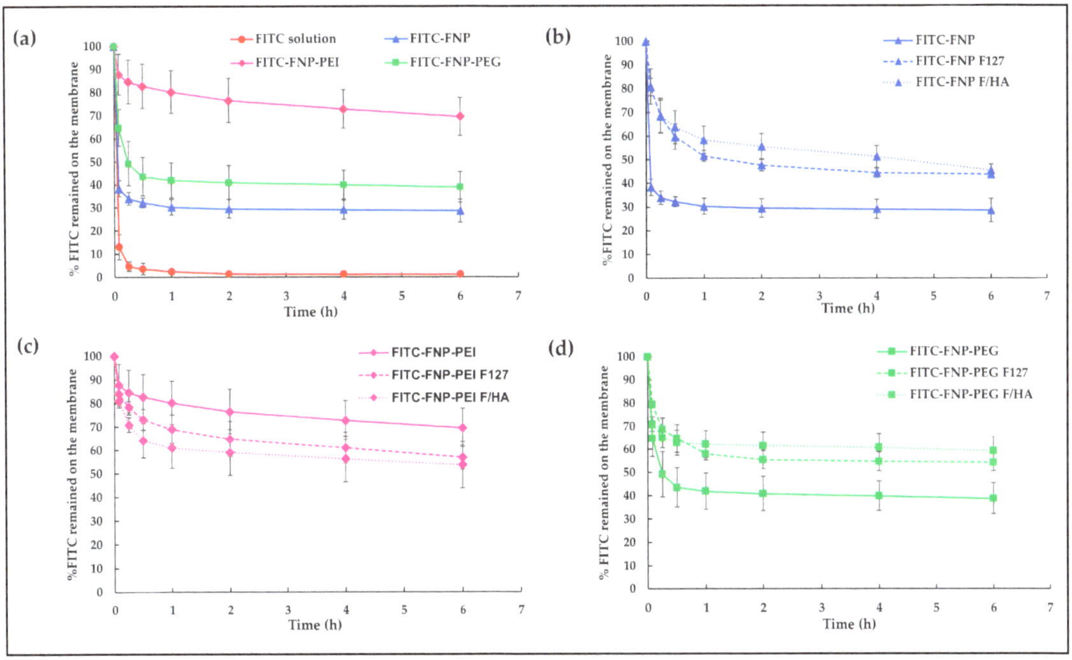

Figure 4. The in vitro mucoadhesive properties of FITC-labeled FNPs and FITC-FNP-ISG. (**a**) %FITC remaining on the mucus membrane of FITC solution compared with three types of FITC-FNP dispersion; (**b**) %FITC remaining of FITC-FNP dispersion; (**c**) %FITC remaining of FITC-FNP-PEI dispersion; and (**d**) %FITC remaining of FITC-FNP-PEG dispersion compared with their in situ hydrogel composites.

Interestingly, FITC-FNP-PEI exhibited a higher remaining percentage (69%) than FITC-FNP-PEG (39%), and FITC-FNP (29%) after 6 h continuous fluid flow. The nature of the polymer coating on the particles could be attributed as the primary factors influencing mucoadhesion. Several studied reported that the mucoadhesive properties of the nanoparticles could be enhanced by coating the particles with hydrophilic or cationic polymers via the interaction between the polymer and mucin chain [44]. PEI could increase the mucoadhesion on the mucus membrane through ionic interaction between the positively charged PEI and the negatively charged mucin. On the other hand, PEG, being a hydrophilic polymer with abundant hydroxyl contents, facilitates the penetration of the polymer chain into the mucus layer and engages in hydrogen bonding with mucin [45]. Based on these results, FNP-PEI and FNP-PEG exhibited strong adhesion on the mucus membrane when compared with uncoated FNP. Moreover, mucoadhesive properties of FNPs and FNPs-ISG formulations were compared. As expected, the FITC-FNP-ISG (Figure 4b) and FITC-FNP-PEG-ISG (Figure 4d) exhibited significantly greater FITC remaining than their particles only. This result could be attributed to the increasing viscosity of the gelling system. Therefore, the combination of in situ gelling and nanoparticles could prolong the drug retention on the ocular surface. However, the FITC-FNP-PEI in situ hydrogel (Figure 4c) showed lower FITC remaining than FITC-FNP-PEI particles. This result might be due to the hydrogel network retarding the interaction between positively charged PEI and negatively charged mucin.

3.5. Ex Vivo Mucoadhesive Study

To confirm the mucoadhesion of the prepared formulations on the corneal surface, FITC-FNP-PEG in situ hydrogel and FITC-FNP-PEG were chosen for ex vivo mucoadhesive study. Similar to the in vitro mucoadhesive results, the FITC-FNP-PEG (Figure 5e–h), FITC-FNP-PEG-F127 ISG (Figure 5i–l), and FITC-FNP-PEG-F127/HA ISG (Figure 5m–p) showed good adhesion to the porcine cornea up to 6 h, while the FITC solution (Figure 5a–d) exhibited low intensity of green fluorescence at 30 min and was completely cleared away at 2 h. As expected, FITC-FNP-PEG-F127 in situ hydrogels showed higher intensity of green fluorescence for all time points than FITC-FNP-PEG under fluid flow. These results confirmed that the combination of FNP and in situ hydrogels could enhance the adhesion ability of the particles on the ocular surface.

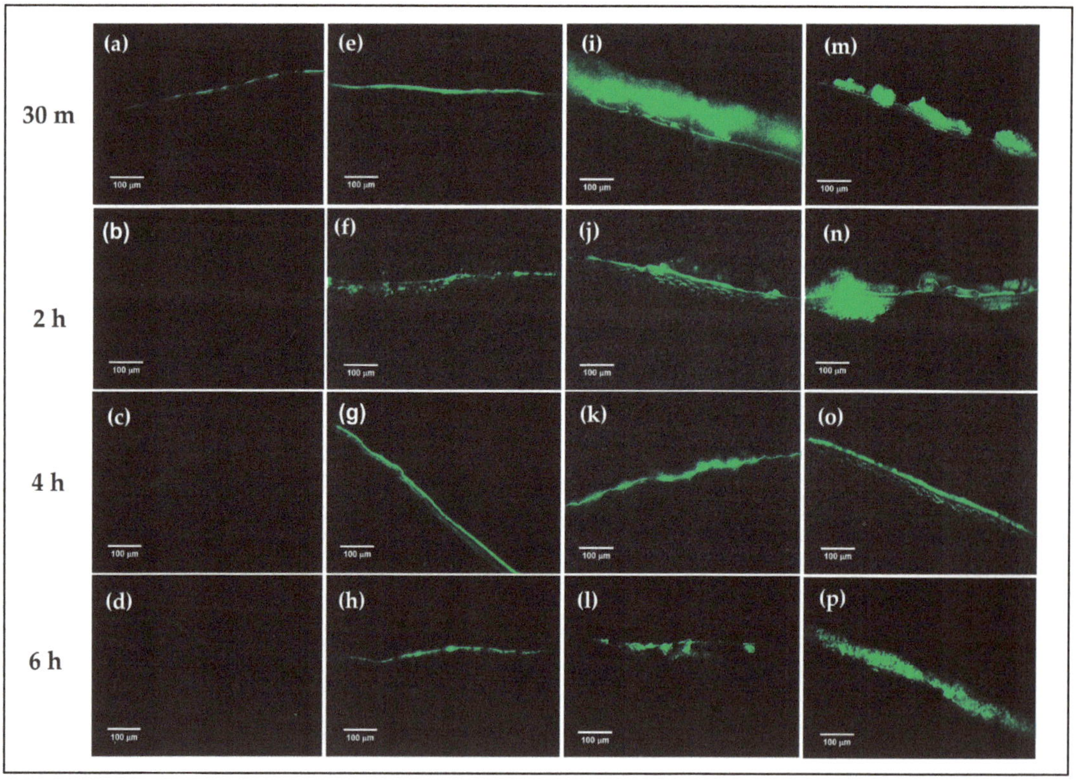

Figure 5. Ex vivo mucoadhesive studies of the nanoparticles and in situ hydrogel compared with nanodispersion and solution formulations. The fluorescence images of the remaining fluorescence on the cross-sectional porcine cornea after treatment with FITC solution (**a–d**), FITC-labeled FNP-PEG (**e–h**), FNP-PEG-F127 ISG (**i–l**), and FNP-PEG-F127/HA ISG (**m–p**) under continuous flow of STF at different time points (200×).

3.6. In Vitro Eye Irritation Studies

During drug development, the ocular irritation potential and toxicity of the ocular formulation must be tested to ensure the safety and biocompatibility of the product before clinical trial in humans. The in vitro cell model is one of the most applicable to eye irritation assessment because it is inexpensive, simple, and quick to implement compared to in vivo testing. In this study, the cytotoxicity study was carried out using the HCE cell line and modified following the STE protocol, as suggested for the assessment of eye irritation potential rather than animal testing [34,46]. The toxicity was accessed using MTT reagent based on mitochondrial activity, which is proportional to the number of viable cells, and the sample demonstrating cell viability higher than 70% was classified as a non-irritant [34].

The effect of AmB-loaded FNPs on HCE cell viability is shown in Figure 6a. The exposure of HCE cells to AmB deoxycholate containing 5, 15, and 150 μg/mL of AmB exhibited cell viability of 76%, 12%, and 0% respectively, demonstrating their severe irritation effect. These toxic effects were associated with the aggregation state of AmB in this formulation, which was confirmed by the absorption spectra from our previous report [27]. Several studies have reported that the aggregated AmB form could bind to both ergosterol in fungal cells and cholesterol in mammalian cells, resulting in leakage of metabolites and ions from the cell membrane, and eventually leading to cell death [47]. In addition, sodium deoxycholate acted as a surfactant in this formulation, which could enhance cell membrane permeability of both mammalian and fungal cells, leading to cell damage [48,49].

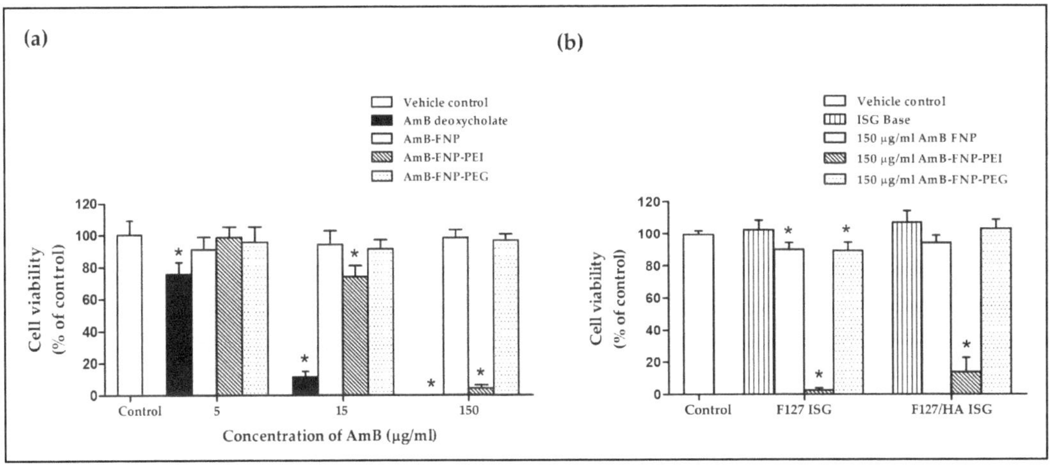

Figure 6. In vitro cytotoxicity study of prepared formulations. (**a**) Percentage of HCE cell viability after 12 h incubation with AmB deoxycholate and AmB-FNP formulations (equivalent concentration of AmB at 5, 15, and 150 μg/mL). (**b**) Percentage of HCE cell viability after 12 h incubation with AmB-FNPs embedded in F127 and F127/HA in situ hydrogel (formulation dose at 150 μg/mL of AmB). (mean ± SD, $n = 9$, * $p < 0.05$).

Interestingly, all concentrations of AmB-FNP and AmB-FNP-PEG exhibited cell viability > 90%, and could therefore be categorized as a non-irritant and safe for the eye. The lower toxicity of both formulations could be explained by the AmB encapsulated in silk fibroin nanoparticles, which could reduce molecular aggregation of AmB, as characterized by the absorbance ratio of the first to fourth peaks from UV–Vis spectroscopy, indicating a higher specificity to ergosterol than cholesterol [27]. Moreover, AmB-FNP and AmB-FNP-PEG, which are composed of fibroin and PEG 400, are biocompatible with ocular tissue; consequently, these formulations exhibited cell viability similar to that of the control and greater than that of AmB deoxycholate.

Unfortunately, AmB-FNP-PEI showed a toxic effect to HCE cells at the high concentration even though this formulation exhibited the partial aggregation of AmB. The exposure of HCE cells to AmB-FNP-PEI containing 5, 15, and 150 μg/mL of AmB possessed cell viability of 98%, 74%, and 4%, respectively, which indicated classification as a potential irritant at the high dose. This toxic effect was related to the amount of PEI because the blank FNP-PEI also showed the toxicity to HCE cells in a dose-dependent manner (data not shown). Although PEI is capable of binding with the negative charge of mucin on the ocular surface, which enhances the precorneal retention time, it is also toxic like other cationic polymers. Fischer et al. reported that the high charge density of branching PEI can interact with the negative charge of the cell membrane, leading to weakening of the plasma membrane integrity and causing cell death [50]. In addition, they also found that the PEI affected the metabolic activity, and the severity of cytotoxic effects depends on the exposure time and concentration of the polymer.

Figure 6b demonstrates the cell viability of the HCE cells after exposure to ISG bases, AmB-FNPs F127 ISG, and AmB-FNPs F127/HA ISG. Both ISG bases showed % cell viability having no significant difference from the control, and thus indicating no irritation. The HCE cells exposed to AmB-FNP-ISG and AmB-FNP-PEG ISG showed viability of more than 80%, suggesting no irritation and that they are safe for ocular uses. However, the exposure of HCE cells to AmB-FNP-PEI ISG exhibited cell viability of less than 20%, indicating a toxic effect due to PEI as described above.

Furthermore, the morphology and density of HCE cells after treatment were assessed via crystal violet staining. The morphology of HCE cells treated with AmB-FNPs and AmB-FNP-ISG are shown in Figures 7 and 8, respectively. When compared with untreated cells and the vehicle control, the cells exposed with AmB-FNP, AmB-FNP-PEG, and their AmB-FNP in situ hydrogels revealed no differences in cell number and morphology from the control cells. These results confirmed that AmB-FNP-ISG and AmB-FNP-PEG-ISG have good biocompatibility and are safe for topical ocular application. In contrast, the cells exposed to AmB deoxycholate, which is the marketed formulation, demonstrated a small amount of cell debris remaining on the well plate, indicating significant cell death and detachment from the surface. Additionally, the cell exposure to AmB-FNP-PEI and its hydrogel revealed cell debris and changes in morphology, indicating cell damage and unsuitability for ocular application.

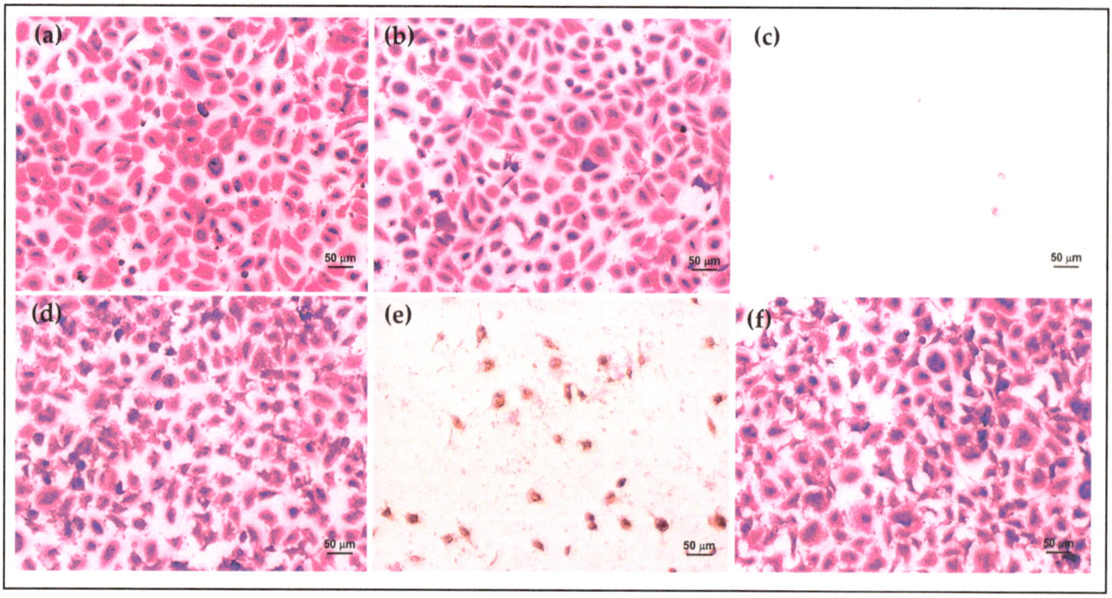

Figure 7. Morphology of HCE cells after exposure to AmB-FNP dispersions compared with the controls for 12 h: (**a**) untreated cells; (**b**) vehicle control; (**c**) 150 µg/mL AmB deoxycholate; (**d**) 150 µg/mL AmB-FNP; (**e**) 150 µg/mL AmB-FNP-PEI; and (**f**) 150 µg/mL AmB-FNP-PEG (200×).

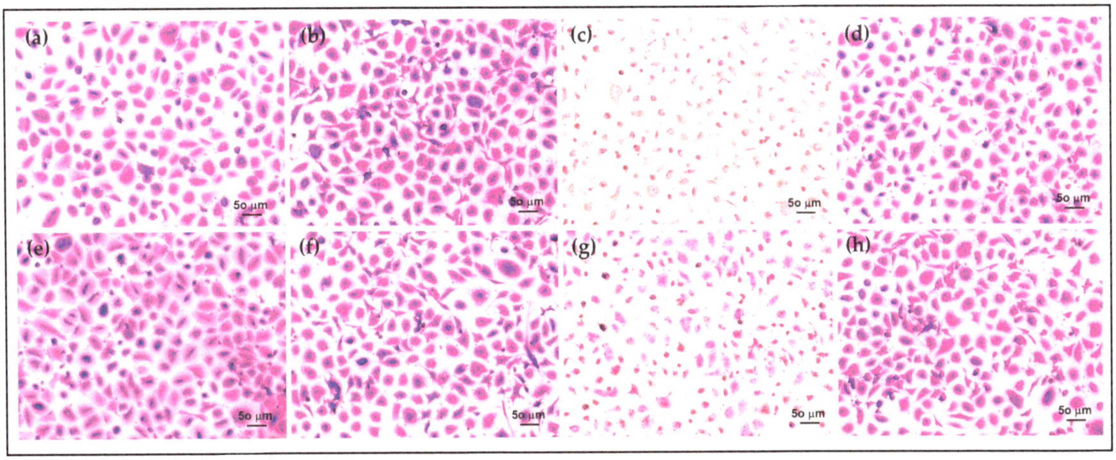

Figure 8. Morphology of HCE cells after exposure to AmB-FNPs ISG compared with their hydrogel bases for 12 h: (**a**) F127 ISG base; (**b**) AmB-FNP-F127; (**c**) AmB-FNP-PEI F127; (**d**) AmB-FNP-PEG F127; (**e**) F127/HA ISG base; (**f**) AmB-FNP-F127/HA; (**g**) AmB-FNP-PEI-F127/HA; and (**h**) AmB-FNP-PEG-F127/HA (200×).

4. Conclusions

Our current study demonstrates the efficacy and safety of the combination system of AmB-FNPs-ISG as mucoadhesive ophthalmic eye drops for FK treatment. The optimized AmB-FNPs demonstrated a spherical shape with a mean particle size of 215 nm and high entrapment efficiency of 71%. The developed thermosensitive AmB-FNP in situ hydrogel formulations displayed satisfactory gelling capacity, a translucent homogeneous solution, pH ~7, osmolality of ~323–348 mOsmol/kg, and %transmittance > 90%. All composite formulations illustrated optimal viscosity and pseudoplastic behavior, which are suitable for ocular application. Furthermore, the nanoparticle in situ hydrogel formulations showed good physiochemical stability under storage conditions for 90 days. The combined systems of AmB-FNPs-ISG exhibited an effective antifungal effect against *C. albicans* similar to that of commercial AmB, and they showed a greater antifungal effect than the single AmB-FNP as a result of the synergistic effect of the Pluronic surfactant in the hydrogel. As expected, in vitro and ex vivo mucoadhesive results of the combined system showed higher fluorescence intensity than the solution and nanodispersion, which confirmed that the combination of FNPs with in situ hydrogels could enhance the retention time of the particles on the corneal tissue for more than 6 h. However, AmB-FNP-PEI ISG demonstrated toxicity to the HCE cells depending on the PEI content in the particles, whereas AmB-FNP-ISG, AmB-FNP-PEG-ISG, and their FNP counterparts exhibited significantly less toxicity on HCE cells than commercial AmB, thus making them more suitable for ocular application. Thus, the smart AmB-FNP-PEG-ISG has potential as ready-to-use AmB eye drops for FK treatment.

Author Contributions: Conceptualization, W.T.; data curation, P.C.; formal analysis, P.C.; funding acquisition, P.C.; investigation, P.C. and P.S.; methodology, P.C. and P.S.; project administration, W.T.; supervision, N.S., S.L. and W.T.; validation, P.C. and W.T.; visualization, P.C. and W.T.; writing—original draft preparation, P.C.; writing—review and editing, N.S., S.L. and W.T. All authors have read and agreed to the published version of the manuscript.

Funding: This research was funded by the National Research Council of Thailand (NRCT) under the Royal Golden Jubilee Ph.D. program, grant number PHD/0160/2560.

Institutional Review Board Statement: Not applicable.

Data Availability Statement: The data presented in this study are available on request from the corresponding author.

Acknowledgments: The authors are thankful to the Centre of Excellence in Fungal Research and Department of Microbiology and Parasitology, Faculty of Medical Science and Faculty of Pharmaceutical Sciences, Naresuan University for providing necessary facilities.

Conflicts of Interest: The authors declare no conflicts of interest.

References

1. Atta, S.; Perera, C.; Kowalski, R.P.; Jhanji, V. Fungal Keratitis: Clinical Features, Risk Factors, Treatment, and Outcomes. *J. Fungi* **2022**, *8*, 962. [CrossRef] [PubMed]
2. Huang, J.F.; Zhong, J.; Chen, G.P.; Lin, Z.T.; Deng, Y.; Liu, Y.L.; Cao, P.Y.; Wang, B. A Hydrogel-Based Hybrid Theranostic Contact Lens for Fungal Keratitis. *ACS Nano* **2016**, *10*, 6464–6473. [CrossRef] [PubMed]
3. Sun, R.L.; Jones, D.B.; Wilhelmus, K.R. Clinical characteristics and outcome of Candida keratitis. *Am. J. Ophthalmol.* **2007**, *143*, 1043–1045. [CrossRef] [PubMed]
4. Petrillo, F.; Sinoca, M.; Fea, A.M.; Galdiero, M.; Maione, A.; Galdiero, E.; Guida, M.; Reibaldi, M. Candida Biofilm Eye Infection: Main Aspects and Advance in Novel Agents as Potential Source of Treatment. *Antibiotics* **2023**, *12*, 1277. [CrossRef] [PubMed]
5. Ghosh, A.K.; Rudramurthy, S.M.; Gupta, A.; Choudhary, H.; Singh, S.; Thakur, A.; Jatana, M. Evaluation of Liposomal and Conventional Amphotericin B in Experimental Fungal Keratitis Rabbit Model. *Transl. Vis. Sci. Technol.* **2019**, *8*, 35. [CrossRef] [PubMed]
6. Mahmoudi, S.; Masoomi, A.; Ahmadikia, K.; Tabatabaei, S.A.; Soleimani, M.; Rezaie, S.; Ghahvechian, H.; Banafsheafshan, A. Fungal keratitis: An overview of clinical and laboratory aspects. *Mycoses* **2018**, *61*, 916–930. [CrossRef] [PubMed]
7. Raj, N.; Vanathi, M.; Ahmed, N.H.; Gupta, N.; Lomi, N.; Tandon, R. Recent Perspectives in the Management of Fungal Keratitis. *J. Fungi* **2021**, *7*, 907. [CrossRef]
8. Serrano, D.R.; Ruiz-Saldaña, H.K.; Molero, G.; Ballesteros, M.P.; Torrado, J.J. A novel formulation of solubilised amphotericin B designed for ophthalmic use. *Int. J. Pharm.* **2012**, *437*, 80–82. [CrossRef]
9. Chennell, P.; Yessaad, M.; Abd El Kader, F.; Jouannet, M.; Wasiak, M.; Bouattour, Y.; Sautou, V. Do Ophthalmic Solutions of Amphotericin B Solubilised in 2-Hydroxypropyl-γ-Cyclodextrins Possess an Extended Physicochemical Stability? *Pharmaceutics* **2020**, *12*, 786. [CrossRef]
10. Chen, H.; Jin, Y.; Sun, L.; Li, X.; Nan, K.; Liu, H.; Zheng, Q.; Wang, B. Recent Developments in Ophthalmic Drug Delivery Systems for Therapy of Both Anterior and Posterior Segment Diseases. *Colloids Interface Sci. Commun.* **2018**, *24*, 54–61. [CrossRef]
11. Bachu, R.D.; Chowdhury, P.; Al-Saedi, Z.H.F.; Karla, P.K.; Boddu, S.H.S. Ocular Drug Delivery Barriers-Role of Nanocarriers in the Treatment of Anterior Segment Ocular Diseases. *Pharmaceutics* **2018**, *10*, 28. [CrossRef] [PubMed]
12. Dubald, M.; Bourgeois, S.; Andrieu, V.; Fessi, H. Ophthalmic Drug Delivery Systems for Antibiotherapy—A Review. *Pharmaceutics* **2018**, *10*, 10. [CrossRef] [PubMed]
13. Jumelle, C.; Gholizadeh, S.; Annabi, N.; Dana, R. Advances and limitations of drug delivery systems formulated as eye drops. *J. Control. Release.* **2020**, *321*, 1–22. [CrossRef] [PubMed]
14. Zhou, W.; Wang, Y.; Jian, J.; Song, S. Self-aggregated nanoparticles based on amphiphilic poly(lactic acid)-grafted-chitosan copolymer for ocular delivery of amphotericin B. *Int. J. Nanomed.* **2013**, *8*, 3715–3728. [CrossRef]
15. Chhonker, Y.S.; Prasad, Y.D.; Chandasana, H.; Vishvkarma, A.; Mitra, K.; Shukla, P.K.; Bhatta, R.S. Amphotericin-B entrapped lecithin/chitosan nanoparticles for prolonged ocular application. *Int. J. Biol. Macromol.* **2015**, *72*, 1451–1458. [CrossRef]
16. Fu, T.; Yi, J.; Lv, S.; Zhang, B. Ocular amphotericin B delivery by chitosan-modified nanostructured lipid carriers for fungal keratitis-targeted therapy. *J. Liposome Res.* **2017**, *27*, 228–233. [CrossRef]
17. Lakhani, P.; Patil, A.; Wu, K.W.; Sweeney, C.; Tripathi, S.; Avula, B.; Taskar, P.; Khan, S. Optimization, stabilization, and characterization of amphotericin B loaded nanostructured lipid carriers for ocular drug delivery. *Int. J. Pharm.* **2019**, *572*, 118771. [CrossRef]
18. Morand, K.; Bartoletti, A.C.; Bochot, A.; Barratt, G.; Brandely, M.L.; Chast, F. Liposomal amphotericin B eye drops to treat fungal keratitis: Physico-chemical and formulation stability. *Int. J. Pharm.* **2007**, *344*, 150–153. [CrossRef]
19. Albadr, A.A.; Tekko, I.A.; Vora, L.K.; Ali, A.A.; Laverty, G.; Donnelly, R.F.; Thakur, R.R.S. Rapidly dissolving microneedle patch of amphotericin B for intracorneal fungal infections. *Drug Deliv. Transl. Res.* **2022**, *12*, 931–943. [CrossRef]
20. Göttel, B.; Lucas, H.; Syrowatka, F.; Knolle, W.; Kuntsche, J.; Heinzelmann, J.; Viestenz, A.; Mäder, K. In situ Gelling Amphotericin B Nanofibers: A New Option for the Treatment of Keratomycosis. *Front. Bioeng. Biotechnol.* **2020**, *8*, 600384. [CrossRef]
21. Dannert, C.; Stokke, B.T.; Dias, R.S. Nanoparticle-Hydrogel Composites: From Molecular Interactions to Macroscopic Behavior. *Polymers* **2019**, *11*, 275. [CrossRef] [PubMed]
22. Almeida, H.; Amaral, M.H.; Lobao, P.; Silva, A.C.; Loboa, J.M. Applications of polymeric and lipid nanoparticles in ophthalmic pharmaceutical formulations: Present and future considerations. *J. Pharm. Pharm. Sci.* **2014**, *17*, 278–293. [CrossRef] [PubMed]
23. Mazet, R.; Yaméogo, J.B.G.; Wouessidjewe, D.; Choisnard, L.; Gèze, A. Recent Advances in the Design of Topical Ophthalmic Delivery Systems in the Treatment of Ocular Surface Inflammation and Their Biopharmaceutical Evaluation. *Pharmaceutics* **2020**, *12*, 570. [CrossRef] [PubMed]
24. Faustino, C.; Pinheiro, L. Lipid Systems for the Delivery of Amphotericin B in Antifungal Therapy. *Pharmaceutics* **2020**, *12*, 29. [CrossRef]

25. Patel, A.; Cholkar, K.; Agrahari, V.; Mitra, A.K. Ocular drug delivery systems: An overview. *World J. Pharmacol.* **2013**, *2*, 47–64. [CrossRef]
26. Hsu, X.-L.; Wu, L.-C.; Hsieh, J.-Y.; Huang, Y.-Y. Nanoparticle-Hydrogel Composite Drug Delivery System for Potential Ocular Applications. *Polymers* **2021**, *13*, 642. [CrossRef]
27. Chomchalao, P.; Saelim, N.; Tiyaboonchai, W. Preparation and characterization of amphotericin B-loaded silk fibroin nanoparticles-in situ hydrogel composites for topical ophthalmic application. *J. Mater. Sci.* **2022**, *57*, 12522–12539. [CrossRef]
28. Almeida, H.; Lobão, P.; Frigerio, C.; Fonseca, J.; Silva, R.; Quaresma, P.; Lobo, J.M.S.; Amaral, M.H. Development of mucoadhesive and thermosensitive eyedrops to improve the ophthalmic bioavailability of ibuprofen. *J. Drug Deliv. Sci. Technol.* **2016**, *35*, 69–80. [CrossRef]
29. Chomchalao, P.; Nimtrakul, P.; Pham, D.T.; Tiyaboonchai, W. Development of amphotericin B-loaded fibroin nanoparticles: A novel approach for topical ocular application. *J. Mater. Sci.* **2020**, *55*, 5268–5279. [CrossRef]
30. Balouiri, M.; Sadiki, M.; Ibnsouda, S.K. Methods for in vitro evaluating antimicrobial activity: A review. *J. Pharm. Anal.* **2016**, *6*, 71–79. [CrossRef]
31. Rodríguez-Tudela, J.L.; Cuenca-Estrella, M.; Díaz-Guerra, T.M.; Mellado, E. Standardization of antifungal susceptibility variables for a semiautomated methodology. *J. Clin. Microbiol.* **2001**, *39*, 2513–2517. [CrossRef] [PubMed]
32. Yan, J.; Chen, X.; Yu, S.; Zhou, H. Comparison of different in vitro mucoadhesion testing methods for hydrogels. *J. Drug Deliv. Sci. Technol.* **2017**, *40*, 157–163. [CrossRef]
33. Chaiyasan, W.; Srinivas, S.P.; Tiyaboonchai, W. Mucoadhesive chitosan-dextran sulfate nanoparticles for sustained drug delivery to the ocular surface. *J. Ocul. Pharmacol. Ther.* **2013**, *29*, 200–207. [CrossRef] [PubMed]
34. Takahashi, Y.; Koike, M.; Honda, H.; Ito, Y.; Sakaguchi, H.; Suzuki, H.; Nishiyama, N. Development of the short time exposure (STE) test: An in vitro eye irritation test using SIRC cells. *Toxicol. Vitr.* **2008**, *22*, 760–770. [CrossRef] [PubMed]
35. Jung, Y.-s.; Park, W.; Park, H.; Lee, D.-K.; Na, K. Thermo-sensitive injectable hydrogel based on the physical mixing of hyaluronic acid and Pluronic F-127 for sustained NSAID delivery. *Carbohydr. Polym.* **2017**, *156* (Suppl. C), 403–408. [CrossRef] [PubMed]
36. Elhabal, S.F.; Ghaffar, S.A.; Hager, R.; Elzohairy, N.A.; Khalifa, M.M.; Mohie, P.M.; Gad, R.A.; Omar, N.N. Development of thermosensitive hydrogel of Amphotericin-B and Lactoferrin combination-loaded PLGA-PEG-PEI nanoparticles for potential eradication of ocular fungal infections: In-vitro, ex-vivo and in-vivo studies. *Int. J. Pharm. X* **2023**, *5*, 100174. [CrossRef] [PubMed]
37. Curti, C.; Lamy, E.; Primas, N.; Fersing, C.; Jean, C.; Bertault-Peres, P.; Vanelle, P. Stability studies of five anti-infectious eye drops under exhaustive storage conditions. *Pharmazie* **2017**, *72*, 741–746. [CrossRef] [PubMed]
38. Loza, K.; Epple, M.; Maskos, M. Stability of Nanoparticle Dispersions and Particle Agglomeration. In *Biological Responses to Nanoscale Particles: Molecular and Cellular Aspects and Methodological Approaches*; Gehr, P., Zellner, R., Eds.; Springer International Publishing: Cham, Switzerland, 2019; pp. 85–100.
39. Schreier, S.; Malheiros, S.V.P.; de Paula, E. Surface active drugs: Self-association and interaction with membranes and surfactants. Physicochemical and biological aspects. *Biochim. Biophy. Acta—Biomembr.* **2000**, *1508*, 210–234. [CrossRef]
40. Bubonja-Šonje, M.; Knežević, S.; Abram, M. Challenges to antimicrobial susceptibility testing of plant-derived polyphenolic compounds. *Arh. Hig. Rada Toksikol.* **2020**, *71*, 300–311. [CrossRef]
41. Hossain, M.L.; Lim, L.Y.; Hammer, K.; Hettiarachchi, D.; Locher, C. A Review of Commonly Used Methodologies for Assessing the Antibacterial Activity of Honey and Honey Products. *Antibiotics* **2022**, *11*, 975. [CrossRef]
42. Pham, D.T.; Saelim, N.; Cornu, R.; Béduneau, A.; Tiyaboonchai, W. Crosslinked Fibroin Nanoparticles: Investigations on Biostability, Cytotoxicity, and Cellular Internalization. *Pharmaceuticals* **2020**, *13*, 86. [CrossRef] [PubMed]
43. Grassiri, B.; Zambito, Y.; Bernkop-Schnürch, A. Strategies to prolong the residence time of drug delivery systems on ocular surface. *Adv. Colloid Interface Sci.* **2021**, *288*, 102342. [CrossRef] [PubMed]
44. Ludwig, A. The use of mucoadhesive polymers in ocular drug delivery. *Adv. Drug Deliv. Rev.* **2005**, *57*, 1595–1639. [CrossRef] [PubMed]
45. Shaikh, R.; Raj Singh, T.R.; Garland, M.J.; Woolfson, A.D.; Donnelly, R.F. Mucoadhesive drug delivery systems. *J. Pharm. Bioallied Sci.* **2011**, *3*, 89–100. [CrossRef] [PubMed]
46. Sakaguchi, H.; Ota, N.; Omori, T.; Kuwahara, H.; Sozu, T.; Takagi, Y.; Takahashi, Y.; Tanigawa, K. Validation study of the Short Time Exposure (STE) test to assess the eye irritation potential of chemicals. *Toxicol. In Vitr.* **2011**, *25*, 796–809. [CrossRef] [PubMed]
47. Hartsel, S.C.; Bauer, E.; Kwong, E.H.; Wasan, K.M. The effect of serum albumin on amphotericin B aggregate structure and activity. *Pharm. Res.* **2001**, *18*, 1305–1309. [CrossRef]
48. Alvarez, C.; Shin, D.H.; Kwon, G.S. Reformulation of Fungizone by PEG-DSPE Micelles: Deaggregation and Detoxification of Amphotericin B. *Pharm. Res.* **2016**, *33*, 2098–2106. [CrossRef]
49. Adhikari, K.; Buatong, W.; Thawithong, E.; Suwandecha, T.; Srichana, T. Factors Affecting Enhanced Permeation of Amphotericin B Across Cell Membranes and Safety of Formulation. *AAPS PharmSciTech.* **2016**, *17*, 820–828. [CrossRef]
50. Fischer, D.; Li, Y.; Ahlemeyer, B.; Krieglstein, J.; Kissel, T. In vitro cytotoxicity testing of polycations: Influence of polymer structure on cell viability and hemolysis. *Biomaterials* **2003**, *24*, 1121–1131. [CrossRef]

polymers

Article

Biopolymers as Seed-Coating Agent to Enhance Microbially Induced Tolerance of Barley to Phytopathogens

Aizhamal Usmanova [1], Yelena Brazhnikova [1,2,*], Anel Omirbekova [1,2], Aida Kistaubayeva [1], Irina Savitskaya [1] and Lyudmila Ignatova [1,2]

1 Faculty of Biology and Biotechnology, Al-Farabi Kazakh National University, Almaty 050038, Kazakhstan; aizhamalduszhanovna@mail.ru (A.U.)
2 Scientific Research Institute of Biology and Biotechnology Problems, Al-Farabi Kazakh National University, Almaty 050038, Kazakhstan
* Correspondence: polb_4@mail.ru

Abstract: Infections of agricultural crops caused by pathogen ic fungi are among the most widespread and harmful, as they not only reduce the quantity of the harvest but also significantly deteriorate its quality. This study aims to develop unique seed-coating formulations incorporating biopolymers (polyhydroxyalkanoate and pullulan) and beneficial microorganisms for plant protection against phytopathogens. A microbial association of biocompatible endophytic bacteria has been created, including *Pseudomonas flavescens* D5, *Bacillus aerophilus* A2, *Serratia proteamaculans* B5, and *Pseudomonas putida* D7. These strains exhibited agronomically valuable properties: synthesis of the phytohormone IAA (from 45.2 to 69.2 μg mL^{-1}), antagonistic activity against *Fusarium oxysporum* and *Fusarium solani* (growth inhibition zones from 1.8 to 3.0 cm), halotolerance (5–15% NaCl), and PHA production (2.77–4.54 g L^{-1}). A pullulan synthesized by *Aureobasidium pullulans* C7 showed a low viscosity rate (from 395 Pa·s to 598 Pa·s) depending on the concentration of polysaccharide solutions. Therefore, at 8.0%, *w*/*v* concentration, viscosity virtually remained unchanged with increasing shear rate, indicating that it exhibits Newtonian flow behavior. The effectiveness of various antifungal seed coating formulations has been demonstrated to enhance the tolerance of barley plants to phytopathogens.

Keywords: polyhydroxyalkanoate; polysaccharides; dynamic viscosity; seed coating; beneficial microorganisms; biocontrol

Citation: Usmanova, A.; Brazhnikova, Y.; Omirbekova, A.; Kistaubayeva, A.; Savitskaya, I.; Ignatova, L. Biopolymers as Seed-Coating Agent to Enhance Microbially Induced Tolerance of Barley to Phytopathogens. *Polymers* **2024**, *16*, 376. https://doi.org/10.3390/polym16030376

Academic Editors: Masoud Ghaani and Stefano Farris

Received: 28 November 2023
Revised: 24 January 2024
Accepted: 26 January 2024
Published: 30 January 2024

1. Introduction

According to the estimates of the Food and Agriculture Organization (FAO), 20–40% of global crop losses are related to plant diseases, with 42% of those attributed to infections caused by pathogenic fungi. *Fusarium* fungi are common pathogens of cereals, including barley, and can cause diseases such as Fusarium head blight, seedling blight, root rot, and Fusarium crown rot throughout their life cycle. In addition, many species of *Fusarium* are capable of producing mycotoxins (deoxynivalenol, nivalenol, HT2/T2, zearalenone), even in some cases in the absence of severe disease symptoms [1].

The use of microorganisms and their metabolites as bio-control agents is one of the most promising methods for the effective and safe protection of plants. The widespread use of antibiotics in the food industry, agriculture, and medicine leads to an increase in antibiotic resistance of pathogenic microorganisms. In this regard, endophytes have advantages over other biocontrol agents, as they are producers of many biologically active metabolites, such as phenolic acids, alkaloids, quinones, steroids, saponins, tannins, and terpenoids. Microbiological strategies for protecting agricultural crops are based on the plant growth-promoting properties of these strains.

Biopolymers can also be used in the development of plant protection products against phytopathogens. They are non-toxic and biodegradable and can be obtained from renewable sources, making them suitable for use in organic farming. Additionally, they can interact with many hydrophobic and hydrophilic compounds in more complex formulations. Biopolymers play a protective role for plants against pathogenic fungi through several mechanisms [2]. Polymers can directly interact with fungi, suppressing spore germination and mycelium growth, as demonstrated, for example, with chitosan [3,4]. They can act as effective elicitors, inducing the plant immune system to fight pathogens [5]. They can also be used as carriers for active ingredients with controlled release [2].

Among biopolymers, the most used are carboxymethyl cellulose, chitosan, xanthan gum, gum arabic, polyvinyl alcohol, starch, gelatin, polyacrylamide, and alginates. These polymers are used for treating the seeds of turnips, tomatoes, chickpeas, corn, beans, eggplants, okra, chili peppers, guar, pumpkins, cucumbers, lupine, clover, soybeans, and wheat [6–8].

One of the promising microbial polymers for seed coating is pullulan. It is a water-soluble, low-viscosity polysaccharide that has the property of biodegrading under the action of microorganisms. Pullulan has oxygen barrier properties, excellent moisture retention, and also prevents the growth of pathogens. Additionally, pullulan is a prebiotic—a substance that stimulates the growth and development of microorganisms [9,10].

Polyhydroxyalkanoates (PHAs) are of great interest as well, as they are non-toxic, biodegradable, and biocompatible polymers [11]. Previous studies have reported the use of PHA with the addition of polycaprolactone to obtain biodegradable films for rice seed germination [12]. Furthermore, some PHAs exhibit antagonistic activity against bacteria [13,14]. In our previous studies, we demonstrated that PHA produced by the strain *Pseudomonas fluorescens* D5 has pronounced antifungal activity against *Fusarium graminearum*, *Fusarium solani*, *Fusarium oxysporum* [15], and *Penicillium expansum* [16].

However, there is no information about the use of a mixture of PHA with pullulan in the composition of seed coatings. Therefore, this work is aimed at developing unique compositions for seed treatment, including effective microorganisms and biopolymers (PHA and pullulan) as seed coating agents to enhance the microbially induced tolerance of plants to phytopathogenic fungi.

The main objectives of the present study are as follows: (1) formation of a microbial association with agronomically valuable properties, (2) investigation of the rheological properties of polysaccharide solutions, and (3) study of various seed coating types for barley tolerance to *Fusarium*.

This study is significant for a better understanding of the effect of seed coating on the microbially induced tolerance of barley to phytopathogens. In this research, a new opportunity is proposed for the use of pullulan as a seed coating agent, expanding the areas of application for microbial polymers. The excellent gelling and thickening properties, as well as the biodegradability and non-toxicity of the investigated biopolymers, pullulan, and PHA, make them promising for use in antifungal formulations for seed treatments.

2. Materials and Methods

The following strains were used in the present study:

Bacillus aerophilus A2 (accession number OQ569360) isolated from leaves of peppermint (*Mentha piperita*);
Pseudomonas flavescens D5 (accession number OP642636) isolated from flowers of common chicory (*Cichórium intybus*);
Serratia proteamaculans B5 (accession number OR858823) isolated from the leaves of Iris;
Bacillus simplex B9 (accession number OR864231) isolated from the roots of wormwood (*Artemisia absinthium*);
Pseudomonas putida D7 (accession number OR863903) isolated from the roots of Echinacea (*Echinacea purpurea*);
Aureobasidium pullulans C7 (accession number OR864236) isolated from dark chestnut soil;
Bacillus thuringiensis C8 (accession number OR858828) isolated from the surface of apples.

2.1. Production of IAA

To determine the amount of indole-3-acetic acid (IAA) produced by microorganisms, a colorimetric method was employed. Isolates were cultivated in nutrient broth for 48 h at 28 °C. After incubation, the culture was centrifuged at 6000× *g* for 20 min. The supernatant, with a volume of 1 mL, was mixed with 2 mL of Salkowski reagent. The optical density was measured at 530 nm. The concentration of IAA was expressed in μg mL^{-1} [17].

2.2. The Antifungal Properties of Microorganisms

The antifungal activity was determined using the agar disk diffusion method. Bacteria were cultured for 48 h in nutrient broth at 28 °C with aeration. Bacterial cultures were spread on the surface of nutrient agar as a continuous lawn in a volume of 100 μL, and after 48 h of growth, 5 mm diameter disks were cut. Disks with bacterial culture were placed on Petri dishes previously inoculated with a continuous lawn of phytopathogenic test cultures (*Fusarium solani*, *Fusarium oxysporum*) at a concentration of 10^6 spores mL^{-1}. A nutrient agar disk served as a control. The Petri dishes were incubated at 28 °C for 72 h. The antifungal activity was assessed by measuring the diameter of the growth inhibition zone of the tested phytopathogens [18].

2.3. Determination of Microbial Halotolerance

To assess the halotolerance of bacteria, nutrient agar medium supplemented with NaCl at concentrations of 5%, 10%, 15%, and 25% was used. Microorganisms were inoculated using the streak method. Strains capable of cultivation at different salt concentrations were selected based on the research results.

2.4. PHA Production Assay

Strains producing PHA were cultivated in liquid MSM medium at 28 °C for 48 h at 150 rpm. The medium composition (g·L^{-1}) was as follows: $MgSO_4 \cdot 7H_2O$—0.1; KH_2PO_4—0.68; K_2HPO_4—1.73; NaCl—4.0; NH_4NO_3—1.0; $FeSO_4 \cdot 7H_2O$—0.03; $CaCl_2 \cdot 2H_2O$—0.02; and glucose—5.0 [19]. Subsequently, the suspension was centrifuged at 6000× *g* for 10 min, the supernatant was decanted, and PHA was extracted from the residue. Sodium hypochlorite and hot chloroform were added to the residue at a 1:1 ratio, and the mixture was kept at 30 °C for 1 h. The suspension was then centrifuged at 6000× *g* for 15 min, and the upper and middle layers were removed. The residue was precipitated with a 1:1 mixture of ethanol and acetone, dried at 35 °C, and weighed [20].

2.5. Microbe–Microbe In Vitro Compatibility Test

Five bacterial strains (*Pseudomonas flavescens* D5, *Bacillus aerophilus* A2, *Serratia myotis* B5, *Bacillus simplex* B9, and *Pseudomonas putida* D7) were used for in vitro compatibility test. The agar diffusion method was selected to determine biocompatibility. Bacterial strains were separately grown on nutrient agar at 28 °C for 24 h. Then, colonies were transferred to nutrient broth and incubated overnight at 28 °C at 160 rpm [21].

A volume of 100 μL of the test microorganism (0.5 McFarland) was spread on the surface of nutrient agar. Sterile filter paper disks (d = 5 mm) were inoculated with the overnight bacterial culture adjusted to a concentration of 0.5 McFarland. Inoculated disks were placed on Petri dishes (4 disks per each) with the test microorganism, and each was incubated in darkness at 28 °C for 4 days. Experiments were conducted with three replicates.

2.6. Extraction of Polysaccharide

For the extraction of polysaccharides, the 4-day-old culture of *A pullulans* C7 and 3-day-old culture of *B. thuringiensis* C8 were centrifuged for 15 min at 10,000× *g*. The fungal polysaccharide was precipitated with a double volume of 96% ethanol and the bacterial polysaccharide with a triple volume of alcohol. The yield coefficient for biomass (P/X) was calculated as a ratio of production of EPS to the dry biomass and expressed in percent. The yield coefficient for substrate (P/S) was calculated as a ratio of production of EPS to the utilized glucose and expressed in percent [22].

2.7. Measurement of Dynamic Viscosity

The dynamic viscosity of the polymer's solution with different concentrations—2, 4, 6, 8, 10, and 12% *w*/*v*—was performed using a rotational Ametek Brookfield DVPlus viscometer with a ULA spindle, at different shear rates, ranging from 0.1 to 500 s^{-1} at 25 °C [23]. The viscosimetric analyses of the samples were performed at 25 °C.

2.8. Development of Various Options for Processing Barley Seeds

Various antifungal formulations for seed treatments were developed, including (1) bacterial strain suspension, (2) polymer mixture, and (3) bacterial strain suspension + polymer mixture.

Bacterial strains were separately cultured in nutrient broth for 48 h at 180 rpm and 28 °C. The cultures were then centrifuged at 6000× *g* for 10 min and resuspended in a phosphate-buffered saline (PBS (g L^{-1}), 8.0 NaCl, 0.2 KCl, 1.44 Na_2HPO_4, and 0.24 KH_2PO_4). The optical density of each bacterial suspension was adjusted to 10^8 CFU mL^{-1} Strain suspensions were mixed in equal proportions.

The polymer mixture was prepared using PHA at a concentration of 0.05% and pullulan at a concentration of 2% (wt./vol.), incorporated into phosphate-buffered saline.

For coating of seeds simultaneously in a bacterial suspension and a polymer blend, PHA at a concentration of 0.05% and pullulan at a concentration of 2% (wt./vol.) were introduced into a mixture of bacterial suspensions.

2.9. Pot Experiments

To conduct the research, barley seeds sterilized in a 5% sodium hypochlorite solution were used. Subsequently, the seeds were rinsed with sterile water and sown on nutrient agar medium [22] to ensure the absence of bacteria on the seed surface.

Phytopathogenic load conditions were simulated by introducing a suspension of the phytopathogenic fungus *F. oxysporum* into the soil at a titer of 10^8 spores mL^{-1}, with 2 mL of the suspension per 100 g of soil.

Experiment options:

T1—Untreated seeds;
T2—Untreated seeds + *Fusarium oxysporum;*
T3—Seed treatment with bacterial suspension + *Fusarium oxysporum;*
T4—Seed coating with polymeric mixture + *Fusarium oxysporum;*
T5—Simultaneous seed coating with bacterial suspension and polymeric mixture of *Fusarium oxysporum.*

Pre-sterilized seeds were immersed in various antifungal formulations, followed by transferring the seeds to 0.1 M $CaCl_2$. After coating, the seeds were dried for 20 min before planting.

In each pot containing 300 g of sterile soil, 10 barley seeds were planted. The experiment was conducted under sterile conditions with three replicates. The plants were grown for 12 days.

2.10. Determination of Free Proline Concentration

The content of free proline was determined using a non-heated acidic ninhydrin reagent prepared as follows: (1.25 g ninhydrin + 30 mL glacial acetic acid + 20 mL 6 M H_3PO_4). A portion of fresh plant tissue from a leaf blade (200 mg) was homogenized in 10 mL of a 3% aqueous solution of sulfosalicylic acid and left for 1 h in a water bath at 100 °C.

Subsequently, 1.5 mL of glacial acetic acid, 1.5 mL of ninhydrin reagent, and 1.5 mL of the prepared extract were poured into a clean test tube. The samples were incubated for 1 h in a water bath at 100 °C and then rapidly cooled to room temperature. After artificial cooling (using cold water or ice), the optical density of the reaction products was measured at a wavelength of 520 nm using a spectrophotometer. Proline content values were calculated using a calibration curve, constructed using chemically pure proline [24].

2.11. Determination of Chlorophyll Concentration

To obtain an ethanolic extract, 2 g of leaves were sliced and thoroughly ground in a mortar, gradually adding 96% ethanol in small portions (a total of 10 mL). The extract was centrifuged for 15 min at 6000× *g* [25]. Photocolorimetry was carried out using a spectrophotometer at wavelengths of 665 and 649 nm in a cuvette with an optical path length of 1 cm. The comparison cuvette was filled with 96% ethanol. The pigment concentration was determined using the following formula:

$$Chl_a = 13.95 \times A_{665} - 6.88 \times A_{649}$$

$$Chl_b = 24.96 \times A_{649} - 7.32 \times A_{665}$$

2.12. Preparation of the Extract for the Determination of Antioxidant Enzymes

Antioxidant enzyme activity was determined spectrophotometrically based on the rate of NADH oxidation using the method [26]. For this purpose, plant material (1.5–2 g) was homogenized with an extracting medium containing 50 mM K-phosphate buffer (pH 7.5), 1 mM EDTA, 0.3%, 1 mM ascorbic acid, filtered and centrifuged (15 min, 8000× *g*). The obtained supernatant was used to determine the activity of the enzymes.

2.12.1. Investigation of Catalase Activity

Catalase activity was determined using H_2O_2 according to the method [26]. The reaction mixture consisted of 15 mM H_2O_2, 100 mM K-phosphate buffer (pH 7.0), and 0.1 mL of the sample. Changes in optical density were measured at 240 nm, and activity was calculated using the extinction coefficient ε = 0.03 mM^{-1} cm^{-1}. All experiments were conducted in triplicate and expressed in units per milligram of protein.

2.12.2. Investigation of Ascorbate Peroxidase Activity

The activity of ascorbate peroxidase was determined in a medium with the following composition: 50 mM K-phosphate buffer pH 7.0, 0.5 mM ascorbate, and 0.2 mM H_2O_2. The reaction was initiated by adding 0.1 mL of the sample [27]. Changes in optical density were measured at 290 nm. Enzyme activity was calculated using the extinction coefficient ε = 2.8 mM^{-1} cm^{-1} and expressed as 1 mmol of ascorbate min^{-1} per mg protein.

2.12.3. Investigation of Guaiacol Peroxidase Activity

Guaiacol peroxidase activity was determined using a spectrophotometric method, considering absorption due to guaiacol oxidation [28]. The reaction mixture consisted of 50 mM phosphate buffer (pH 7), 9 mM guaiacol, 10 mM H_2O_2, and 0.2 mL of the sample. Optical density was measured at 470 nm for 1 min, and enzyme activity was calculated using the extinction coefficient ε = 26.6 mM^{-1} cm^{-1}, expressed as 1 mmol of ascorbate min^{-1} per mg protein.

2.13. Statistical Analysis

All the data are presented as the mean ± standard deviation (SD) of three replicates. The data were processed by the standard methods of one-way analysis of variance (ANOVA) using the software Statistica version 10.0 (TIBCO Software Inc., Palo Alto, CA, USA). Tukey's honestly significant difference (HSD) test ($p < 0.05$) was performed for multiple comparisons to estimate significant differences between means.

3. Results

3.1. Characterization of the Biological Activity of Endophytic Bacteria

For the application of microorganisms both to enhance plant growth and to protect them from adverse factors, a crucial step is the selection of strains possessing a set of beneficial properties. In the first stage of the research, the agronomically valuable properties of five endophytic bacterial strains were investigated (Table 1).

Table 1. Plant growth-promoting properties of bacterial strains.

Strains	Properties					
	IAA, μg mL^{-1}	Zone of Inhibition of Phytopathogen Growth, cm		Halotolerance		PHA Production, g L^{-1}
		Fusarium solani	*Fusarium oxysporum*	5% NaCl	15% NaCl	
Pseudomonas flavescens D5	45.2 ± 2.1 a	-	3.0 ± 0.1 b	+		2.77 ± 0.07 a
Bacillus aerophillus A2	52.4 ± 2.1 b	-	-	+		4.54 ± 0.08 b
Serratia proteamaculans B5	62.7 ± 2.1 c	2.6 ± 0.1 b	-	+		
Bacillus simplex B9	-	2.1 ± 0.05 a	1.8 ± 0.05 a	+		
Pseudomonas putida D7	69.2 ± 3.1 c	-	-	+	+	

Values are given as the mean ± SD. Values represented by the same letter are not significantly different according to the Tukey test ($p \leq 0.05$).

One of the well-known mechanisms for improving and regulating plant growth by microorganisms is their ability to synthesize various phytohormones. The stimulation of plant growth resulting from the application of microorganisms is predominantly associated with their ability to synthesize auxins, primarily IAA [29]. All examined bacteria demonstrated the ability to produce IAA (Table 1), except for the *Bacillus simplex* B9 strain. The highest concentration of IAA was found in the *Pseudomonas putida* D7 strain (Table 1). The amount of produced IAA varied between 45.2 and 69.2 μg mL^{-1} depending on the strain, which is similar to or significantly higher than that observed in other endophytic bacterial strains [30,31].

The next criterion for assessing the biological activity of the strains was the evaluation of their resistance to adverse environmental factors.

Among the adverse factors of biotic nature, phytopathogenic microflora plays a key role. Infections of agricultural crops caused by pathogenic fungi are among the most widespread and harmful, as they not only reduce the quantity of the harvest but also significantly degrade its quality due to the accumulation of mycotoxins [32]. One of the positive effects of bacteria on crops is their ability to protect plants from phytopathogens through direct and indirect mechanisms [33].

The study of the antagonistic activity of bacterial strains against *Fusarium solani* and *Fusarium oxysporum* showed that three out of five strains inhibit the growth of phytopathogens (Figure 1). The zones of growth suppression ranged from 1.8 to 3.0 cm (Table 1).

Salinization of soils is one of the most crucial abiotic stress factors that negatively impact plant privity [34]. The application of salt-tolerant growth-promoting bacteria may contribute to stress alleviation and enhance the resilience of crops grown in saline soils [34].

In the study of halotolerance, it was shown that all strains were resistant to a salt concentration of 5%, and one strain, *Pseudomonas putida* D7, demonstrated the ability to grow in a medium with 15% NaCl (Table 1, Figure 2). According to the classification, the investigated strains are moderately halophilic, exhibiting optimal growth at NaCl concentrations ranging from 3% to 15% (~0.5–2.7 M). Halophilic bacteria have several advantages compared to other microorganisms, as they possess high metabolic activity, allowing them to grow in extreme conditions and produce a variety of valuable biologically active compounds, including those with antimicrobial properties [35].

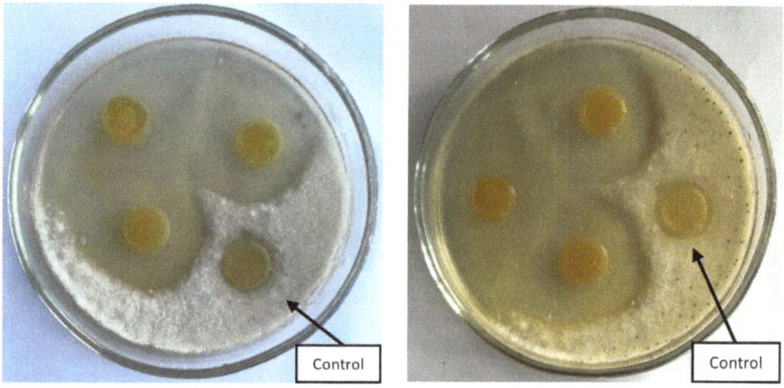

(**a**) (**b**)

Figure 1. Inhibition of the growth of the phytopathogenic fungus *Fusarium solani* by strains (**a**) *Bacillus simplex* B9 and (**b**) *Serratia proteamaculans* B5.

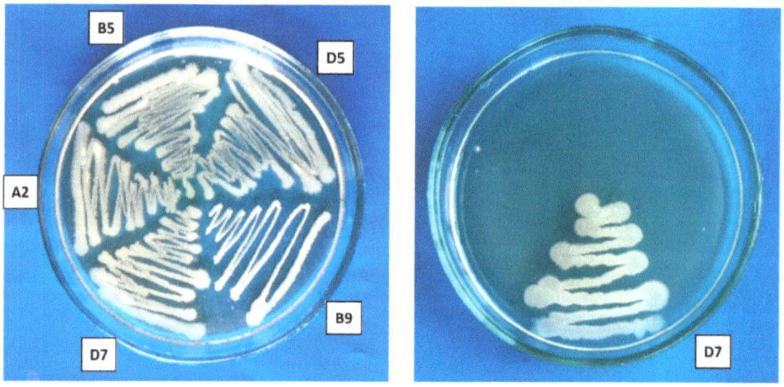

(**a**) (**b**)

Figure 2. Growth of bacterial strains on medium containing (**a**) 5% NaCl and (**b**) 15% NaCl. D5—*Pseudomonas flavescens* D5, A2—*Bacillus aerophillus* A2, B5—*Serratia proteamaculans* B5, B9—*Bacillus simplex* B9, D7—*Pseudomonas putida* D7.

PHA is a class of polyesters of various hydroxyalkanoic acids, which are synthesized by many Gram-positive and Gram-negative bacteria and accumulate intracellularly [33]. In the present study, the strains *Ps. flavescens* D5 and *B. aerophillus* A2 demonstrated the ability to produce PHA (Table 1).

3.2. Biocompatibility Assessment of Strains

Currently, the advantages of preparations based on microbial consortia over monocultures are convincingly confirmed, as the biotechnological potential of microorganisms in such preparations is more fully realized. There are several advantages of multi-component preparations: multiplicity of action, synergistic effect, increased stability and adaptability to different agro-climatic conditions, the ability to utilize inhomogeneous substrates in composition, and more complete utilization of the functional capabilities of microorganisms [36–38].

In the development of multi-strain inoculants, it is crucial to consider the type of relationships between microorganisms and the possibility of their combination. Therefore, the next stage of the research was the in vitro testing of the selected strains for compatibility during their co-cultivation on a solid nutrient medium (Table 2).

Table 2. Pairwise compatibility among bacterial strains.

Strain	*Pseudomonas flavescens* **D5**	*Bacillus aerophillus* **A2**	*Serratia proteamaculans* **B5**	*Bacillus simplex* **B9**	*Pseudomonas putida* **D7**
Pseudomonas flavescens D5					
Bacillus aerophillus A2	+				
Serratia proteamaculans B5	+	+			
Bacillus simplex B9	–	–	+		
Pseudomonas putida D7	+	+	+	–	

«+»—compatible; «–»—incompatible.

It was shown that during co-cultivation, four out of five strains did not suppress the growth and development of each other (Table 2), indicating their compatibility and the possibility of including them in the composition of a multi-strain inoculant. The identified biocompatibility of the studied strains indicates the absence of competition between them and insensitivity to the produced extracellular metabolites with antagonistic properties. An exception was the *B.simplex* B9 strain, which demonstrated pronounced incompatibility with most of the investigated bacterial strains (Table 2). Thus, for seed treatment in subsequent experiments, four out of five strains that showed compatibility were used.

3.3. Biosynthesis of Microbial Exopolysaccharides and Their Rheological Properties

In addition to plant-beneficial microorganisms, such ingredients of seed coating as binders that help to release a suitable amount of plant-beneficial microorganisms in physiologic conditions and ensure the adherence and cohesion of the material on the seed surface and keep the ingredients active are used [37,39,40]. The microbial polymer solution should be water-soluble with a low viscosity for complete atomization of the liquid onto seeds [40].

Earlier, we isolated strains *Aureobasidium pullulans* C7 [41] and *Bacillus thuringiensis* C8, which showed the ability to biosynthesize exopolysaccharide (EPS). The *A. pullulans* C7 strain synthesized 12.53 ± 0.48 g L^{-1} exoglycan on the 4th day of fermentation, and the yield coefficient for biomass was 349.02% (Table 3). This indicates that in this medium the substrate is utilized to a greater extent for the formation of EPS than for the formation of cell mass. The amount of polysaccharide accumulated by the studied strain is comparable with the data of other researchers [42,43].

Table 3. Production of exopolysaccharides by strains *A. pullulans* C7 and *B. thuringiensis* C8 in presence of glucose.

Strain	The Dry Weight of Cells, g L^{-1} (X)	Production of EPS, g L^{-1} (P)	Utilized Glucose, g L^{-1} (S)	The yield Coefficient for Biomass P/X, %	The yield Coefficient for Substrate P/S, %
A. pullulans C7	3.59 ± 0.13	12.53 ± 0.48	17.12 ± 0.81	349.02	73.19
B. thuringiensis C8	1.86 ± 0.06	3.97 ± 0.15	10.15 ± 0.61	213.44	39.11

Values are given as the mean ± SD.

The strain *B. thuringiensis* C8 produced 3.97 g L^{-1} of exoglycan (Table 3). The yield coefficient for bacterial biomass indicates the potential of this strain as a producer of EPS. The ability of strains of the genus *Bacillus*, including *B. thuringiensis*, to produce EPS is confirmed in the works of other researchers [44,45].

Further, measurements of dynamic viscosity were made for polymer solutions obtained by cultivation of *A. pullulans* C7 and *B.thuringiensis* C8.

The dynamic viscosity of each concentration solution produced by *B. thuringiensis* C8 obviously decreased with the increase in shear rate (Figure 3), showing a shear-thinning behavior, which means that this kind of polysaccharide solution belongs to non-Newtonian fluid (or pseudoplastic flow behavior). Solutions of polymers in water and at the same concentrations can sometimes have oppositional behaviors, i.e., Newtonian or non-Newtonian fluids, depending on their structures, molecular weights, and polymer microbial producers [46,47].

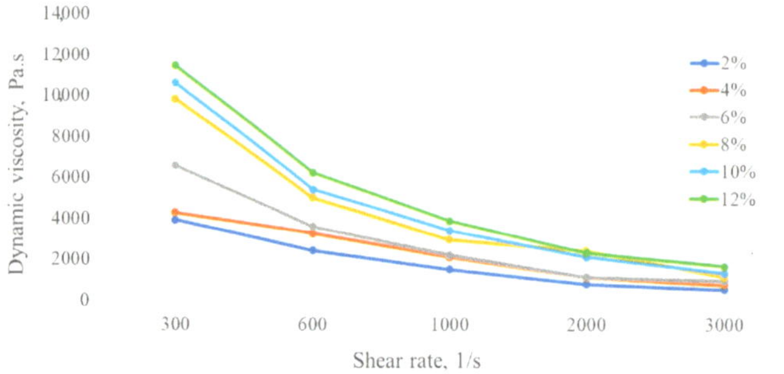

Figure 3. Viscosity of exopolysaccharide solutions produced by *B. thuringiensis* C8 in different concentrations.

It was also noted that the viscosity increased with increasing polymer solution concentration; however, at lower concentrations, the rheological measurements became erratic. The shear-thinning phenomenon could be due to the rate of formation of new entanglements lower than the externally imposed disruption rate with an increase in shear rate. Another distinguishing feature is that the microbial solution showed comparatively high viscosity rates at all dilute concentrations.

The measurement of the dynamic viscosity of the polymer solution obtained by the cultivation of *A. pullulans* C7 showed the dependence on dynamic viscosity at the shear rate at different concentrations ranging from 2% to 12% (w/v) at 25 °C (Figure 4). The shear rate increased with increasing polymer concentration, thereby demonstrating that viscosity strongly depends on concentration.

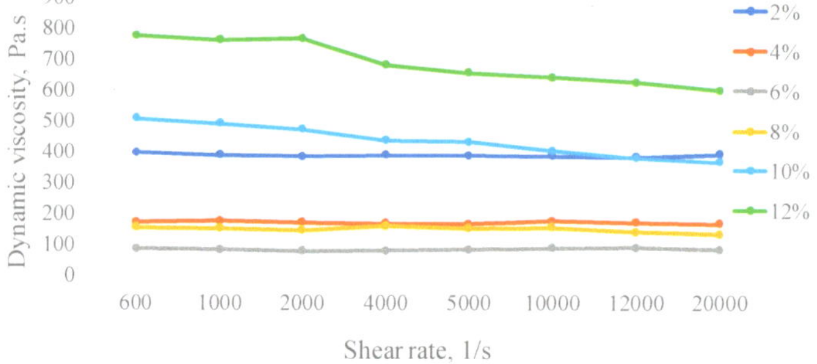

Figure 4. Viscosity of pullulan solutions produced by *A. pullulans* C7 in different concentrations.

At lower concentrations (<8.0%, w/v), viscosity virtually remained unchanged with increasing shear rate, thus suggesting that the pullulan aqueous solution exhibits Newtonian flow behavior. It is known that Newtonian fluid viscosity is constant no matter the shear rate or applied shear stress experienced by the fluid. However, with increasing concentration to 8% w/v, the flow behavior was changed to pseudoplastic. Such flow behavior of pullulan can happen because of the separation of exopolysaccharides from each other or the alignment of them with the shear field and thereby a decrease in viscosity up to an approximately constant value [48]. Also, it is known that the viscosity is dependent on the structure and concentration of the polymer, its molecular weight and distribution, the conformation of macromolecules in the solution and its interaction with solvents, the type of intermolecular and intramolecular aggregation, and the flexibility of the chains with temperature. A similar change in flow behavior with increasing concentration was reported for pullulan and other polysaccharides [49,50].

The data obtained allow us to suggest the microbial polymer, pullulan, produced by *A. pullulans* C7 as a potential seed coating binder due to rheological characteristics. It is already known as an excellent film former and is functional for a variety of applications, including for use as an adhesive, binder, and thickener to modify or maintain the texture of food.

It has also been reported that pullulan has considerable mechanical strength and other functional properties such as adhesiveness, film and fiber formability, and enzymatically mediated degradability [51]. High flexibility and a lack of crystallinity provide pullulan with the capacity to form thin layers, electrospun nanofibers, nanoparticles, flexible coatings, stand-alone films, and three-dimensional objects [51,52]. Due to its peculiar characteristics, pullulan is extensively used in different sectors, the three main realms of application pertaining to the pharmaceutical, biomedical, and food fields.

3.4. The Use of Various Antifungal Formulations for Seed Treatments in Pot Experiments

Seed coating is a method that involves applying exogenous materials to the surface of seeds to enhance their properties and/or deliver active components (such as plant growth regulators, nutrients, and microbial inoculants). This process can protect seeds from phytopathogens, increase germination rates, improve plant resistance to stress factors, and enhance overall plant growth [37,52–54].

In the next stage of the research, various antifungal formulations for seed coating were developed, and their impact on barley growth under phytopathogenic conditions was assessed (Figure 5).

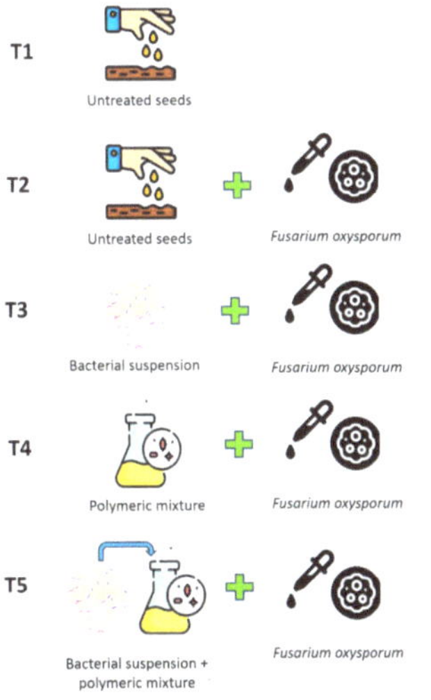

Figure 5. Scheme of pot experiments.

As active components, a microbial inoculant consisting of a suspension of four compatible strains was used: *Ps. flavescens* D5, *B. aerophilus* A2, *S. proteamaculans* B5, and *Ps. putida* D7. As polymer components, PHA produced by the strain *Ps. flavescens* D5, and pullulan, produced by the yeast strain *A. pullulans* C7, were used. PHA was included in the mixture due to its antifungal properties, as previously identified in earlier studies [15,16].

Uniform seed emergence and early crop development are crucial aspects for achieving high crop yields. Seed coating is an effective method that improves seed-sowing qualities and activates the internal resources of the seed material [52]. In the conducted research, pre-sowing seed treatment, in most cases, enhanced their germination energy and germination capacity. The greatest effect was observed when applying a bacterial suspension in combination with a polymer mixture, where germination energy and germination capacity reached 95% and 97%, respectively (Figure 6).

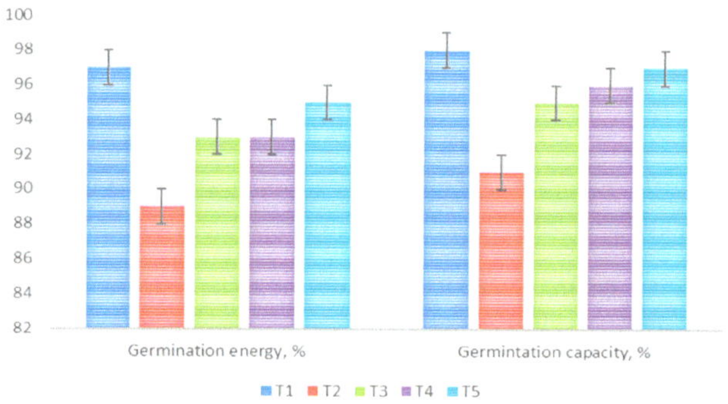

Figure 6. Influence of various pre-sowing treatments on barley seed germination energy and germination capacity.

The pre-sowing treatment of seeds demonstrated a pronounced growth-stimulating effect on barley plants, as evidenced by a significant ($p < 0.05$) increase in morphometric indicators (Table 4).

Table 4. Influence of various pre-sowing seed treatments on growth parameters of barley.

Treatment Variants	Dry Mass of stem, g	Dry Mass of Root, g	Length of Stem, cm	Length of Root, cm
T1	1.2 ± 0.03 d	0.9 ± 0.04 a	22.0 ± 0.9 c	11.5 ± 0.5 b
T2	0.6 ± 0.02 a	0.8 ± 0.03 a	15.0 ± 0.7 a	8.5 ± 0.3 a
T3	1.1 ± 0.04 c	0.9 ± 0.04 a	22.5 ± 0.9 c	12.5 ± 0.2 c
T4	0.9 ± 0.03 b	1.3 ± 0.02 c	20.8 ± 0.8 b	11.0 ± 0.5 b
T5	1.0 ± 0.03 c	1.2 ± 0.03 b	23.8 ± 0.9 c	13.6 ± 0.5 d

Values are given as the mean ± SD. Values represented by the same letter are not significantly different according to the Tukey test ($p \leq 0.05$).

The barley's response varied depending on the type of treatment. Root length is a crucial morphometric indicator as roots are in contact with soil and soil microflora, absorbing water with mineral compounds. The greatest root elongation was observed in variant T5 with the application of a bacterial suspension and a polymer mixture (1.6 times), followed by treatment T3, where root elongation was noted at 1.5 times. Stem length is also a significant characteristic when assessing the plant's response to different pre-sowing seed treatments. Treated variants showed an increase in stem length by 20–53%. The greatest increase in stem length was observed in variants T3 and T5 (Figure 7). It is shown that the stem mass of treated plants was more than 1.5–1.8 times greater, and root mass was 1.1–1.6 times greater compared to the untreated control (Table 4).

In the conducted research, the observed stimulating effect on the growth parameters of barley can be attributed to several reasons. One of the mechanisms of the positive influence on plants is the ability of the strains included in the composition to produce the phytohormone IAA, which regulates cell division and elongation, their proliferation and differentiation, as well as the development of vascular tissues and apical dominance [55]. Another mechanism for improving morphometric plant parameters under conditions of phytopathogenic stress is the biocontrol properties of strains and the protective role of biopolymers.

(a) (b) (c)

Figure 7. Barley growth under phytopathogenic stress with different seed treatment variants: (**a**) T2—untreated, (**b**) T3—treatment with bacterial suspension, (**c**) T5—seed coating in polymer mixture with bacterial suspension.

The state of the photosynthetic apparatus is an indicator of the physiological condition of plants. One of the primary characteristics of photosynthetic activity is the content of chlorophyll pigments [56]. In previous studies, fluorescence visualization analysis of chlorophyll was applied to assess the condition of the plant photosynthetic system under the influence of biotic [56–58] and abiotic [59–61] stress. High chlorophyll content may indicate potentially high agricultural productivity [56].

In the present study, under conditions of biotic stress induced by the phytopathogen *F. oxysporum*, the total chlorophyll content in barley leaves decreased by 2.7 times compared to the indicator for plants grown under normal conditions, reaching 1.03 ± 0.03 mg g^{-1} (Table 5). This likely indicates changes in the pigment–protein complexes of light-harvesting antennae and reaction centers of photosystems. Seed treatment had a positive effect on the photosynthetic activity of barley under phytopathogenic stress. This positive effect was to increase the content of chlorophyll *a* in leaves by 1.4–2.1 times, chlorophyll *b* by 2–2.4 times, and the total content of chlorophyll (a + b) by 1.6–2.2 times. The maximum effect was achieved in variant T5 with the application of a bacterial suspension and a polymer mixture (Table 5). The observed differences in pigment content may be associated with the production of certain compounds by the studied bacteria, influencing the biosynthesis and/or degradation processes of chlorophylls, as well as creating more favorable growth conditions for plants under stress.

Table 5. Influence of different pre-sowing seed treatments on proline and chlorophyll content in barley.

Treatment Variant	Proline Content, mg g^{-1}	Chlorophyll a Content, mg g^{-1}	Chlorophyll b Content, mg g^{-1}	Total Chlorophyll Content (a + b), mg g^{-1}
T1	0.94 ± 0.03 a	1.89 ± 0.07 e	0.94 ± 0.02 e	2.83 ± 0.1 e
T2	1.70 ± 0.07 e	0.69 ± 0.02 a	0.34 ± 0.01 a	1.03 ± 0.04 a
T3	1.4 ± 0.03 d	1.23 ± 0.04 c	0.57 ± 0.02 b	1.8 ± 0.07 b
T4	1.22 ± 0.04 c	0.94 ± 0.03 b	0.68 ± 0.03 c	1.62 ± 0.05 c
T5	1.1 ± 0.03 b	1.42 ± 0.03 d	0.81 ± 0.04 d	2.23 ± 0.07 d

Values are given as the mean ± SD. Values represented by the same letter are not significantly different according to the Tukey test ($p \leq 0.05$).

It is known that stress factors lead to a disruption in the balance between the generation of reactive oxygen species (ROS) and their neutralization. Among the essential mechanisms of plant tolerance mediated by bacteria is the involvement of these microorganisms in detoxifying ROS through the modulation of the natural antioxidant defense systems of plants—both non-enzymatic (proline, ascorbic acid, glutathione, cysteine, flavonoids, carotenoids, and tocopherol) and enzymatic (superoxide dismutase, peroxidase, catalase, ascorbate peroxidase, guaiacol peroxidase, and glutathione reductase), all components of which are in complex functional interaction [62,63].

The increase in proline content is one of the characteristic responses of plants to various types of stress, including biotic stress, providing the first stage of plant adaptation. Proline serves multiple functions, including the regulation of cytosolic acidity, minimization of lipid peroxidation by scavenging free radicals, and stabilization of subcellular components and structures (proteins and membranes) [64]. A higher level of proline in barley leaves was observed when plants were grown in soil with an elevated infectious background compared to untreated plants in sterile soil (Table 5). In the untreated variant under phytopathogenic stress, the proline concentration was 1.7 mg g^{-1}, exceeding this indicator in plants grown in favorable conditions by 1.8 times. In treated plants, the proline content was lower. The most noticeable decrease in proline was observed in the variant with simultaneous seed coating in a bacterial suspension and a polymer mixture (Table 5). The obtained results indicate a reduction in the stress experienced by plants due to the pre-sowing seed treatment. Similar to our findings, a reduction in proline levels in various plant species under the influence of microbial treatment has been demonstrated in several studies [65,66].

In the conducted studies, an increase in the activity of antioxidant enzymes was observed when untreated plants were grown under conditions of phytopathogenic stress compared to plants grown in stress-free conditions (Table 6). The obtained data indicate that in response to the action of stress factors, there is an activation of the plant's defense system.

Table 6. Influence of various pre-sowing seed treatments on the activity of antioxidant enzymes in barley.

Treatment Variant	Catalase, mol min^{-1} mg of Protein $^{-1}$	Ascorbate Peroxidase, mol min^{-1} mg of Protein^{-1}	Guaiacol Peroxidase, mol min^{-1} mg of Protein^{-1}
T1	0.12 ± 0.005 a	9.6 ± 0.3 a	4.8 ± 0.2 a
T2	0.23 ± 0.004 b	12.47 ± 0.2 c	7.14 ± 0.3 b
T3	0.34 ± 0.007 d	12.34 ± 0.5 c	6.7 ± 0.3 b
T4	0.35 ± 0.007 d	10.03 ± 0.5 b	6.9 ± 0.3 b
T5	0.29 ± 0.006 c	30.37 ± 0.9 d	19.2 ± 0.7 c

Values are given as the mean ± SD. Values followed by the same letter do not differ according to the Tukey test ($p \leq 0.05$).

The pre-sowing seed treatment (T3–T5) led to an increase in catalase activity by 1.3–1.5 times under stress conditions. For ascorbate peroxidase and guaiacol peroxidase, an increase in enzyme activity was observed under phytopathogenic stress in the seed treatment with the bacterial suspension and polymer mixture (T5)—by 2.4 times and 2.7 times, respectively (Table 6). Similarly to the obtained data, previous studies have reported an increase in the activity of antioxidant enzymes in plants when inoculated with bacteria as one of the defense mechanisms of plants when grown under stressful conditions [67–69].

Thus, it was shown that when seeds were treated with the T5 composition, plant growth parameters (weight and length of roots) significantly increased compared to the T3 variant with a bacterial suspension. In addition, the use of the T5 composition contributed more to the attenuation of plant stress caused by phytopathogens compared to the use of microorganisms only (Tables 5 and 6). This indicates that the addition of biopolymers to formulations for seed treatments enhances microbe-induced plant tolerance to phytopathogens.

4. Conclusions

As a result of the research, a microbial association of bio-compatible endophytic bacteria has been created, possessing agronomically valuable properties such as the synthesis of the phytohormone IAA, antagonistic activity against *Fusarium oxysporum* and *Fusarium solani*, halotolerance, and PHA production. The study of the rheological properties of polysaccharide solutions showed that pullulan produced by *Aureobasidium pullulans* C7 can be used as seed coating binder at a low concentration of the polymer solution characterized by low viscosity ratio and exhibits Newtonian flow behavior. The effectiveness of various seed coating treatments including biopolymers (PHA and pullulan) and beneficial microorganisms in enhancing the resistance of barley plants to phytopathogens has been demonstrated. The innovative, eco-friendly antifungal seed treatments provide protection for barley against Fusarium diseases, significantly improving seed germination and plant growth in the field. In addition, these polymers will be a new progressive material with the possibility of use in medicine in the form of capsules for prolonged action of drugs, as absorbable suture threads, and dressings. In the form of a film material, the obtained microbial polymers can be used for packaging and storage of food products.

Author Contributions: Conceptualization, L.I. and Y.B.; Methodology, A.U., A.K. and I.S.; Software, A.O.; Validation, L.I. and Y.B.; Formal Analysis, Y.B., A.K. and I.S.; Investigation, L.I., A.O., Y.B. and A.U.; Resources, L.I. and A.O.; Data Curation, L.I., A.K. and I.S.; Writing—Original Draft Preparation, L.I., A.U., A.O. and Y.B.; Writing—Review and Editing, L.I. and Y.B.; Visualization, A.U. and A.O.; Supervision, L.I. and Y.B.; Project Administration, L.I.; Funding Acquisition, L.I. All authors have read and agreed to the published version of the manuscript.

Funding: This research was funded by the Ministry of Science and Higher Education of the Republic of Kazakhstan, grant number AP19679444.

Institutional Review Board Statement: Not applicable.

Informed Consent Statement: Not applicable.

Data Availability Statement: The data that support the findings of this study are available upon request from the corresponding author.

Acknowledgments: The Institute of Polymer Materials and Technologies (Almaty, Kazakhstan) is greatly acknowledged for performing rheological studies of polymers.

Conflicts of Interest: The authors declare no conflicts of interest.

References

1. Karlsson, I.; Persson, P.; Friberg, H. Fusarium Head Blight From a Microbiome Perspective. *Front. Microbiol.* **2021**, *12*, 628373. [CrossRef]
2. Korbecka-Glinka, G.; Piekarska, K.; Wi'sniewska-Wrona, M. The Use of Carbohydrate Biopolymers in Plant Protection against Pathogenic Fungi. *Polymers* **2022**, *14*, 2854. [CrossRef]

3. Lopez-Moya, F.; Suarez-Fernandez, M.; Lopez-Llorca, L.V. Molecular mechanisms of chitosan interactions with fungi and plants. *Int. J. Mol. Sci.* **2019**, *20*, 332. [CrossRef] [PubMed]
4. Xing, K.; Zhu, X.; Peng, X.; Qin, S. Chitosan antimicrobial and eliciting properties for pest control in agriculture. *Agron. Sustain. Dev.* **2015**, *35*, 569. [CrossRef]
5. Zheng, F.; Chen, L.; Zhang, P.F.; Zhou, J.Q.; Lu, X.F.; Tian, W. Carbohydrate polymers exhibit great potential as effective elicitors in organic agriculture: A review. *Carbohydr. Polym.* **2020**, *230*, 115637. [CrossRef] [PubMed]
6. Chin, J.M.; Lim, Y.Y.; Ting, A.S.Y. Biopolymers for biopriming of Brassica rapeseeds: A study on coating efficacy, bioagent viability and seed germination. *J. Saudi Soc. Agric. Sci.* **2021**, *20*, 198. [CrossRef]
7. Jurado, M.M.; Suárez-Estrella, F.; Toribio, A.J.; Martínez-Gallardo, M.R.; Estrella-González, M.J.; López-González, J.A.; López, M.J. Biopriming of cucumber seeds using actinobacterial formulas as a novel protection strategy against *Botrytis cinerea*. *Front. Sustain. Food Syst.* **2023**, *7*, 1158722. [CrossRef]
8. Ren, X.X.; Chen, C.; Ye, Z.H.; Su, X.Y.; Xiao, J.J.; Liao, M.; Cao, H.Q. Development and Application of Seed Coating Agent for the Control of Major Soil-Borne Diseases Infecting Wheat. *Agronomy* **2019**, *9*, 413. [CrossRef]
9. Singh, R.S.; Kaur, N.; Kennedy, J.F. Pullulanproductionfromagro-industrialwasteanditsapplicationsinfoodindustry: A review. *Carbohydr. Polym.* **2019**, *217*, 46. [CrossRef]
10. Mahmoud, Y.A.; El-Naggar, M.E.; Abdel-Megeed, A.; El-Newehy, M. Recent Advancements in Microbial Polysaccharides: Synthesis and Applications. *Polymers* **2021**, *13*, 4136. [CrossRef]
11. Behera, S.; Priyadarshanee, M.; Vandana, D.S. Polyhydroxyalkanoates, the bioplastics of microbial origin. Properties. Biochemical synthesis and their applications. *Chemosphere* **2022**, *294*, 133723. [CrossRef]
12. Nor Azillah, F.O.; Sarala, S.; Noriaki, S. Biodegradable dual-layer Polyhydroxyalkanoate (pha)/Polycaprolactone (pcl) mulch film for agriculture. *Energy Nexus* **2022**, *8*, 100137.
13. Ma, L.; Zhang, Z.; Li, J.; Yang, X.; Fei, B.; Leung, P.H.M.; Tao, X.M. A New Antimicrobial Agent: Poly (3-Hydroxybutyric Acid) Oligomer. *Macromol. Biosci.* **2019**, *19*, 1800432. [CrossRef]
14. Abou-Aiad, T.H.M. Morphology and Dielectric Properties of Polyhydroxybutyrate (PHB)/Poly(Methylmethacrylate)(PMMA) Blends with Some Antimicrobial Applications. *Polym. Plast. Technol. Eng.* **2007**, *46*, 435. [CrossRef]
15. Ignatova, L.; Usmanova, A.; Brazhnikova, Y.; Omirbekova, A.; Egamberdieva, D.; Mukasheva, T.; Kistaubayeva, A.; Savitskaya, I.; Karpenyuk, T.; Goncharova, A. Plant Probiotic Endophytic Pseudomonas flavescens D5 Strain for Protection of Barley Plants from Salt Stress. *Sustainability* **2022**, *14*, 15881. [CrossRef]
16. Ignatova, L.; Brazhnikova, Y.; Omirbekova, A.; Usmanova, A. Polyhydroxyalkanoates (PHAs) from Endophytic Bacterial Strains as Potential Biocontrol Agents against Postharvest Diseases of Apples. *Polymers* **2023**, *15*, 2184. [CrossRef] [PubMed]
17. Abulfaraj, A.; Jalal, R. Use of plant growth-promoting bacteria to enhance salinity stress in soybean (*Glycine max* L.) plants. *Saudi J. Biol. Sci.* **2021**, *28*, 3823. [CrossRef]
18. Sharma, N. Polyhydroxybutyrate (PHB) Production by Bacteria and Its Application as Biodegradable Plastic in Various Industries. *Acad. J. Polym. Sci.* **2019**, *2*, 3. [CrossRef]
19. Mostafa, Y.S.; Alrumman, S.A.; Otaif, K.A.; Alamri, S.A.; Mostafa, M.S.; Sahlabji, T. Production and Characterization of Bioplastic by Polyhydroxybutyrate Accumulating *Erythrobacter aquimaris* Isolated from Mangrove Rhizosphere. *Molecules* **2020**, *25*, 179. [CrossRef]
20. Mahitha, G.; Jaya, M.R. Microbial polyhydroxybutyrate production by using cheap raw materials as substrates. *Indian J. Pharm. Biol. Res.* **2016**, *4*, 57. [CrossRef]
21. Siddiqui, I.A.; Shaukat, S.S. Combination of Pseudomonas Aeruginosa and Pochonia Chlamydosporia for Control of Root-Infecting Fungi in Tomato. *J. Phytopathol.* **2003**, *151*, 215. [CrossRef]
22. Goksungur, Y.; Uzunogullari, P.; Dagbagli, S. Optimization of pullulan production from hydrolysed potato starch waste by response surface methodology. *Carbohydr. Polym.* **2011**, *83*, 1330. [CrossRef]
23. Oliveira, V.C. Study of the Molecular Weight of Pullulan Produced by *Aureobasidium pullulans* from Industrial Waste. *Mater. Res.* **2023**, *26*, e20230060. [CrossRef]
24. Bates, L.E.; Waldre, R.P.; Teare, I.D. Rapid determination of free proline for water stress studies. *Plant Soil.* **1973**, *39*, 205. [CrossRef]
25. Lichtestaller, H.K. Determination of total carotenoids and chlorophylls a and b of leaves extracts in different solvents. *Biochem. Soc. Trans.* **1983**, *11*, 591.
26. Al-Taweel, K.; Iwaki, T.; Yabuta, Y. A bacterial transgene for catalase protects translation of D1 protein during exposure of salt-stressed tobacco leaves to strong light. *Plant Physiol.* **2007**, *145*, 258–265. [CrossRef] [PubMed]
27. Khor, S.; Rahmad, Z.; Sreeramanan, S. Ascorbate Peroxidase Activity of Aranda Broga Blue Orchid Protocorm-like Bodies (PLBs) In Response to PVS2 Cryopreservation Method. *Trop. Life Sci. Res.* **2016**, *1*, 139. [CrossRef] [PubMed]
28. Mika, A.; Lüthje, S. Properties of Guaiacol Peroxidase Activities Isolated from Corn Root Plasma Membranes. *Plant Physiol.* **2003**, *132*, 1489. [CrossRef] [PubMed]
29. Afzal, I.; Shinwari, Z.K.; Sikandar, S.; Shahzad, S. Plant Beneficial Endophytic Bacteria: Mechanisms, Diversity, Host Range and Genetic Determinants. *Microbiol. Res.* **2019**, *221*, 36. [CrossRef] [PubMed]
30. Dhungana, S.A.; Itoh, K. Effects of Co-Inoculation of Indole-3-Acetic Acid-Producing and -Degrading Bacterial Endophytes on Plant Growth. *Horticulturae* **2019**, *5*, 17. [CrossRef]

31. Khianngam, S.; Meetum, P.; Chiangmai, P.N.; Tanasupawat, S. Identification and Optimisation of Indole-3-Acetic Acid Production of Endophytic Bacteria and Their Effects on Plant Growth. *Trop. Life Sci. Res.* **2023**, *34*, 219. [CrossRef] [PubMed]
32. Yu, J.; Pedroso, I.R. Mycotoxins in Cereal-Based Products and Their Impacts on the Health of Humans. Livestock Animals and Pets. *Toxins* **2023**, *15*, 480. [CrossRef] [PubMed]
33. Newitt, J.T.; Prudence, S.M.M.; Hutchings, M.I.; Worsley, S.F. Biocontrol of Cereal Crop Diseases Using Streptomycetes. *Pathogens* **2019**, *8*, 78. [CrossRef] [PubMed]
34. Behera, T.K.; Krishna, R.; Ansari, W.A.; Aamir, M.; Kumar, P.; Kashyap, S.P.; Pandey, S.; Kole, C. Approaches Involved in the Vegetable Crops Salt Stress Tolerance Improvement: Present Status and Way Ahead. *Front. Plant Sci.* **2022**, *12*, 787292. [CrossRef] [PubMed]
35. Corral, P.; Amoozegar, M.A.; Ventosa, A. Halophiles and Their Biomolecules. Recent Advances and Future Applications in Biomedicine. *Mar. Drugs* **2019**, *18*, 33. [CrossRef] [PubMed]
36. Thomloudi, E.E.; Tsalgatidou, P.C.; Douka, D.; Spantidos, T.N.; Dimou, M.; Venieraki, A.; Katinakis, P. Multistrainversus Single-Strain Plant Growth Promoting Microbial Inoculants. The Compatibility Issue. *Hell. Plant Prot. J.* **2019**, *12*, 61.
37. Paravar, A.; Piri, R.; Balouchi, H.; Ma, Y. Microbial seed coating: An attractive tool for sustainable agriculture. *Biotechnol. Rep.* **2023**, *37*, e00781. [CrossRef]
38. Zhubanova, A.A.; Ernazarova, A.K.; Kaiyrmanova, G.K.; Zayadan, B.K.; Savitskaya, I.S.; Abdieva, G.Z.; Kistaubaeva, A.S.; Akimbekov, N.S. Construction of cyanobacterial-bacterial consortium on the basis of axenic cyanobacterial cultures and heterotrophic bacteria cultures for bioremediation of oil-contaminated soils and water ponds. *Russ. J. Plant Physiol.* **2013**, *60*, 555. [CrossRef]
39. Ma, Y. Seed coating with beneficial microorganisms for precision agriculture. *Biotechnol. Adv.* **2019**, *37*, 107423. [CrossRef]
40. Sohail, M.N.; Tahira, P.; Charles, H. Recent Advances in Seed Coating Technologies: Transitioning Toward Sustainable Agriculture. *Green Chem.* **2022**, *5*, 225. [CrossRef]
41. Brazhnikova, Y.V.; Mukasheva, T.D.; Ignatova, L.V. Shtamm Drojjepodobnogo Griba Aureobasidium pullulans C7—Producent Ekzopolisaharida i Indoliluksusnoi Kisloti [Strain of Yeast-like Fungus Aureobasidium pullulans C7—PRODUCER of Exopolysaccharide and Indolylacetic Acid]. RK Patent No. 32992, 6 August 2018.
42. Duan, X.; Chi, Z.; Li, H. High pullulan yield is related to low UDP-glucose level and high pullulan-related synthases activity in *Aureobasidium pullulans* Y68. *Ann. Microbiol.* **2007**, *57*, 243. [CrossRef]
43. Chi, Z.; Wang, F.; Chi, Z.; Yue, L.; Liu, G.; Zhang, T. Bioproducts from *Aureobasidium pullulans*, a biotechnologically important yeast. *Appl. Microbiol. Biotechnol.* **2009**, *82*, 793. [CrossRef] [PubMed]
44. Wang, Z.; Sheng, J.; Tian, X.; Wu, T.; Liu, W.; Shen, L. Optimization of the production of exopolysaccharides by Bacillus thuringiensis 27 in sand biological soil crusts and its bioflocculant activity. *Afr. J. Microbiol. Res.* **2011**, *5*, 207.
45. Malick, A.; Khodaei, N.; Benkerroum, N.; Karboune, S. Production of exopolysaccharides by selected Bacillus strains: Optimization of media composition to maximize the yield and structural characterization. *Int. J. Biol. Macromol.* **2017**, *102*, 539. [CrossRef]
46. Lee, Y.K.; Jung, S.K.; Chang, Y.H. Rheological properties of a neutral polysaccharide extracted from maca (*Lepidium meyenii* Walp.) roots with prebiotic and anti-inflammatory activities. *Int. J. Biol. Macromol.* **2020**, *152*, 757. [CrossRef]
47. Hosseini, E.; Mozafari, H.; Hojjatoleslamy, M.; Rousta, E. Influence of temperature, pH and salts on rheological properties of bitter almond gum. *Food Sci. Technol.* **2017**, *37*, 437–443. [CrossRef]
48. Haghighatpanah, N.M. Optimization and characterization of pullulan produced by a newly identified strain of *Aureobasidium pullulans*. *Int. J. Biol. Macromol.* **2020**, *152*, 305–313. [CrossRef]
49. TeránHilares, R.; Resende, J.; Orsi, C.A.; Ahmed, M.A.; Lacerda, T.M.; Santos, J.C. Exopolysaccharide (pullulan) production from sugarcane bagasse hydrolysate aiming to favor the development of biorefineries. *Int. J. Biol. Macromol.* **2019**, *15*, 169.
50. Farris, S.; Unalan, I.U.; Introzzi, L.; Fuentes-Alventosa, J.M.; Cozzolino, C.A. Pullulan-based films and coatings for food packaging: Present applications, emerging opportunities, and future challenges. *J. Appl. Polym. Sci.* **2014**, *131*, 40539. [CrossRef]
51. Li, R. Electrospinning pullulan fibers from salt solutions. *Polymers* **2017**, *9*, 32. [CrossRef]
52. Munhuweyi, K.; Mpai, S.; Sivakumar, D. Extension of avocado fruit postharvest 591 quality using non-chemical treatments. *Agronomy* **2020**, *10*, 212. [CrossRef]
53. Pirzada, T.; Farias, B.V.; Mathew, R.; Guenther, R.H.; Byrd, M.V.; Sit, T.L.; Pal, L.; Opperman, C.H.; Khan, S.A. Recent Advances in Biodegradable Matrices for Active Ingredient Release inCrop Protection: Towards Attaining Sustainability in Agriculture. *Curr. Opin. Colloid Interface Sci.* **2020**, *48*, 121. [CrossRef] [PubMed]
54. Rocha, I.; Ma, Y.; Souza-Alonso, P.; Vosátka, M.; Freitas, H.; Oliveira, R.S. Seed Coating: A Tool for Delivering Beneficial Microbes to Agricultural Crops. *Front. Plant Sci.* **2019**, *10*, 1357. [CrossRef] [PubMed]
55. Gouda, S.; Kerry, R.G.; Das, G.; Paramithiotis, S.; Shin, H.S. Revitalization of plant growth promoting rhizobacteria for sustainable development in agriculture. *Microbiol. Res.* **2018**, *206*, 131. [CrossRef] [PubMed]
56. Moustaka, J.; Moustakas, M. Early-Stage Detection of Biotic and Abiotic Stress on Plants by Chlorophyll Fluorescence Imaging Analysis. *Biosensors* **2023**, *13*, 796. [CrossRef] [PubMed]
57. Stamelou, M.L.; Sperdouli, I.; Pyrri, I.; Adamakis, I.D.S.; Moustakas, M. Hormetic responses of photosystem II in tomato to Botrytis cinerea. *Plants* **2021**, *10*, 521. [CrossRef] [PubMed]
58. Suárez, J.C.; Vanegas, J.I.; Contreras, A.T.; Anzola, J.A.; Urban, M.O.; Beebe, S.E.; Rao, I.M. Chlorophyll fluorescence imaging as a tool for evaluating disease resistance of common bean lines in the western Amazon region of Colombia. *Plants* **2022**, *11*, 1371. [CrossRef]

59. Legendre, R.; Basinger, N.T.; van Iersel, M.W. Low-cost chlorophyll fluorescence imaging for stress detection. *Sensors* **2021**, *21*, 2055. [CrossRef]
60. Asfi, M.; Ouzounidou, G.; Panajiotidis, S.; Therios, I.; Moustakas, M. Toxicity effects of olive-mill wastewater on growth, photosynthesis and pollen morphology of spinach plants. *Ecotoxicol. Environ. Saf.* **2012**, *80*, 69. [CrossRef]
61. Guidi, L.; Calatayud, A. Non-invasive tool to estimate stress-induced changes in photosynthetic performance in plants inhabiting Mediterranean areas. *Environ. Exp. Bot.* **2014**, *103*, 42. [CrossRef]
62. Zandi, P.; Schnug, E. Reactive Oxygen Species, Antioxidant Responses and Implications from a Microbial Modulation Perspective. *Biology* **2022**, *11*, 155. [CrossRef]
63. Sahu, P.K.; Jayalakshmi, K.; Tilgam, J.; Gupta, A.; Nagaraju, Y.; Kumar, A.; Hamid, S.; Singh, H.V.; Minkina, T.; Rajput, V.D.; et al. ROS generated from biotic stress: Effects on plants and alleviation by endophytic microbes. *Front. Plant Sci.* **2022**, *13*, 1042936. [CrossRef]
64. Liang, X.; Zhang, L.; Natarajan, S.K.; Becker, D.F. Proline mechanisms of stress survival. *Antioxid. Redox Signal.* **2013**, *19*, 998. [CrossRef]
65. Ma, Y.; Rajkumar, M.; Oliveira, R.S.; Zhang, C.; Freitas, H. Potential of plant beneficial bacteria and arbuscular mycorrhizal fungi in phytoremediation of metal-contaminated saline soils. *J. Hazard. Mater.* **2019**, *379*, 120813. [CrossRef]
66. Singh, R.P.; Jha, P.N. The PGPR Stenotrophomona smaltophilia SBP-9 augments resistance against biotic and abiotic stress in wheat plants. *Front. Microbiol.* **2017**, *8*, 1945. [CrossRef]
67. Piri, R.; Moradi, A.; Balouchi, H.; Salehi, A. Improvement of cumin (*Cuminum cyminum*) seed performance under drought stress by seed coating and biopriming. *Sci. Hortic.* **2019**, *257*, 21. [CrossRef]
68. Bhattacharyya, C.; Banerjee, S.; Acharya, U. Evaluation of plant growth promotion properties and induction of antioxidative defense mechanism by tea rhizobacteria of Darjeeling. *Sci. Rep.* **2020**, *10*, 15536. [CrossRef]
69. Neshat, M.; Abbasi, A.; Hosseinzadeh, A.; Sarikhani, M.R.; Dadashi Chavan, D.; Rasoulnia, A. Plant growth promoting bacteria (PGPR) induce antioxidant tolerance against salinity stress through biochemical and physiological mechanisms. *Physiol. Mol. Biol. Plants.* **2022**, *28*, 347. [CrossRef]

polymers

MDPI

Article

Accelerated Weathering Testing (AWT) and Bacterial Biodegradation Effects on Poly(3-hydroxybutyrate-co-3-hydroxyvalerate) (PHBV)/Rapeseed Microfiber Biocomposites Properties

Madara Žiganova *, Remo Merijs-Meri, Jānis Zicāns, Agnese Ābele, Ivan Bochkov and Tatjana Ivanova

Institute of Chemistry and Chemistry Technology, Faculty of Natural Sciences and Technology, Riga Technical University, 3 Paula Valdena Street, LV-1048 Riga, Latvia; remo.merijs-meri@rtu.lv (R.M.-M.); janis.zicans@rtu.lv (J.Z.); agnese.abele_1@rtu.lv (A.Ā.); ivans.bockovs@rtu.lv (I.B.); tatjana.ivanova@rtu.lv (T.I.)

* Correspondence: madara.ziganova@rtu.lv

Abstract: In the context of sustainable materials, this study explores the effects of accelerated weathering testing and bacterial biodegradation on poly(3-hydroxybutyrate-co-3-hydroxyvalerate) (PHBV)/rapeseed microfiber biocomposites. Accelerated weathering, simulating outdoor environmental conditions, and bacterial biodegradation, representing natural degradation processes in soil, were employed to investigate the changes in the mechanical, thermal and morphological properties of these materials during its post-production life cycle. Attention was paid to the assessment of the change of structural, mechanical and calorimetric properties of alkali and N-methylmorpholine N-oxide (NMMO)-treated rapeseed microfiber (RS)-reinforced plasticized PHBV composites before and after accelerated weathering. Results revealed that accelerated weathering led to an increase in stiffness, but a reduction in tensile strength and elongation at break, of the investigated PHBV biocomposites. Additionally, during accelerated weathering, the crystallinity of PHBV biocomposites increased, especially in the presence of RS, due to both the hydrolytic degradation of the polymer matrix and the nucleating effect of the filler. It has been observed that an increase in PHBV crystallinity, determined by DSC measurements, correlates with the intensity ratio $I_{1225/1180}$ obtained from FTIR-ATR data. The treatment of RS microfibers increased the biodegradation capability of the developed PHBV composites, especially in the case of chemically untreated RS. All the developed PHBV composites demonstrated faster biodegradation in comparison to neat PHBV matrix.

Keywords: rapeseed microfibers; alkali treatment; N-methylmorpholine N-oxide treatment; poly(3-hydroxybutyrate-co-3-hydroxyvalerate); biocomposite; biodegradation; accelerated weathering

Citation: Žiganova, M.; Merijs-Meri, R.; Zicāns, J.; Ābele, A.; Bochkov, I.; Ivanova, T. Accelerated Weathering Testing (AWT) and Bacterial Biodegradation Effects on Poly(3-hydroxybutyrate-co-3-hydroxyvalerate) (PHBV)/Rapeseed Microfiber Biocomposites Properties. *Polymers* **2024**, *16*, 622. https://doi.org/10.3390/polym16050622

Academic Editors: Stefano Farris and Masoud Ghaani

Received: 1 December 2023
Revised: 19 February 2024
Accepted: 21 February 2024
Published: 24 February 2024

1. Introduction

Finite resources and climate issues require moving from a 'take-make-dispose' society to a carbon-neutral, environmentally sustainable, toxin-free and fully circular economy by 2050 [1]. Polyhydroxyalkanoate (PHA)-group polymers can be thermoplastically recycled, depolymerized to oligomers or monomers for use as renewable feedstock or converted to biomass by industrial or home composting [2]. Thus, PHA stands as a sustainable polymer for food packaging applications and beyond due to it nontoxicity, biocompatibility, recyclability and biodegradability. Each PHA molecule contains monomer units of certain (R)-hydroxy fatty acid connected each to other by ester bonds. However, applications of PHA often are limited due to it poor mechanical properties, such as low impact resistance, reduced elongation at break, and fragility. This is especially characteristic of the simplest member of the PHA group—polyhydroxy-3-butyrate (PHB). To overcome these issues to some extent, the length of the side chain may be increased. In general, PHAs may be categorized into different types on the basis of their structural chain length: short-chain-length PHAs, such as PHB, consist of monomeric building blocks of 3–5 carbons;

medium-chain-length polymers, such as poly(3-hydroxyoctanoate), consist of 6–14 carbons; long-chain-length PHAs have monomeric building blocks with 15 or more carbon atoms, for example, poly(3-hydroxypentadecanoate) [3]. PHA copolymers also may be used instead of homopolymers. Building PHA structural chains from different monomeric units, in various proportions, can improve the properties of the material. One of the most studied PHB-based copolymers is polyhydroxy-3-butyrate-co-3-hydroxyvalerate (PHBV). An increase in the 3-hydroxyvalerate fraction in the copolymer reduces the melting point and significantly increases flexibility due to higher chain mobility [4]. PHA copolymers, especially at higher co-monomer content, however, possess a low modulus and thermal resistance. To improve stiffness, thermal resistance and dimension stability, polymers can be modified with anisodiametric particles as reinforcing phase. Modification of PHA with natural fibers is a good option for obtaining sustainable environmentally friendly materials for various applications without compromising its biodegradability [5]. Better properties of the composites usually are achieved by performing modification either of the fibers or the polymer matrix for increased interaction in the interfacial region. Furthermore, polymer reinforcement with low-cost fibers obtained from agricultural or forestry residues can reduce final costs of the materials. According to Xiaoying Zhao et al. [6], to reduce brittleness and costs, microfibers of invasive plants—hanarygrass (HG) and honeysuckle (HS)—may be used as reinforcing agents for PHBV at 5–20 wt % fiber content. The authors tested the interfacial bond shear strength (IBSS) of the PHBV composites containing untreated and treated microfibers. As expected, the IBSS for the untreated microfiber composites (1.9 ± 0.4 MPa; HS—2.3 ± 0.3 MPa) was smaller than that for the composite containing 2% NaOH-treated fibers (HG—3.3 ± 0.3 MPa; HS—2.7 ± 0.3 MPa). Berthet M. A. et al. [7] found that PHBV composite with wheat microfibers in the range 0–30 wt % possessed a significantly increased water vapor transmission rate (from 11 up to 110 g m^{-2} day^{-1}); at the same time, Young's modulus Young's modulus was not significantly affected, which could fulfil the requirement for the packaging of respiring fresh food products, such as strawberries, thus enabling their preservation in a better way than currently used polyolefins. To compensate for the processability drop and decreased flexibility of composites, one of the most common approaches is plasticization. A broad range of PHA plasticizers have been examined, with citrate-based compounds as one of more effective ones. Thus, plasticization of PHB with triethyl citrate in the range 0–30 wt % led to an increase in the ultimate elongation from 5.8 to 6.9% [8]. MarieAlix Berthert et al. [9] compared wheat straw, brewers spent grains and olive mill particles' potential use as fillers, with a mass content of 0–50% in PHBV for food-packaging applications. According to water vapor permeability results, PHBV composites with 20% wheat straw fibers demonstrated a 3.5-fold increment, being in consent with the requirements of respiring food products, whereas PHBV composites with 20% olive mills, demonstrating a 2.5-fold increment, could be more adapted for water-sensitive products. W. Frącz, G. Janowski, R. Smusz and M. Szumski [10] found that by adding up to 30% of various natural fibers with aspect ratio 10 to PHBV matrix, the tensile strength of the composite increased from 35 MPa to 42.9 MPa for hemp fibers and 40.18 MPa for flax fibers, but decreased to 30.68 MPa for wood fibers due to less regular distribution of the fibers within the matrix. Similar results were observed by Singh S. et. al. [11] for PHBV composites with wood a fiber content of 10–40 wt % at a somewhat lower aspect ratio (4–5). The tensile strength of this PHBV composite decreased from 21.42 MPa to 16.75 MPa when it was loaded with 40 wt % of wood fibers. The authors also found that tensile modulus of the investigated composite improved by 167% at the highest wood fiber content.

Usually, higher interaction, and hence properties, are achieved in the case of alkali-pretreated fibers [12,13]. Alkali pretreatment, however, requires the use of corrosive chemical compounds and creates large amount of alkaline wastewater. To overcome this issue, other fiber pretreatment methods, such as hydrothermal treatment, i.e., boiling in water [14,15] or steam explosion [16,17], may be used. In this aspect, N-methylmorpholine N-oxide (NMMO) may be considered as a non-toxic and biodegradable solvent for lignocellulosic fiber pre-treatment to reduce the amount of waxes and other undesirable impurities. Although NMMO has been used for wet spinning of Lyocel fibers, there is almost no evidence on the use of NMMO for lignocellulosic fiber pre-treatment aiming at improvement of interfacial interaction in the composites.

Rosenau et al. [18] found that by modifying cellulostic fibers using an aqueous solution of NMMO in an "organic spinning process", the mechanical properties of the obtained Lyocell fibers considerably improved, demonstrating very high dry and wet tensile strength, due to the formation of cellulose II. According to our previous research [19], NMMO treatment has a remarkable influence on the RS fiber chemical composition. It has been observed that the hemicellulose content in fibers decreased by 15% after NMMO treatment, thus providing better reinforcing capability.

Only a few research teams have investigated the influence of the composition of bio-based and biodegradable polymer composites with natural fibers on its weathering resistance and biodegradability. K.C. Batista et al. [20] found that with an increase in the peach palm particle (PPp) mass content from 10 to 25 wt %, the thermal stability of a PHBV composite decreased from 298 °C to 278 °C because of biodegradation in soil due to the increased hydrophylity of the composite. Micro-cavities, caused by the introduction of higher contents of PPp in the polymer matrix, allowed easier entering of water and microorganisms into the biocomposite matrix, resulting in enhanced degradation. In addition to that, plant residue could provide some nutrients such as polysaharides, proteins and minerals, which create favorable conditions for microorganisms [21].

However, to the authors' knowledge, there is no sufficient evidence on the assessment of the change in the same PHBV composite's properties with ligncellulosic fibers during accelerated aging in combination with biodegradation studies. Consequently, in the current research, attention is paid to the comparative evaluation of the effects of alkali and NMMO treatments of RS on the weathering resistance and biodegradability of PHBV/RS composites. It is expected that the study could provide more information on the performance of PHBV composites with lignocellulosic fibers during its life cycle and beyond.

2. Materials and Methods

2.1. Materials

Poly(3-hydroxybutyrate-co-3-hydroxyvalerate is a commercial product (PHBV, Ningbo City, China, TianAn Biopolymer: ENMAT Y1000) with 1 mol % HV 3-(hydroxyvalerate) content. Triethyl citrate (TEC, Burlington, MA, USA, Sigma Aldrich, Mw = 276.28 $g \cdot mol^{-1}$, ρ = 1.137 $g \cdot L^{-1}$) was used as a plasticizer. Winter rapeseed straw (RS) was collected as biomass waste from local farms Braslini, Pasiles and Susuri. RS were ground with a Retsch SM300 rotary grinder (Retsch GmbH, located in Haan, Germanya) a speed of 700 rpm using a 0.25 mm sieve. The obtained fibers were dried in an oven for 24 h at 60 °C and stored in closed plastic bags until further use. Sodium hydroxide pellets EMSURE were supplied by Sigma Aldrich (Sigma-Aldrich Chemie GmbH, located in Taufkirchen, Germany). A 50 wt % aqueous solution of NMMO was supplied by Merck (Biotecha Latvia, Riga, Latvia). Propyl gallate was purchased from Biosynth Carbosynth (Biosynth Carbosynth Ltd., located in Compton, Berkshire, United Kingdom).

2.2. Fiber Alkali Treatment (Mercerization)

According to our previous research [22], the most optimal conditions for treatment of RS were as follows. A certain amount (50 g) of the ground RS fibers was immersed in aqueous solution with a NaOH concentration of 2%, and the obtained suspension was mixed for 30 min. At the end of the mercerization, the fibers were washed several times using distilled water until neutral reaction. The treated fibers were dried in an oven at 60 °C for about 24 h and stored in closed plastic bags until further use.

2.3. Fiber NMMO Treatment

Firstly, the commercial NMMO solution (50 wt % in H_2O) was concentrated to an 85% (*w*/*w*) solution by vacuum evaporation. Secondly, RS fibers were added to the prepared NMMO solution to achieve the optimal NMMO/water/cellulose ratio of 74:10:14. To prevent thermos-oxidative degradation, 1% (*w*/*w*) of propyl gallate was added. The vessel was placed in an oil bath at 90 °C under continuous mixing for different periods of time (1.3 h, 5 h and 30 h) to allow the treatment to occur. At the end of the treatment, the fibers were washed using distilled water until pH = 7. The treated fibers were dried in an oven at 60 °C for about 24 h and stored in closed plastic bags until further use.

2.4. Preperation of Composites

According to the previous research [23], the optimal amount of the TEC plasticizer was 20%, and the conditions for preparation of the composites with microfibers were as follows. The biopolymer and the microfibers were dried at 60 °C in a vacuum oven for 24 h. As shown in Table 1, systems with unmodified and two different types of modified RS microfibers were melt-compounded using a two-roll mill LRM-S-110/3E from Lab Tech Engineering Company Ltd., Phraeksa, Thailand. The mixing time was 3 min and the roll temperatures were 165 °C and 175 °C. Furthermore, the systems were milled at room temperature and 700 rpm using a Retsch cutting mill (SM300 Retsch GmbH, Haan, Germany) with a 6 mm sieve. The obtained flakes, with average dimensions of 3 mm × 2 mm, were used for manufacturing test specimens using compression molding as follows: 3 min pre-heating, 3 min pressing under 5 MPa pressure, 5 min cooling under pressure. Test specimens for mechanical property tests were cut from ~0.5 mm thick plates with dimensions of 60 mm × 100 mm obtained by hot pressing at 190 °C. Samples for bacterial biodegradation were cut from 1 mm thick plates with dimensions of 60 mm × 100 mm, similarly obtained by hot pressing.

Table 1. Compositions of the obtained composite systems.

Code	PHBV (wt %)	RS (wt %)	RSa (wt %)	RS_{NMMO} (h)	TEC (wt %)
PHBV	100	–	–	–	0
PHBV20	80	–	–	–	20
PHBV20/2RS	78	2	–	–	20
PHBV20/2RSa	78	–	2	–	20
PHBV20/5RSa	75	–	5	–	20
PHBV20/10RSa	70	–	10	–	20
PHBV20/2$RS_{NMMO1.2h}$	78	–	–	1 h 20 min	20
PHBV20/2RS_{NMMO5h}	78	–	–	5 h	20
PHBV20/2$RS_{NMMO30h}$	78	–	–	30 h	20

RS—rapeseed straw microfibers; RSa—alkali treated rapeseed straw microfibers; RS_{NMMO}—rapeseed straw microfibers treated with NMMO, TEC—triethyl citrate.

2.5. Composite Characterization

2.5.1. Weathering

A QUV accelerated weathering tester (Q-Lab Co., Westlake, OH, USA) equipped with fluorescent UVA-340 lamps (Q-Lab Co., USA) was used to simulate outdoor conditions in accordance with ISO 4892–3:2016 [24]. The test cycle consisted of ultraviolet irradiation (8 h) at black panel temperature 50 ± 3 °C and irradiance 0.75 W/m^2, spray (15 min), and condensation (3 h and 45 min at 50 °C). A pre-determined amount of tensile test specimens was inserted into the QUV weathering chamber and exposed to the previously mentioned accelerated weathering conditions until 250 h and 500 h were reached.

2.5.2. Fourier Transform Infrared Spectroscopy (FTIR)

FTIR spectra were obtained using a Thermo Fisher Scientific Nicolet 6700 spectrometer (Thermo Fisher Scientific, Waltham, Massachusetts, United States) via the attenuated total reflectance (ATR) technique. All spectra were recorded in the range of 650 to 4000 cm^{-1} with a resolution of 4 cm^{-1}.

2.5.3. Colorimetry

The color analysis of the prepared compositions was evaluated using a Ci7600 Sphere Benchtop Spectrophotometer (x-rite Pantone, Kentwood, MI, USA). Three parallel measurements were done using a total transmittance aperture of 6 mm, a wavelength range of 360–750 nm, a photometric resolution of 0.01%, and a white paper background due to the transparency of the samples. Lightness or luminance (L^*) and chromaticity coordinates or base color parameters a^* and b^* (a^*—the green-red component, b^*—the blue-yellow component), saturation (C^*) and hue angle (h°) were measured for ten replicate samples. The total color change (ΔE) was calculated using Equation (1) according to ASTM D 2244-02 [25].

$$\Delta E = \sqrt{\Delta L^{*2} + \Delta a^{*2} + \Delta b^{*2}} \tag{1}$$

where ΔL^*, Δa^* and Δb^* are the differences between the initial and final values for L^*, a^* and b^*, respectively.

2.5.4. Differential Scanning Calorimetry (DSC)

Melting/crystallization behavior was evaluated using a Mettler/Toledo differential scanning calorimeter DSC 1/200W (Mettler-Toledo International Inc., Columbus, Ohio, United States). The specimen of approximately 10 mg was sealed in an aluminum pan and subjected to the following regime: the first heating run from 25 °C to 200 °C at a rate of 10 °C/min, ending with a holding at the target temperature for 5 min; the cooling run from 200 °C to −50 °C at a rate of 10 °C/min, ending with a holding at the target temperature for 5 min; and the second, final, heating run from −50 °C to 200 °C at a rate of 10 °C/min under a nitrogen atmosphere.

The degree of crystallinity (χ) was calculated using the following equation:

$$\chi = \frac{\Delta H_C}{\Delta H_m^{\circ}(1 - w)} \times 100 \tag{2}$$

where ΔHc is the measured specific melting/crystallization enthalpy of PHBV phase and ΔH_m° is the melting enthalpy of the 100% crystalline PHB = 146 J/g [26].

2.5.5. Tensile Properties

Tensile stress–strain characteristics were determined at a temperature of 20 °C in accordance with EN ISO 527 [27] using Zwick Roell material testing equipment, a BDO—FB020TN (Zwick Roell Group, Ulm, Germany), equipped with pneumatic grips. Type 5A test specimens were stretched at a constant deformation speed of 50 mm/min. Demonstrated values represent the averaged results of the measurements performed on 10 test specimens.

2.5.6. Bacterial Biodegradation

Biodegradation at 58 ± 2 °C and 42% of soil humidity was carried out to characterize the biodegrabalility of the developed compositions by the rate of mass loss as a function of time. Compost soil (swamp peat) with a pH of 6.64 was acquired from a local distributor of UAB "Juknevičiaus Kompostas" (Vilnius, Lithuania). Similarly to Martins Nabels-Sneiders et al. [28], the samples were cut from a pressed plate with a thickness of 1300 μm. The samples were circular samples with d = 25.7 mm, and were encased between sieves and deposited in the soil at a depth of 1.5 cm using closed plastic containers. Samples were regularly recovered, dried in a vacuum oven at 60 °C for 4 h, weighed, photographed, and inserted back into the soil. The test was finished when the mass loss of the sample reached 50%.

3. Results and Discussion

3.1. Accelerated Weathering Impact on Properties

3.1.1. Surface Chemistry (FTIR)

FTIR spectra in the absorption region of carbonyl bonds of unaged and aged PHBV and its composites with RS are shown in Figure S1a–i in Supplementary Materials. Liqing Wei et al. have previously reported that carbonyl absorption band intensities in the 1680–1800 cm^{-1} region may be related to the crystalline and amorphous parts in the polymer, i.e., absorption intensities of PHBV with 33 mol % of HV groups at 1720 cm^{-1} and 1740 cm^{-1} have been related to crystalline and amorphous parts in the polymer, allowing the calculation of the crystallinity index. In the case of the unirradiated PHBV with a low amount of HV units used in this research, intensities related to the crystalline and amorphous parts in the polymer may be observed at ca. 1718 cm^{-1} and ca. 1735 cm^{-1}, respectively. The peak intensity ratio for unexposed PHBV was observed to be 3.4. After accelerated weathering, this ratio changed insignificantly. Kann et al. have minutely investigated the determination of crystallinity of PHA group polymers by means of FTIR, DSC and X-ray analysis [29]. The authors of this research have suggested the FTIR method for evaluation of the crystallinity of PHAs as a simpler and easier method in comparison to X-ray diffraction or scattering. They have also found correlations between FTIR results and DSC measurements. In their contribution, they suggested to determine the crystallinity index of PHAs between absorption bands 1230 cm^{-1} and 1184–1186 cm^{-1}, related to the conformational band at helical chains, observed only in the crystalline phase and C-O-C asymmetric stretching, pronounced in the amorphous phase, respectively. In the current research, the respective absorption bands for crystalline and amorphous parts of PHBV have been observed at ca. 1220–1230 cm^{-1} and 1180 cm^{-1} (Figure S2a–i in Supplementary Materials). By calculating the crystallinity ratio from these absorption bands (see Table 2), it may be observed that by increasing the exposure time, the crystallinity of PHBV is increased, which confirms with findings from DSC measurements (Section 3.1.3). By evaluating the effect of weathering time, it may also be observed that the intensity of the small peak at ca 1685 cm^{-1} increased with increasing weathering time, which is most probably attributed to partial hydrolysis of PHBV and the formation of carboxylic acids during the aging process. It may also be observed that the chosen method of the fiber surface treatment does not considerably affect the position of the carbonyl peak.

Another common indicator of aging of polymers is the carbonyl index, which was calculated from the FTIR spectra by comparing the main carbonyl absorbance peak at 1718 cm^{-1} with the peak relatively stable under the conditions of accelerated weathering, such as the peak at 1379 cm^{-1}, which is attributable to $-CH_3$ group fluctuations. It is observed that $I_{1718/1379}$ for all the investigated composites decreases with increasing weathering time. Similar results have been obtained by Ana Antuanes et al. [30], who have observed degradation in both the crystalline and amorphous phases of PHBV nanocomposites with TiO_2. It is expected that decrease in $I_{1718/1379}$ occurs because of hydrolytic degradation of PHBV during accelerated weathering. Iggui et al. [31] have demonstrated that thedecrease in the carbonyl index because of hydrolytic aging is clearly related to the reduced molecular weight of PHBV.

Table 2. Carbonyl index and crystalline/amorphous ratios for PHBV and the PHBV biocomposites, both before and after weathering.

Code	Carbonyl Index $I_{1718/1379}$	Crystallinity Index $I_{1718/1735}$	Crystallinity Index $I_{1225/1180}$
PHBV 0 h	3.3	3.4	1.3
PHBV 250 h	3.3	3.1	1.2
PHBV 500 h	3.2	3.1	1.2

Table 2. *Cont.*

Code	Carbonyl Index $I_{1718/1379}$	Crystallinity Index $I_{1718/1735}$	Crystallinity Index $I_{1225/1180}$
PHBV20 0 h	3.8	3.3	1.0
PHBV20 250 h	3.6	3.4	1.3
PHBV20 500 h	3.2	3.2	1.2
PHBV20/2RS 0 h	3.5	2.6	0.9
PHBV20/2RS 250 h	3.0	3.2	1.2
PHBV20/2RS 500 h	3.0	3.1	1.2
PHBV20/2RSa 0 h	4.2	3.0	0.8
PHBV20/2RSa 250 h	3.1	3.2	1.2
PHBV20/2RSa 500 h	2.6	2.9	1.1
PHBV20/5RSa 0 h	4.3	3.1	0.8
PHBV20/5RSa 250 h	3.0	3.2	1.1
PHBV20/5RSa 500 h	3.0	3.2	1.2
PHBV20/10RSa 0 h	4.0	2.9	0.9
PHBV20/10RSa 250 h	3.3	3.3	1.2
PHBV20/10RSa 500 h	3.1	3.2	1.1
PHBV20/2RS$_{NMMO1.2h}$ 0 h	4.2	3.0	0.8
PHBV20/2RS$_{NMMO1.2h}$ 250 h	3.3	3.0	1.2
PHBV20/2RS$_{NMMO1.2h}$ 500 h	3.4	3.5	1.2
PHBV20/2RS$_{NMMO5h}$ 0 h	4.3	2.9	0.8
PHBV20/2RS$_{NMMO5h}$ 250 h	3.0	3.3	1.1
PHBV20/2RS$_{NMMO5h}$ 500 h	3.0	2.9	1.2
PHBV + 20/2RS$_{NMMO30h}$ 0 h	3.9	3.7	0.8
PHBV20/2RS$_{NMMO30h}$ 250 h	3.4	3.4	1.2
PHBV20/2RS$_{NMMO30h}$ 500 h	2.4	2.6	1.1

3.1.2. Colorimetry

Examination of the color fastness of the PHBV and its composites led to obtaining the photometric images shown in Table 3, whereas the colorimetric parameters are summarized in Figure 1 and Supplementary Materials Table S1. As one can see, the fibers' treatment with alkali or modification with TEC didn't considerably change lightness (L^* changes not more than 2 units in comparison to that of neat polymer matrix) of the developed PHBV composites. The effect of NMMO treatment is greater; however, it may be connected with longer treatment times, which were 3 to 60 times longer in comparison to those used during mercerization. In general, NMMO-treated fibers yielded a darker surface color of PHBV composites, especially if treated for longer time. As a result of UV irradiation and increased moisture during accelerated weathering, the surface of the PHBV composite samples became rougher, most probably because of partial thermal and hydrolytic degradation of the biopolymer matrix. In addition, all the investigated systems during aging tended to fade out, as shown by increasing L^* values. This is characteristic for natural fiber-reinforced biopolymer composites, as may be confirmed by results obtained by Aneta Tor-Świątek et al. [32] for linen fiber–biopolymer composites or A.T. Michel for hemp fiber-reinforced PHB composites. As may be expected, a larger color change ΔE was observed for the investigated PHBV composites with higher amounts of microfibers.

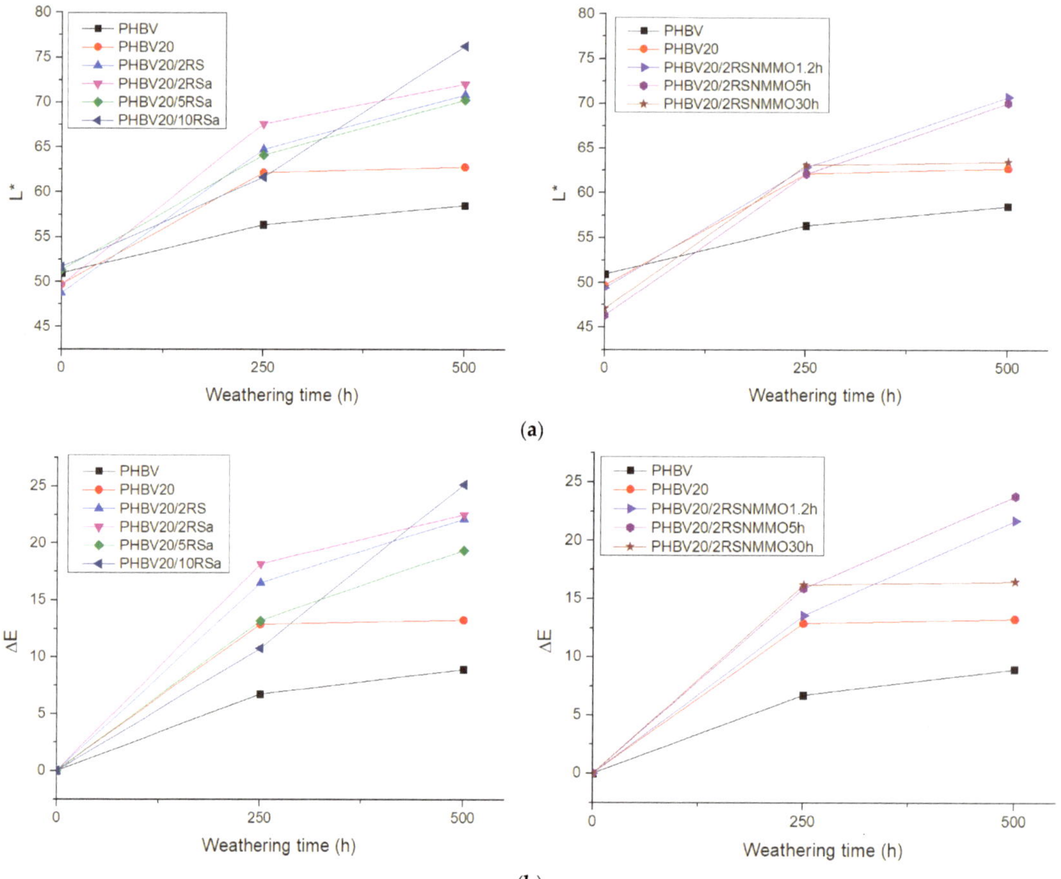

Figure 1. Change of colorimetric parameters L^* (**a**) and ΔE (**b**) with accelerated weathering time.

Table 3. Photometric images of color change of PHBV and its plasticized composites during accelerated weathering for 0 h, 250 h, 500 h.

Sample Code	**0 – 500 h**		
PHBV			
PHBV20			
PHBV20/2RS			

Table 3. *Cont.*

Sample Code	0 – 500 h
PHBV20/2RSa	
PHBV20/5RSa	
PHBV20/10RSa	
PHBV20/2$RS_{NMMO1.2h}$	
PHBV20/2RS_{NMMO5h}	
PHBV20/2$RS_{NMMO30h}$	

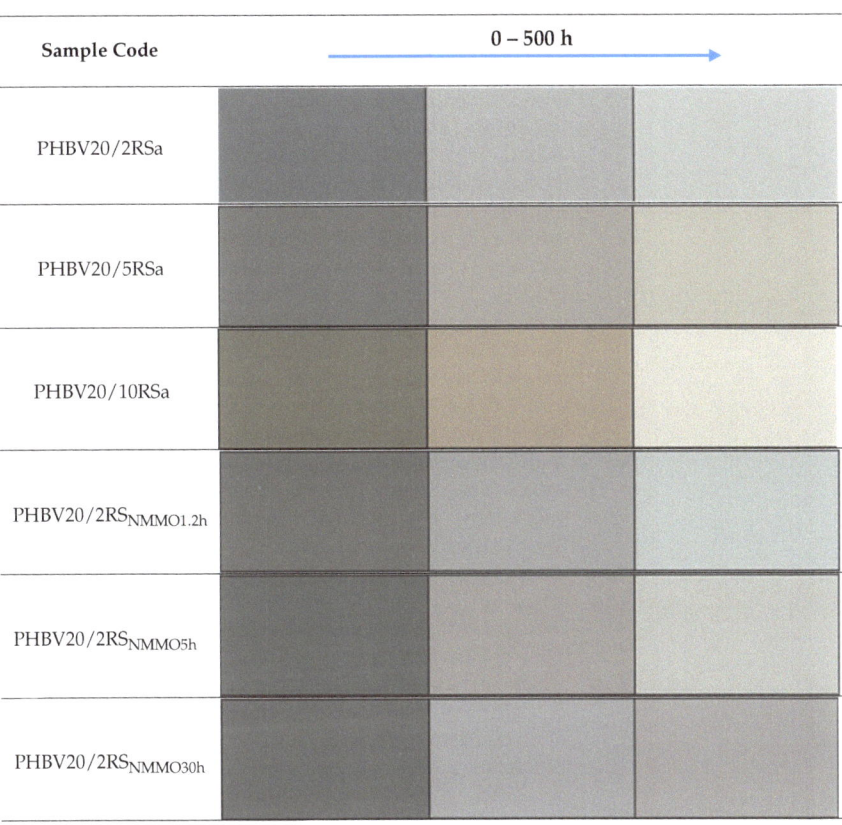

3.1.3. Thermal Properties (DSC)

DSC analysis was used to determine changes in the melting/crystallization behavior of PHBV and its plasticized composites before and after accelerated weathering. The results of DSC analysis are summarized in Figure 2 and Table 4. From the first-run DSC thermograms of plasticized PHBV and its composite with RS, depicted in Figure 2, it is possible to determine the cold-crystallization and melting regions of the PHBV crystalline phase. Glass transition relaxation of the PHBV phase is hardly detectable from DSC measurements and is therefore not analyzed. It should be mentioned, however, that the glass transition temperature of PHBV and its composites usually fluctuates within a range of 0–27 °C, depending on the measurement method [23,33]. As demonstrated in Figure 3, as soon as RS is added, cold crystallization of the PHBV phase may be observed. This is most probably related to the effect of RS on the crystallization behavior of PHBV from melt during compression moulding of the test specimens. After exposure to 250 h of accelerated weathering, the cold crystallization region of PHBV + 2RS composite was lost, which may be explained by increased temperature during the test run (60 °C during the UV irradiation cycle and 50 °C during the condensation cycle). This may result in a more complete structural arrangement of the PHBV phase. In addition, because of accelerated weathering, the melting peak of the PHBV crystalline phase is shifted towards higher temperatures by ca. 10 °C. This may be due to structural rearrangement of the PHBV phase within both the plasticized matrix without and with RS. Further increment of exposure time does not considerably change the maximum temperature of the PHBV melting peak. However, the shape of the plasticized PHBV melting peak after 500 h of accelerated aging is changed due to certain degradation of the PHBV phase with the formation of a greater variety of crystalline structures (please see Figure S3 in Supplementary Materials). First-run DSC thermograms give an important information about the investigated materials. A more detailed layout of the calorimetric data from the first-run DSC experiments is given in Table 5, where, besides relaxation peak maximum temperatures, corresponding onset and offset temperatures are given, accompanied by respective enthalpy values, from which values of crystallinity degree have been calculated. Cold crystallization peak enthalpy was also considered when calculating the crystallinity degree of the investigated systems. However, cold crystallization resulted in a broad flat peak with unexpressed maximum, allowing the assumption that cold crystallization temperature changed insignificantly with addition of the rapeseed microfibers. The maximum melting temperature also changed to a relatively small extent, i.e., within 6 °C. Comparatively larger changes of T_m have been observed for the composites, containing NMMO-modified fibers with the highest treatment time. In general, by increasing the fibers' content, some increase in the onset, maximum and offset temperatures of the melting is observed, resulting in the shift of the melting peaks towards somewhat higher temperatures. The total crystallinity degree of the plasticized PHBV phase, however, is decreased for the composites with fibers. Similarly to PHBV and its composite with untreated rapeseed fibers, after 250 h of exposure, there is an increment in the melting peak temperatures of the composites containing treated fibers. By further increasing exposure time up to 500 h, the changes in these calorimetric parameters are comparatively smaller, denoting that the greatest structural changes have occurred already, prior to the exposure time of 250 h. However, it should be mentioned that first-run DSC thermograms are considerably influenced by thermal history events; therefore, more reliable data, characterizing structural changes of the investigated PHBV composites upon aging, may be found from cooling or second-run DSC thermograms, which are also displayed in Table 5. From the cooling thermograms' data, only one exothermic peak may be observed, which is related to crystallization of the PHBV phase in the composites. As shown, the crystallization temperature maximum is not considerably changed along with the addition of rapeseed microfibers, fluctuating within the range of 7 °C. There are also no considerable differences in crystallization onset and offset temperatures, but crystallization occurs in a somewhat narrower temperature range for rapeseed fiber–containing composites, which may be attributed to promotion of crystallization by the fibers. After 250 h of exposure, crystallization from the melt begins at somewhat higher temperatures, whereas it is more pronounced for the composites, especially in the case of higher RSa fiber content and longer NMMO treatment times of the RS. A similar trend is also observed after 500 h of irradiation, although there is practically no change in the beginning of the crystallization from the melt in comparison to systems exposed for 250 h. Disregarding this, after accelerated weathering, a considerable increase in the crystallinity degree of the composites with fibers is observed, which may be explained by partial hydrolytic degradation of the polymer matrix, resulting in easier crystallizing of macromolecular fragments. It may be concluded that RS microfibers tend to act as nucleants, promoting crystallization of the polymer matrix, especially in the case of PHBV composites. In [30,31], it has been also reported that cellulose particles in a PHBV matrix act like nucleating agent, inducing PHBV to crystallize at lower temperatures because of the reduced energy barrier. Table 5 also contains results from the second DSC run after controlled cooling from the melt. It may be observed that in the second DSC run thermograms, no cold crystallization peaks are observed, most probably because the cooling rate was enough to ensure complete crystallization of PHBV. Consequently, all second heating run DSC thermograms display two melting peaks, the first of which is attributed to melting of less ordered crystallites, formed during controlled cooling. In general, it may be observed that the melting range of the investigated systems after the controlled cooling is shifted to somewhat lower temperatures. In addition, PHBV melting peak temperatures almost do not change with increasing fiber content. More pronounced effects are observed after exposing the investigated compositions to accelerated aging. It may be observed clearly that a shift of the PHBV melting peak towards higher temperatures occurs. In addition, the evolvement of at least three different crystalline structures may be observed along with increasing exposure time. For example, in the case of plasticized PHBV, the lowest melting maxima, observed at ca 155 °C is related to less ordered crystalline structures, whereas its intensity evidently increases with exposure time. The second temperature maxima, observed dominantly for unexposed PHBV is at 165 °C. After exposure for 250 h this temperature maxima is decreased and new temperature maxima at 171 °C appears. Further increment of exposure time up to 500 h results in decrease of this temperature maxima to 168 °C. Similar trends have been observed also in the case of RS fibers containing composites. There is a trend of increased intensity of the highest temperature maxima around 171 °C with increasing both, the RSa fiber content and the RS treatment time with NMMO. This may be related to the fact that larger amount of fibers or more fibrillated fibers more efficiently promote crystallization of PHBV matrix, especially in the case of partially degraded macromolecular fragments at the highest exposure time. As a result crystallinity degree of PHBV has been considerably increased with increasing both RSa fiber content and treatment time of RS fibers by NMMO.

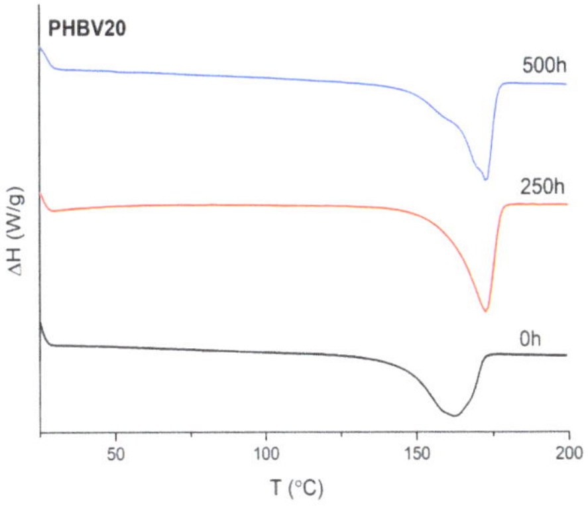

(**a**)

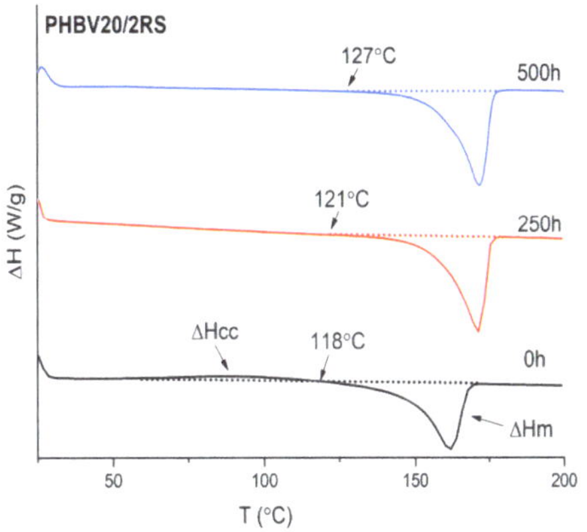

(**b**)

Figure 2. DSC thermograms change of PHBV20 (**a**) and PHBV20/2RS (**b**) during accelerated weathering process.

Table 4. Change of calorimetric properties of PHBV and the plasticized PHBV composites during accelerated weathering process (first heating run).

Calorimetric Parameter–Aging Time / Sample Code	T_{cc}			T_m			χ, %		
	0 h	250 h	500 h	0 h	250 h	500 h	0 h	250 h	500 h
PHBV	-	-	-	175	175	172	64	55	55
PHBV20	-	-	-	163	172	173	57	63	63
PHBV20/2RS	90	-	-	162	171	171	54	65	71
PHBV20/2RSa	91	-	-	162	172	171	55	68	67
PHBV20/5RSa	91	-	-	167	175	172	47	70	66
PHBV20/10RSa	91	-	-	165	172	172	47	70	75
PHBV20/2$RS_{NMMO1.2h}$	99	-	-	165	171	173	47	69	69
PHBV20/2RS_{NMMO5h}	102	-	-	164	170	174	42	68	70
PHBV20/2$RS_{NMMO30h}$	104	-	-	168	176	172	52	67	69

Table 4. *Cont.*

Calorimetric Parameter–Aging Time / Sample Code	(Cooling)					
	T_m			χ, %		
	0 h	**250 h**	**500 h**	**0 h**	**250 h**	**500 h**
PHBV	83	89	91	47	46	38
PHBV20	75	82	73	44	50	47
PHBV20/2RS	76	85	82	45	54	59
PHBV20/2RSa	74	87	77	45	56	56
PHBV20/5RSa	74	91	88	44	58	59
PHBV20/10RSa	77	92	92	47	61	63
PHBV20/2RS$_{NMMO1.2h}$	73	88	82	44	77	59
PHBV20/2RS$_{NMMO5h}$	71	86	77	41	55	57
PHBV20/2RS$_{NMMO30h}$	70	89	81	44	56	59

Calorimetric Parameter–Aging Time / Sample Code	(second heating run)											
	T_{cc}			T_{m1}			T_{m2}			χ, %		
	0 h	**250 h**	**500 h**	**0 h**	**250 h**	**500 h**	**0 h**	**250 h**	**500 h**	**0 h**	**250 h**	**500 h**
PHBV	95	103	99	167	168	166	172	174	172	56	48	49
PHBV20	91	92	91	154	164	156	165	172	168	58	57	52
PHBV20/2RS	90	97	96	154	163	162	165	171	171	54	52	58
PHBV20/2RSa	91	97	90	155	162	160	166	171	170	55	57	57
PHBV20/5RSa	94	100	97	154	166	164	165	172	171	44	62	58
PHBV20/10RSa	95	97	100	155	165	165	165	171	171	49	66	62
PHBV20/2RS$_{NMMO1.2h}$	90	97	94	154	164	163	165	171	170	53	67	64
PHBV20/2RS$_{NMMO5h}$	91	101	91	153	163	161	165	171	170	49	55	59
PHBV20/2RS$_{NMMO30h}$	94	94	93	154	165	162	165	171	170	50	65	65

Table 5. Photos of PHBV and PHBV composites during the biodegradation process in the composting conditions.

Sample Code	Time (Days)									
	0	13	21	27	33	49	58	69	76	92
PHBV										
PHBV20										
PHBV20/2RS										
PHBV20/2RSa										
PHBV20/5RSa										
PHBV20/10RSa										
PHBV20/2RS$_{NMM0_1.2h}$	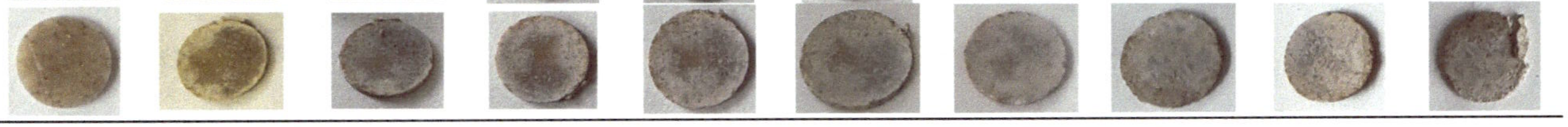									

Table 5. *Cont.*

Sample Code	Time (Days)									
	0	13	21	27	33	49	58	69	76	92
PHBV20/2RS$_{NMM0_5h}$		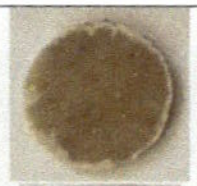								
PHBV20/2RS$_{NMM0_30h}$										

3.1.4. Tensile Properties

The tensile properties of the investigated PHBV composites before and after accelerated weathering are summarized in Figure 3, allowing the assessment of the mechanical property changes of the material during its expected life cycle. As shown in Figure 3a, the tensile modulus E of all the investigated systems increases by increasing the accelerated weathering time. In general, the highest E values, independently from the exposure time, have been observed for neat PHBV. It is not surprising as, by the addition of 20% of TEC as the plasticizer, the stiffness of all the investigated unexposed composites is decreased. If the fibers are added, the modulus of the unexposed plasticized PHBV composites is increased from 956 MPa to 1354 MPa. The highest E value is observed for the composite containing 10% of alkali-treated RS. Larger modulus increment is hindered by the fact that no chemical interaction occurs between the hydrophobic polymer matrix and the hydrophilic fibrous reinforcement, as demonstrated also in the FTIR measurements. By increasing the accelerated weathering time up to 500 h, the E of the plasticized PHBV and its composites increases up to two times, whereas the increase in the ultimate strength of the investigated systems is limited to 250 h. In addition, the reinforcing effect of RS fibers is practically lost after 250 h of accelerated aging. However, after 500 h of accelerated aging, the PHBV composites with 2 wt % of the filler demonstrate ca. 10% higher E values in comparison to the neat biopolymer matrix. Evidently, this is related to the higher crystallinity of PHBV in the composite, as demonstrated by DSC measurements. Unfortunately, the observed increase in E, by increasing the accelerated weathering time, is accompanied with increased brittleness, resulting in rupture of the composites at low ultimate elongation values, even lower than 1.5%. This is due to the limited outdoor exploitation time of the developed composites, where the material is subjected to elevated external temperature, direct water exposure and condensation. This does not exclude short-term application of the material indoors or for a limited time, even in outdoor conditions.

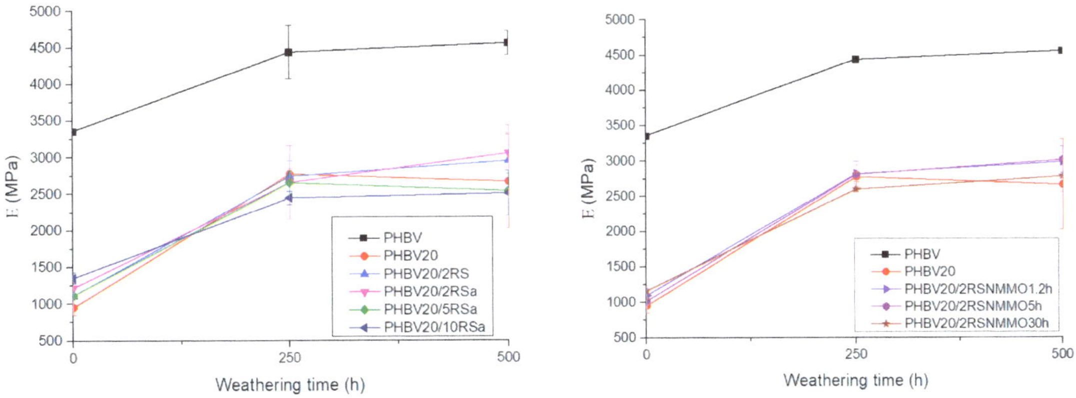

(a)

Figure 3. *Cont.*

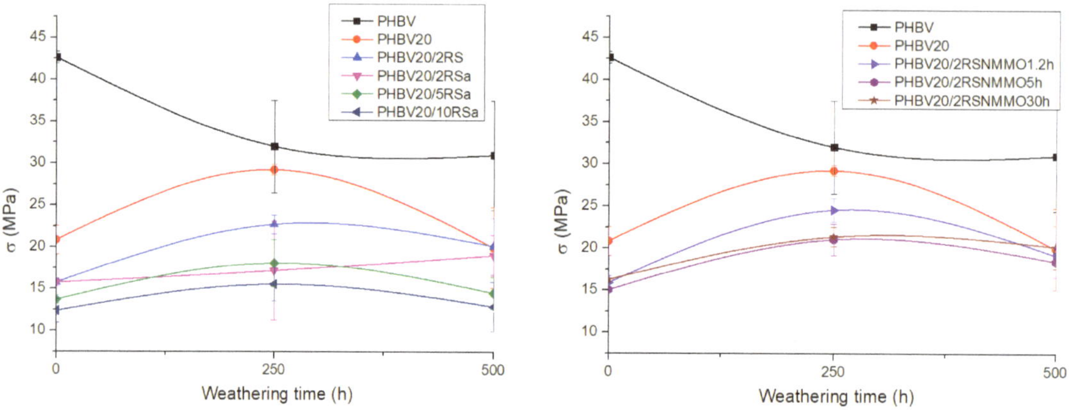

(**b**)

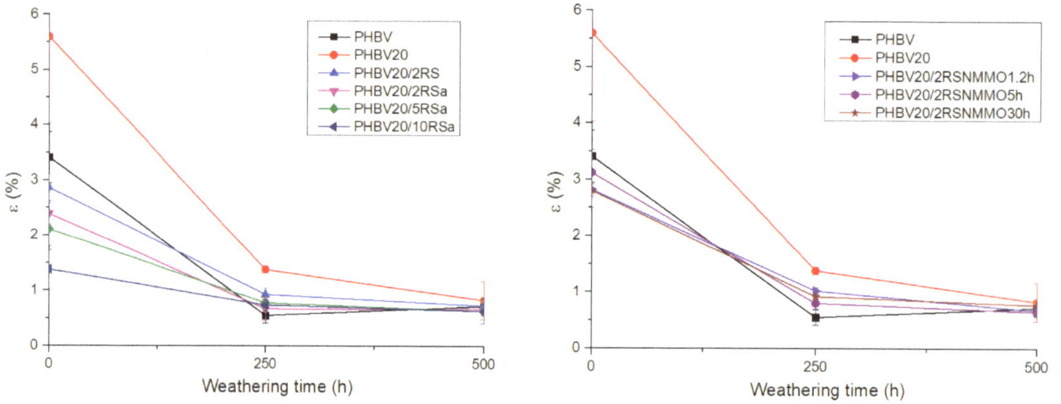

(**c**)

Figure 3. Young's modulus E (**a**), stress at break σ_B (**b**), and ultimate deformation ε_B (**c**), of the PHBV and PHBV biocomposites before accelerated weathering and after 250 h and 500 h of the combined exposure of UV irradiation, temperature, water sprinkling and condensation.

3.2. Biodegradation in Soil

Disregarding the material properties for the expected application, its life cycle is limited. The great advantage of the investigated PHBV composites are their biodegradability, as they totally consist of natural and bio-based constituents. Consequently, in this section, the biodegradability of the developed composites is addressed. Biodegradation is influenced by many factors, such as microbial activity, soil moisture content, temperature, the pH of the environment, the exposed surface area and the composition and molecular weight of the polymer and its crystallinity [34,35]. Rapeseed fibers primarily consist of cellulose, a hydrophilic substance that readily allows water and microorganisms to permeate through its structure to attack the encapsulating polymer matrix, not only via the external environment but also internally. In Table 5, the biodegradation process of the developed PHBV composites in soil is pictured, revealing the evolution of the biodegradation process, accompanied with gradual surface roughening and the development of pits, grooves, cavities and other surface defects, resulting in bulk structure disintegration.

It is shown that biodegradation starts in the PHBV matrix, predominantly in the amorphous phase, revealing the RS fibers, which is especially evident for the PHBV20/10RSa composite. However, except for the plasticized PHBV composite with 2% of untreated RS, biodegradation does not result in fragmentation of the test specimens along with the test run. In the case of the plasticized PHBV composite with untreated RS, fragmentation occurs within 13 days. Disregarding the composition, the investigated test specimens demonstrated ongoing color pattern change during biodegradation in soil. In Table 5, the evolvement of spotted uneven surface images of the test specimens is demonstrated clearly, as biodegradation time is increased.

In Figure 4, optical microscopy images of the investigated compositions are summarized before and after the biodegradation test. It is evident that after the biodegradation test, images become more blurry, which indirectly testifies to the increased surface roughness of the test specimens. The surface of the plasticized PHBV + 2%RS test specimens demonstrate a clearly visible evenly spread multi-fracture network across the surface of the test specimen, testifying to considerable fragmentation. For RSa-containing plasticized PHBV composites surface images do not reveal the development of considerable fractures, except of the composition with the highest RSa content, which may be due to the fact that fracturing initially occurs in the PHBV and RS interface. In the case of NMMO-treated RS-containing composites, fracturing is more visible in comparison to RSa-reinforced PHBV composites, whereas fracturing increases with the increase in NMMO treatment time.

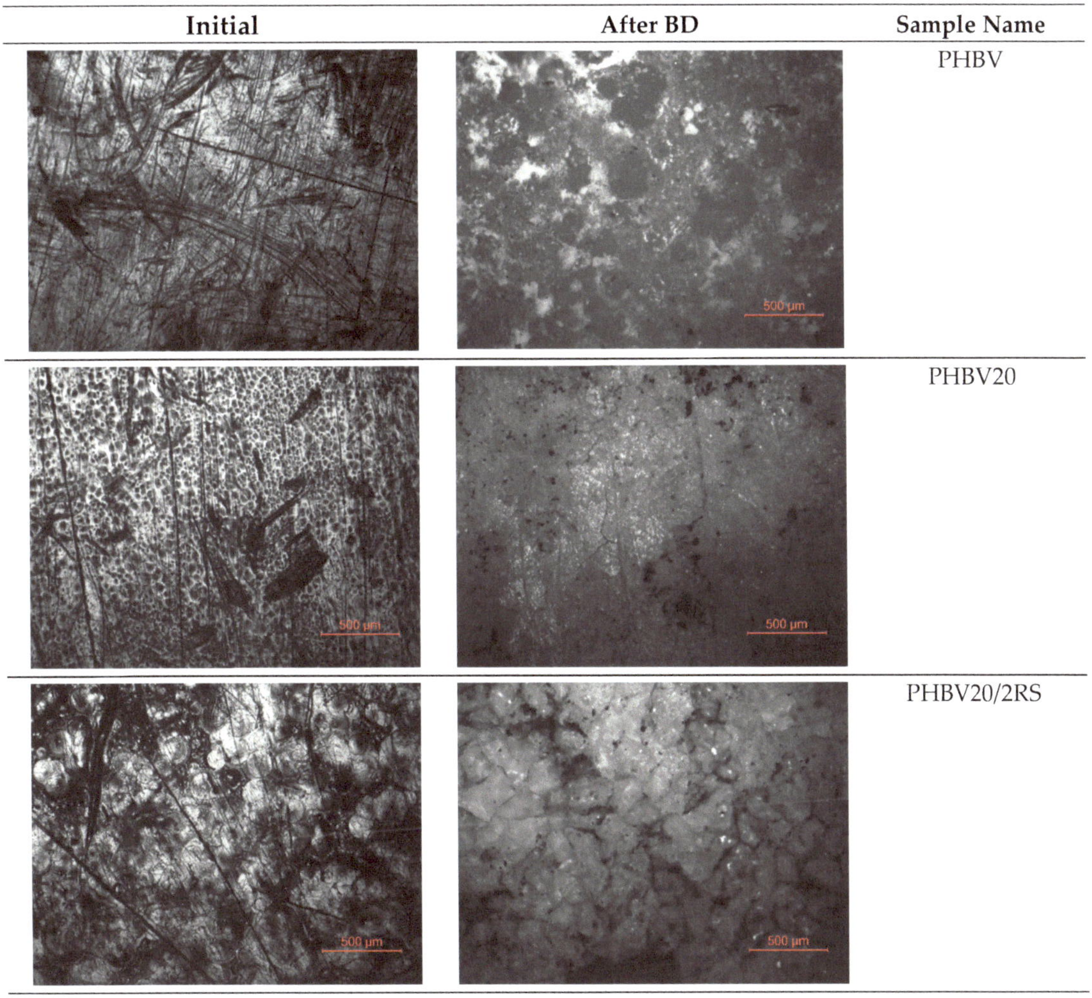

Figure 4. *Cont.*

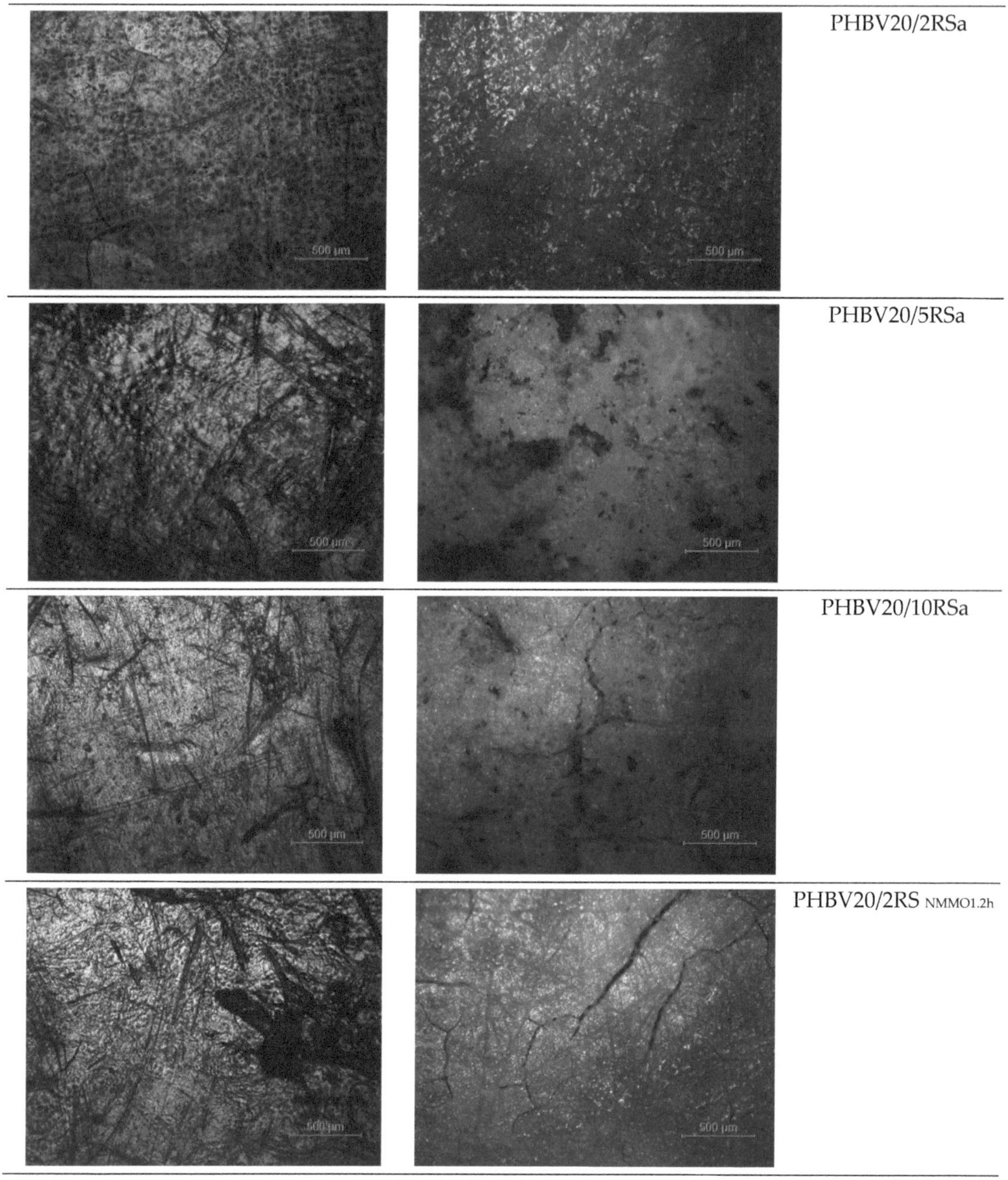

Figure 4. *Cont.*

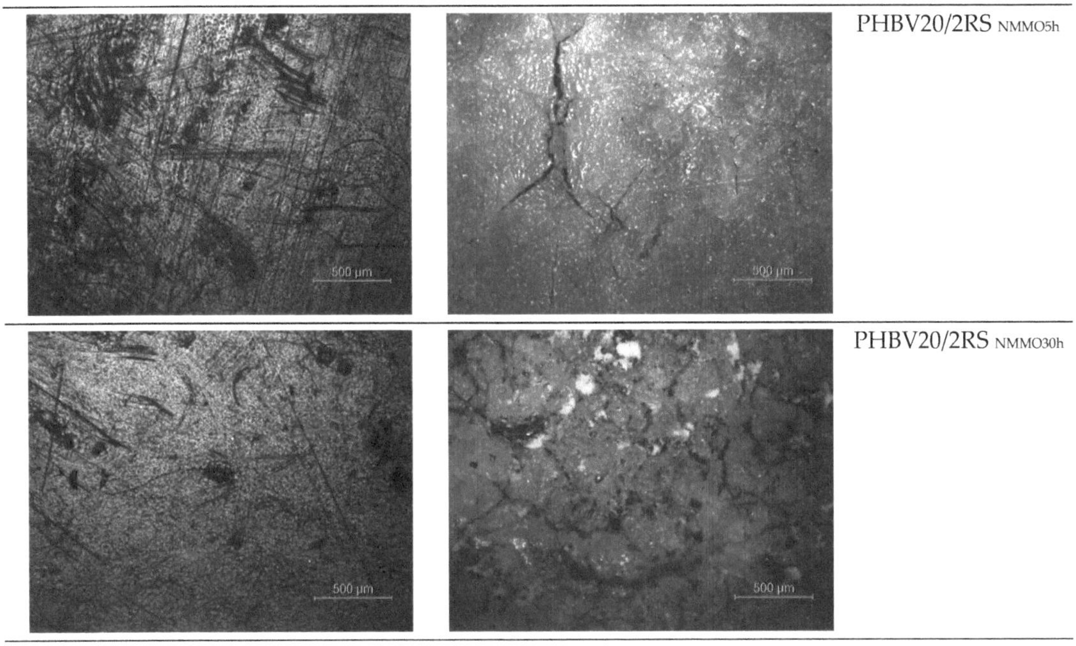

Figure 4. Optical microscopy pictures of PHBV and its plasticized composite samples before and after 3 month biodegradation.

In Figure 5, the mass loss kinetics of the investigated PHBV compositions are shown. It is evident that all the investigated composites demonstrate faster biodegradability in comparison to the neat polymer matrix. This is likely because of the increased hydrophilicity of the composites with the introduction of the fibers in the polymer matrix. TEC also contributes to degradation of PHBV by promoting its hydrolytic degradation. The fastest biodegradation was observed for the PHBV composition with 2 wt % of untreated RS. Surface erosion of the test specimens may have influenced the shape of the biodegradation curves of other PHBV compositions. The reason for faster biodegradation of the composite with untreated RS fibers is most probably lower interfacial adhesion between the polymer matrix and the fiber. Contrarily, due to the fiber treatment, part of the hemicelluloses and lignin are removed along with other impurities, increasing the purified fibers' interaction with the polymer matrix, and hence making biodegradation more difficult. It has been previously reported that fiber treatment influences its structure because of fibrillation and removal of impurities for improved adhesion with polymer [6].

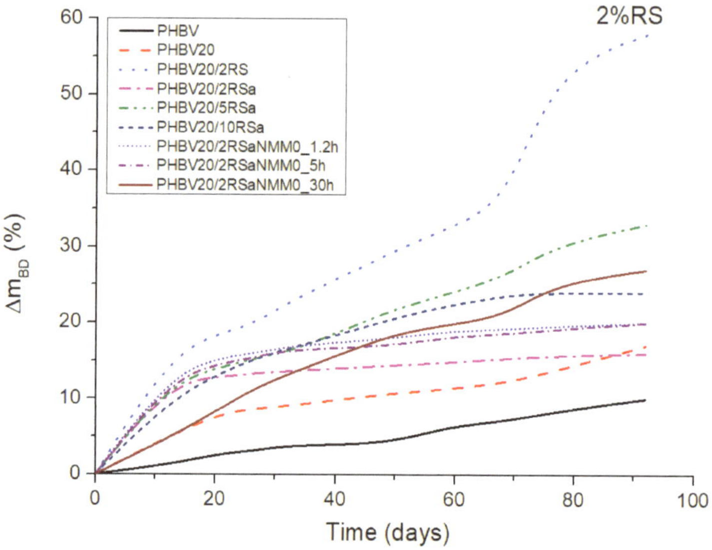

Figure 5. Mass loss during degradation of PHBV and its plasticized composites.

4. Conclusions

This research focuses on the assessment of the structural, mechanical and calorimetric properties change of alkali- and NMMO-treated TEC-plasticized PHBV composites during accelerated weathering and biodegradation in soil. The main results of the study allow the following conclusions. Accelerated weathering during 500 h of the combined exposure of UV irradiation, increased temperature, water sprinkling and condensation led to an increase in stiffness, but a reduction in tensile strength and elongation at break, of the investigated plasticized PHBV biocomposites. It was determined that the largest change in these indicators of mechanical properties occurred prior to 250 h of the exposure, when the reinforcing effect of RS fibers on the modulus of the composites has been lost. The main reasons for the decreased properties were attributed to hydrolytic degradation of PHBV as well as thermal and photochemical destruction of the composites, especially in the presence of RS fibers. Changes in the mechanical properties with increased exposure time were accompanied by increased crystallinity of PHBV, especially in the presence of RS, which was believed to promote crystallization as a nucleant, in addition to the effect of partially degraded PHBV. DSC results correlated with the increment of the crystallinity index, determined as a FTIR-ATR peak intensity ratio between vibrations in crystalline and amorphous-phase $I_{1225/1180}$. Destruction of the PHBV matrix during accelerated weathering was also confirmed by the appearance of multipeak behavior in DSC scans of the UV-exposed composites. In respect to biodegradation, it was observed that all the developed PHBV composites demonstrated faster biodegradation in comparison to neat PHBV matrix. It was suggested that biodegradation primarily occurred in RS-concentrated areas, especially for chemically untreated RS fibers. Contrarily, treatment of RS microfibers delayed biodegradation of the developed PHBV composites due to increased adhesion between the polymer matrix and RS fibers because of partial removal of waxes, lignin and hemicelluloses during the surface treatment.

Supplementary Materials: The following supporting information can be downloaded at: https://www.mdpi.com/article/10.3390/polym16050622/s1, Figure S1: FT-IR spectra at ester carbonyl bond range (1800–1680 cm^{-1}) change during weathering time for neat PHBV (a) and all obtained composites (b–i) according to presented legends; Figure S2: FT-IR spectra at range (1300–1000 cm^{-1}) change during weathering time for neat PHBV (a) and all obtained composites (b–i) according to presented legends; Figure S3: DSC thermograms; Table S1: Colorimetric parameters of PHBV and its plasticized composites during accelerated weathering for 0 h, 250 h, 500 h.

Author Contributions: Conceptualization, R.M.-M. and J.Z.; methodology, M.Ž., R.M.-M., I.B. and A.Ā.; investigation, M.Ž., I.B. and A.Ā.; formal analysis, J.Z.; resources, J.Z. and A.Ā.; data curation, M.Ž., I.B. and A.Ā.; writing—original draft preparation, M.Ž.; writing—review and editing, R.M.-M.; visualization, M.Ž. and T.I.; supervision, R.M.-M.; project administration, T.I.; funding acquisition, J.Z. All authors have read and agreed to the published version of the manuscript.

Funding: This work has been supported by the European Social Fund within the Project No 8.2.2.0/20/I/008 «Strengthening of PhD students and academic personnel of Riga Technical University and BA School of Business and Finance in the strategic fields of specialization» of the Specific Objective 8.2.2 «To Strengthen Academic Staff of Higher Education Institutions in Strategic Specialization Areas» of the Operational Programme «Growth and Employment».

Institutional Review Board Statement: Not applicable.

Data Availability Statement: Data are contained within the article and Supplementary Materials.

Conflicts of Interest: The authors declare no conflicts of interest.

References

1. European Parliament. How the EU Wants to Achieve a Circular Economy by 2050. 2023. Available online: https://www.europarl.europa.eu/news/en/headlines/priorities/circular-economy/20210128STO96607/how-the-eu-wants-to-achieve-a-circular-economy-by-2050 (accessed on 20 October 2023).
2. Vu, D.H.; Åkesson, D.; Taherzadeh, M.J.; Ferreira, J.A. Recycling strategies for polyhydroxyalkanoate-based waste materials: An overview. *Bioresour. Technol.* **2020**, *298*, 122393. [CrossRef] [PubMed]
3. Sharma, V.; Shegal, R.; Reena, G. Polyhydroxyalkanoate (PHA): Properties and Modifications. *Polymers* **2021**, *212*, 123161. [CrossRef]
4. Kaniuk, Ł.; Stachewicz, U. Development and Advantages of Biodegradable PHA Polymers Based on Electrospun PHBV Fibers for Tissue Engineering and Other Biomedical Applications. *ACS Biomater. Sci. Eng.* **2021**, *7*, 5339–5362. [CrossRef] [PubMed]
5. Rivera-Briso, A.L.; Serrano-Aroca, Á. Poly(3-Hydroxybutyrate-co-3-Hydroxyvalerate): Enhancement Strategies for Advanced Applications. *Polymers* **2018**, *10*, 732. [CrossRef] [PubMed]
6. Zhao, X.; Lawal, T.; Rodrigues, M.M.; Geib, T.; Vodovotz, Y. Value-Added Use of Invasive Plant-Derived Fibers as PHBV Fillers for Biocomposite Development. *Polymers* **2021**, *13*, 1975. [CrossRef]
7. Berthet, M.A.; Angellier-Coussy, H.; Chea, V.; Guillard, V.; Gastaldi, E.; Gontard, N. Sustainable food packaging: Valorising wheat straw fibres for tuning PHBV-based composites properties. *Compos. Part A Appl. Sci. Manuf.* **2015**, *72*, 139–147. [CrossRef]
8. Umemura, R.T.; Felisberti, M.I. Plasticization of poly(3-hydroxybutyrate) with triethyl citrate: Thermal and mechanical properties, morphology, and kinetics of crystallization. *J. Appl. Polym. Sci.* **2021**, *138*, 49990. [CrossRef]
9. Bertheta, M.A.; Angellier-Coussya, H.; Machadob, D.; Hillioub, L.; Staeblerc, A.; Vicented, A.; Gontard, N. Exploring the potentialities of using lignocellulosic fibres derived from three food by-products as constituents of biocomposites for food packaging. *Ind. Crops Prod.* **2015**, *69*, 110–122. [CrossRef]
10. Frącz, W.; Janowski, G.; Smusz, R.; Szumski, M. The Influence of Chosen Plant Fillers in PHBV Composites on the Processing Conditions, Mechanical Properties and Quality of Molded Pieces. *Polymers* **2021**, *13*, 3934. [CrossRef] [PubMed]
11. Singh, S.; Mohanty, A.K. Wood fiber reinforced bacterial bioplastic composites: Fabrication and performance evaluation. *Compos. Sci. Technol.* **2007**, *67*, 1753–1763. [CrossRef]
12. Rout, A.K.; Kar, J.; Jesthi, D.K.; Sutar, A.K. Effect of Surface Treatment on the Physical, Chemical, and Mechanical Properties of Palm Tree Leaf Stalk Fibers. *BioResources* **2016**, *11*, 4432–4445. [CrossRef]
13. Hashim, M.Y.; Amin, A.M.; Marwah, O.M.F.; Othman, M.H.; Yunus, M.R.M.; Chuan Huat, N. The effect of alkali treatment under various conditions on physical properties of kenaf fiber. *J. Phys. Conf. Ser.* **2017**, *914*, 1. [CrossRef]
14. Norul Izani, M.A.; Paridah, M.T.; Anwar, U.M.K.; Mohd Nor, M.Y.; H'ng, P.S. Effects of fiber treatment on morphology, tensile and thermogravimetric analysis of oil palm empty fruit bunches fibers. *Compos. Part B Eng.* **2013**, *45*, 1251–1257. [CrossRef]
15. Santos, E.B.C.; Moreno, C.G.; Barros, J.J.P.; de Moura, D.A.; de Carvalho Fim, F.; Ries, A.; Wellen, R.M.R.; da Silva, L.B. Effect of alkaline and hot water treatments on the structure and morphology of piassava fibers. *Mater. Res.* **2018**, *21*, 0365. [CrossRef]
16. Stelte, W. *Steam Explosion for Biomass Pre-Treatment*; Danish Technological Institute: Taastrup, Denmark, 2013; pp. 1–15. [CrossRef]
17. Merijs-Meri, R.; Zicans, J.; Ivanova, T.; Bochkov, I.; Varkale, M.; Franciszczak, P.; Bledzki, A.K.; Danilovas, P.P.; Gravitis, J.; Rubenis, K.; et al. Development and characterization of grain husks derived lignocellulose filler containing polypropylene composites. *Polym. Eng. Sci.* **2019**, *59*, 2467–2473. [CrossRef]
18. Rosenau, T.; Potthast, A.; Sixta, H.; Kosma, P. The chemistry of side reactions and byproduct formation in the system NMMO/cellulose (Lyocell process). *Prog. Polym. Sci.* **2001**, *26*, 1763–1837. [CrossRef]

19. Ābele, A.; Bērziņš, R.; Bērziņa, R.; Merijs-Meri, R.; Žiganova, M.; Zicāns, J. Potential uses of N-methylmorpholine N-oxide for the treatment of agricultural waste biomass. *Proc. Estonian Acad. Sci.* **2023**, *72*, 176–183. [CrossRef]
20. Batista, K.C.; Silva, D.A.K.; Coelho, L.A.F.; Pezzin, S.H.; Pezzin, A.P.T. Soil Biodegradation of PHBV/Peach Palm Particles Biocomposites. *J. Polym. Environ.* **2010**, *18*, 346–354. Available online: https://link.springer.com/article/10.1007/s10924-010-0238-4 (accessed on 5 November 2023). [CrossRef]
21. Wei, L.; Liang, S.; McDonald, A.G. Thermophysical properties and biodegradation behavior of green composites made from polyhydroxybutyrate and potato peel waste fermentation residue. *Ind. Crops Prod.* **2015**, *69*, 91–103. [CrossRef]
22. Žiganova, M.; Merijs-Meri, R.; Zicāns, J.; Ivanova, T.; Bochkov, I.; Kalniņš, M.; Błędzki, A.K.; Danilovas, P.P. Characterisation of Nanoclay and Spelt Husk Microfiller-Modified Polypropylene Composites. *Polymers* **2022**, *14*, 4332. [CrossRef]
23. Žiganova, M.; Merijs-Meri, R.; Zicāns, J.; Bochkov, I.; Ivanova, T.; Vīgants, A.; Ence, E.; Štrausa, E. Visco-Elastic and Thermal Properties of Microbiologically Synthesized Polyhydroxyalkanoate Plasticized with Triethyl Citrate. *Polymers* **2023**, *15*, 2896. [CrossRef]
24. *ISO 4892–3:2016*; Plastics—Methods of Exposure to Laboratory Light Sources—Part 3: Fluorescent UV Lamps. International Organization for Standardization (ISO): Geneva, Switzerland, 2016.
25. *ASTM D 2244-02*; Standard Practice for Calculation of Color Tolerances and Color Differences from Instrumentally Measured Color Coordinates. ASTM: West Conshohocken, PA, USA, 2002.
26. Stanley, A.; Murthy, P.S.K.; Vijayendra, S.V.N. Characterization of Polyhydroxyalkanoate Produced by Halomonas venusta KT832796. *J. Polym. Environ.* **2020**, *28*, 973–983. Available online: https://link.springer.com/article/10.1007/s10924-020-01662-6 (accessed on 5 September 2023). [CrossRef]
27. *EN ISO 527*; Plastics—Determination of Tensile Properties—Part 1: General Principles (ISO 527-1:2019). International Organization for Standardization (ISO): Geneva, Switzerland, 2019.
28. Nabels-Sneiders, M.; Platnieks, O.; Grase, L.; Gaidukovs, S. Lamination of Cast Hemp Paper with Bio-Based Plastics for Sustainable Packaging: Structure-Thermomechanical Properties Relationship and Biodegradation Studies. *J. Compos. Sci.* **2022**, *6*, 246. [CrossRef]
29. Kann, Y.; Shurgalin, M.; Krishnaswamy, R.K. FTIR Spectroscopy for Analysis of Crystallinity of Poly(3-hydroxybutyrate-co-4-hydroxybutyrate) Polymers and Its Utilization in Evaluation of Aging, Orientation and Composition. *Polym. Test.* **2014**, *40*, 218–224. [CrossRef]
30. Antunes, A.; Popelka, A.; Aljarod, O.; Hassan, M.K.; Kasak, P.; Luyt, A.S. Accelerated Weathering Effects on Poly(3-hydroxybutyrate-co-3-hydroxyvalerate) (PHBV) and PHBV/TiO_2 Nanocomposites. *Polymers* **2020**, *12*, 1743. [CrossRef]
31. Iggui, K.; Kaci, M.; Le Moigne, N.; Bergeret, A. Effects of hygrothermal aging on chemical, physical, and mechanical properties of poly(3-hydroxybutyrate-co-3-hydroxyvalerate)/Cloisite 30B bionanocomposite. *Polym. Compos.* **2021**, *42*, 1878–1890. [CrossRef]
32. Tor-Świątek, A.; Garbacz, T. Effect of Abiotic Degradation on the Colorimetric Analysis, Mechanical Properties and Morphology of PLA Composites with Linen Fibers. *Adv. Sci. Technol. Res. J.* **2021**, *15*, 99–109. [CrossRef]
33. Czerniecka-Kubicka, A.; Frącz, W.; Jasiorski, M.; Pilch-Pitera, B.; Pyda, M.; Zarzyka, I. Thermal Properties of poly(3-hydroxybutyrate) Modified by Nanoclay. *J. Therm. Anal. Calorim.* **2017**, *128*, 1513–1526. Available online: https://link.springer.com/article/10.1007/s10973-016-6039-9 (accessed on 13 January 2024). [CrossRef]
34. Lee, S.Y. Plastic bacteria? Progress and prospects for polyhydroxyalkanoate production in bacteria. *Trends Biotechnol.* **1996**, *14*, 431–438. [CrossRef]
35. Boopathy, R. Factors limiting bioremediation technologies. *Bioresour. Technol.* **2000**, *74*, 63–67. [CrossRef]

polymers

Article

Effect of Starch and Paperboard Reinforcing Structures on Insulative Fiber Foam Composites

Gregory M. Glenn [1,*], Gustavo H. D. Tonoli [2], Luiz E. Silva [2], Artur P. Klamczynski [1], Delilah Wood [1], Bor-Sen Chiou [1], Charles Lee [1], William Hart-Cooper [1], Zach McCaffrey [1] and William Orts [1]

1 United States Department of Agriculture, Agricultural Research Service, Western Regional Research Center, Bioproducts Research Unit, 800 Buchanan Street, Albany, CA 94710, USA; artur.klamczynski@usda.gov (A.P.K.); de.wood@usda.gov (D.W.); bor-sen.chiou@usda.gov (B.-S.C.); charles.lee@usda.gov (C.L.); william.hart-cooper@usda.gov (W.H.-C.); zach.mccaffrey@usda.gov (Z.M.); bill.orts@usda.gov (W.O.)

2 Forest Science Department, Federal University of Lavras, Lavras 37203-202, MG, Brazil; gustavotonoli@ufla.br (G.H.D.T.); lesilvaflorestal@gmail.com (L.E.S.)

* Correspondence: greg.glenn@usda.gov

Abstract: Single-use plastic foams are used extensively as interior packaging to insulate and protect items during shipment but have come under increasing scrutiny due to the volume sent to landfills and their negative impact on the environment. Insulative compression molded cellulose fiber foams could be a viable alternative, but they do not have the mechanical strength of plastic foams. To address this issue, a novel approach was used that combined the insulative properties of cellulose fiber foams, a binder (starch), and three different reinforcing paperboard elements (angular, cylindrical, and grid) to make low-density foam composites with excellent mechanical strength. Compression molded foams and composites had a consistent thickness and a smooth, flat finish. Respirometry tests showed the fiber foams mineralized in the range of 37 to 49% over a 46 d testing period. All of the samples had relatively low density (D_d) and thermal conductivity (TC). The D_d of samples ranged from 33.1 to 64.9 kg/m^3, and TC ranged from 0.039 to 0.049 W/mk. The addition of starch to the fiber foam (FF+S) and composites not only increased D_d, drying time (T_d), and TC by an average of 18%, 55%, and 5.5%, respectively, but also dramatically increased the mechanical strength. The FF+S foam and paperboard composites had 240% and 350% higher average flexural strength (σ_{fM}) and modulus (E_f), respectively, than the FF-S composites. The FF-S grid composite and all the FF+S foam and composite samples had equal or higher σ_{fM} than EPS foam. Additionally, FF+S foam and paperboard composites had 187% and 354% higher average compression strength (CS) and modulus (E_c), respectively, than the FF-S foam and composites. All the paperboard composites for both FF+S and FF-S samples had comparable or higher CS, but only the FF+S cylinder and grid samples had greater toughness (Ω_c) than EPS foam. Fiber foams and foam composites are compatible with existing paper recycling streams and show promise as a biodegradable, insulative alternative to EPS foam internal packaging.

Keywords: packaging foam; plastic foam; renewable; compostable; sustainable; starch; plant-based composites

Citation: Glenn, G.M.; Tonoli, G.H.D.; Silva, L.E.; Klamczynski, A.P.; Wood, D.; Chiou, B.-S.; Lee, C.; Hart-Cooper, W.; McCaffrey, Z.; Orts, W. Effect of Starch and Paperboard Reinforcing Structures on Insulative Fiber Foam Composites. *Polymers* **2024**, *16*, 911. https://doi.org/10.3390/polym16070911

Academic Editors: Stefano Farris and Masoud Ghaani

Received: 2 March 2024
Revised: 21 March 2024
Accepted: 24 March 2024
Published: 26 March 2024

1. Introduction

The packaging and distribution of goods is a multi-billion dollar business worldwide, with nearly 500 million packages being transported every day using a myriad of different package configurations [1,2]. In 2022, 161 billion packages were shipped worldwide. That number is expected to reach 256 billion by 2027 [3]. Although there is no meaningful biodegradation of commodity plastics, which are mostly derived from non-renewable resources, they continue to play a major role in the packaging sector. More than 40% of the worldwide production of plastics is used for packaging, much of which is single-use [4,5].

A small percentage of plastics is reused/recycled, but roughly 80% is either landfilled, incinerated, or leaked into the environment [4,6].

For many commercial products such as small appliances, printers, etc., corrugated paperboard is used as an exterior packaging material, while plastic foam primarily from expanded polystyrene (EPS) or polypropylene (EPP) is used as interior packaging/cushioning material [4,7]. EPS foam is one of the preferred internal packaging foams because of its light weight, impact and moisture resistance, low cost, and ability to protect products from temperature extremes [7–10].

Despite its many advantages for internal packaging, EPS foam has become widely recognized for its negative impact on the environment [9,11]. EPS is very resistant to biodegradation [12]. While there are claims that 19–25% of EPS foam is recycled [13], that amount is disputed, in part because there are too few recycling centers available that process it [9]. Furthermore, EPS recycling is expensive partly due to its bulk and resistance to compaction [13–15]. These and other concerns have led several U.S. states and countries to enact legislation to ban single-use EPS foam products in an effort to phase out its use [16–21].

Alternatives to plastic foam packaging are being considered, using bioplastics such as poly(lactic acid) (PLA) and polyhydroxyalkanoate (PHA) as "drop-in" replacements for commodity plastics [11,22]. However, PHAs are still too expensive for single-use packaging, and PLA, like commodity plastics, degrades very slowly in marine and landscape environments [22]. PLA will biodegrade under humid conditions at elevated temperatures, but only a limited number of industrial composting operations are designed to handle PLA products [23]. Furthermore, while PLA can be foamed, the preferred foaming process is expensive due to the complexity and challenges involved in using supercritical fluids [24].

The most successful recycled/reused packaging material is paper and paperboard including corrugated paperboard used for external packaging. The EPA reported that paper and paperboard made up nearly 67% of the recycled municipal solid waste (MSW) materials in the U.S., while paper recycling in Europe exceeds 70% [25,26]. These cellulose-based products are biodegradable, derived from renewable resources, easy to reuse/recycle, and, unlike commodity plastics, will disintegrate and decompose if leaked into waterways or landscape environments [26]. There is growing interest in exploring and expanding the use of cellulose-based materials for internal packaging applications that can supplant EPS foam and be recycled or composted along with paperboard using well-established processing streams.

Interconnecting grid and honeycomb paperboard panels are examples of cellulose-based products designed primarily for internal packaging. These paperboard structures comprise a core consisting of empty square or hexagonal cells constructed in a grid or honeycomb pattern, respectively. The paperboard core may be sandwiched between two face sheets that adhere to the top and bottom surfaces and securely bind/anchor the core [27]. Grid/honeycomb paperboard is lightweight, has excellent compressive strength and shock and vibration resistance and is used extensively as cushioning material for transporting electronic equipment, appliances, furniture, etc., but lacks the insulation properties of EPS foam [27–29]. Abd Kadir et al. (2016) [30] filled the void spaces in the core of a paperboard honeycomb with low-density polyurethane foam to insulate and strengthen the walls of the core. The composite had superior compressive strength, but, like EPS foam, polyurethane foam is not compostable, and such composites would be difficult to recycle.

A foam that is compostable and recyclable can be made from aqueous cellulose fiber suspensions using a foaming agent [31,32]. Recently, a compostable foam made from cellulose fiber with excellent insulative properties was described that could be compression molded into distinct shapes or large panels needed for internal packaging [33]. However, by themselves, these fiber foams do not have the compressive strength or toughness that may be needed for many internal packaging applications [33,34]. Starch has been used to make biopolymer blends with excellent strength and toughness [35]. To our knowledge, there have been no studies reporting the properties of composites made from cellulose fiber foam and paperboard grids or other reinforcing structures that could provide the mechanical strength, toughness, and insulative properties EPS foam packaging provides. The objective of this study was to investigate the physical and mechanical properties of fiber foam/paperboard composites with and without a starch binder and explore their potential as an alternative to EPS foam packaging.

2. Materials and Methods

2.1. Materials

Pulped softwood fiber sheets were obtained from International Paper (Global Cellulose Fibers, Memphis, TN, USA) and produced at their Columbus, MS mill. The fiber was a Southern bleached softwood Kraft with a fiber length ranging from 3.8 to 4.4 mm and an ash content of 0.12%. Reagent-grade sodium dodecyl sulfate (SDS, Cas 151-21-3) was purchased from Thermo Fisher Scientific (Waltham, MA, USA). Paperboard (brown kraft cardboard chipboard (22 point with a thickness of 0.56 mm) was purchased from Magicwater Via GSD (Fontana, CA, USA). Brown kraft paperboard tubes (40 mm diameter × 100 mm length × 0.45 mm thickness) were purchased locally. Polyvinyl alcohol (PVA, Selvol 540, 88% hydrolyzed, 12% acetate, MW = 120,000) was purchased from Sekisui Chemical (Pasadena, TX, USA). Water-soluble pregelatinized waxy corn starch powder (Clearjel, Ingredion, Westchester, IL, USA) containing 0.2% ash, 0.1% protein, and <0.1% fat was obtained from Ingredion (Westchester, IL, USA). Expanded polystyrene (EPS) foam sheets (122 cm × 30.5 cm × 2.62 cm) were purchased locally.

2.2. Paperboard Support Elements

Three different support elements were prepared, consisting of an angular, cylindrical, and interlocking grid design. The angular elements were made by folding paperboard strips (26 mm in width and 52 mm in length) in half to form a 90-degree angle. Cylindrical elements were made by cutting paperboard tubes (40 mm dia.) to a length of 26 mm. The paperboard grid was made by assembling strips 26 mm in width with slots cut every 38 mm along the length into a grid pattern. The paperboard elements were embedded in the fiber foam, as described below.

2.3. Solution Preparation

An aqueous polyvinyl alcohol (PVA) solution (5%, w/w) was made by gradually adding PVA powder to cold water while continuously stirring and then slowly heating (95 °C) until the PVA was solubilized. Water was added to compensate for weight loss due to evaporation. A 29% (w/w) aqueous solution of SDS was made by combining SDS powder and water at room temperature and continuously stirring to achieve dissolution.

2.4. Foam Procedure

A low-moisture fiber foam formulation without starch (FF-S) was developed based on prior research (Table 1) [33]. The pulped fiber was prepared by first weighing the appropriate amount of pulp fiber (Table 1) and placing it in a blender containing approximately 2 L of warm (60 °C) tap water. The fiber was blended for approximately 30 s to disperse and hydrate the fiber. The fiber was allowed to hydrate for approximately 15 min before blending again for 30 s. The fiber mixture was then poured onto a screen (50 mesh) to allow drainage. The fiber was collected from the screen and compressed to expel excess water until the approximate combined weight of the fiber and water was reached for each sample formulation (Table 1). The final combined weight of water and fiber was adjusted by adding water to bring the mixture to the desired final weight.

Table 1. Formulations of fiber foam (FF) samples with and without starch (+S and −S, respectively). The percentage of each ingredient is included in the parentheses.

Sample	FF (−S)	FF (+S)
Fiber	50 g	50 g
	(19.7%)	(15.9%)
Water	100 g	150 g
	(39.4%)	(47.6%)
PVA (5% soln)	100 g	100 g
	(39.4%)	(31.7%)
SDS (29% soln)	4 g	8 g
	(1.57%)	(2.54%)
Starch	0 g	7 g
		(2.22%)

The combined fiber and water sample was added to a 4 L mixing bowl of a planetary mixer (Model KSM 90, KitchenAid, Inc., St. Joseph, MI, USA). For the control sample, additional ingredients were added, as shown in Table 1. The initial weight of the mixing bowl and ingredients was recorded. Water was added occasionally during the mixing step to compensate for weight loss due to evaporation. Mixing started slowly (speed 3) and gradually increased to a speed of 10. A spatula was used to occasionally wipe down the bowl during mixing. The PVA and SDS both facilitated the dispersion of the fiber and prevented aggregation. Once a foam was produced, mixing was paused to measure the wet density (D_w) of the foam and to add water to compensate for any weight loss that occurred due to evaporation. D_w was determined by filling a cup to level with wet foam and recording the weight and volume. The foam was mixed until the desired D_w (Table 2) was achieved.

Table 2. Wet density (D_w), foam volume (Va), drying time (T_d), thickness (T), dry density (D_d), porosity (P), and thermal conductivity (TC) of wet and dry fiber foam with and without starch (FF+S and FF-S, respectively) and composites containing paperboard elements (angle, cylinder, and grid) prepared using a planetary mixer. EPS = expanded polystyrene.

Sample	FF-S	FF+S	FF-S Angle	FF+S Angle	FF-S Cyl.	FF+S Cyl.	FF-S Grid	FF+S Grid	EPS Foam
D_w (kg/m^3)	125 ^{a*}	182 ^{b}	N/A	N/A	N/A	N/A	N/A	N/A	N/A
Va (%)	869 ^{a}	601 ^{b}	N/A	N/A	N/A	N/A	N/A	N/A	N/A
T_d (min)	336 ^{a}	528 ^{c}	333 ^{a}	510 ^{c}	311 ^{a}	529 ^{c}	399 ^{b}	553 ^{c}	N/A
T (cm)	2.66 ^{a}	2.62 ^{a}	2.63 ^{a}	2.63 ^{a}	2.62 ^{a}	2.61 ^{a}	2.70 ^{a}	2.65 ^{a}	2.61 ^{a}
D_d (kg/m^3)	33.1 ^{b}	39.1 ^{c}	35.9 ^{b,c}	44.9 ^{d}	39.1 ^{c}	44.9 ^{d}	57.1 ^{e}	64.9 ^{f}	14.1 ^{a}
P (%)	97.9 ^{a}	97.5 ^{a}	N/A	N/A	N/A	N/A	N/A	N/A	98.6 ^{b}
TC (W/mK)	0.039 ^{a,b}	0.043 ^{a,b,c}	0.042 ^{a,b}	0.044 ^{b,c,d}	0.042 ^{a,b}	0.044 ^{b,c,d}	0.048 ^{c,d}	0.049 ^{d}	0.038 ^{a}

* Mean values within rows followed by a different letter are significantly different ($p < 0.05$).

The mixing procedure for the fiber foam with starch (FF+S) sample was similar to FF-S except for the fact that the water-soluble starch powder was gradually added to the mixing bowl only after the ingredients had started to foam. The starch powder was slowly added to the foam while mixing to ensure that the starch was properly dispersed and solubilized in the foam mixture. Starch tended to reduce the foam volume so higher amounts of water and foaming agent were added to compensate for the reduction in foam volume (Table 1). Notwithstanding the additional amount of water and foaming agent, the final D_w of the foam containing starch was higher than foam without starch (Table 2).

The air uptake volume (*Va*) of the foam was calculated using Equation (1), where V_{system} is the volume of the ingredients before foaming, and V_{air} is the bulk volume of the foamed material. The V_{system} was derived from the specific gravity of each component. Specific gravity values were obtained using a helium gas displacement pycnometer (Micromeritics, model AcuPyc II 1340, Norcross, GA, USA). The specific gravity values (g/cm^3) of the dry ingredients used in calculations included the following: fiber (1.61); PVA (1.30); SDS (1.01); and starch (1.46). The specific gravity of water (1.0 g/cm^3) was used to determine the volume of water added, including in the SDS and PVA solutions. The V_{system} values (cm^3) for the control and starch formulations, as shown in Table 1, were 234 cm^3 and 288 cm^3, respectively. The V_{air} values for the control and starch formulations were 2032 cm^3 and 1731 cm^3, respectively.

$$\text{Va}\ (\%) = \text{V}_{\text{air}}/\text{V}_{\text{system}} \times 100 \tag{1}$$

2.5. Compression Molding

The mold assembly consisted of upper and lower porous platen assemblies, as described previously [33]. The volume of the mold cavity was calculated from dimensional measurements. The weight of foam required to overfill the mold to 135% of the mold volume was calculated from the D_w values. After loading the mold with excess foam, the upper platen was lowered, which compressed the foam, causing it to flow and conform to the mold and form a skin on the upper and lower surfaces that were in contact with the platens. For the fiber foam/paperboard composites, approximately 80% of the foam was added to the mold. A spatula was used to spread the foam uniformly inside the mold. The paperboard elements were then carefully pressed into the foam in a prescribed pattern (Figure 1). The remaining quantity of foam was spread on top of the paperboard elements, and the upper platen was lowered, which compressed the foam, causing it to flow and fill any voids as previously described [33]. The intact platen assembly was placed in an oven for drying.

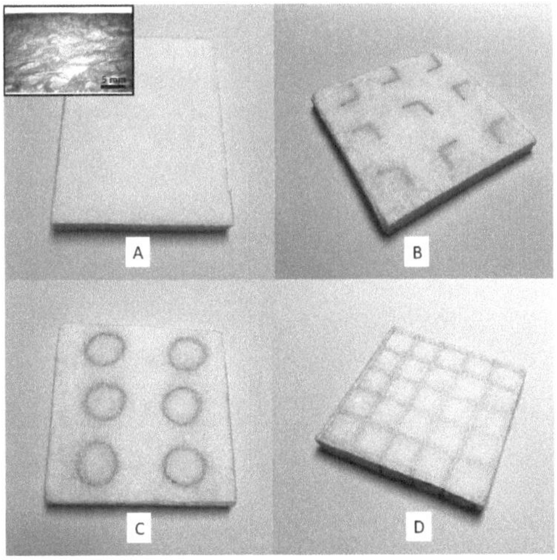

Figure 1. Photographs of fiber foam and paperboard composite samples where the platen assemblies were removed only after the drying process. Samples included the fiber foam (**A**) and composites containing angled (**B**), cylindrical (**C**), and grid (**D**) paperboard elements. Insert in A is a micrograph of cross-sectional view of foam. Scale bar = 5 mm.

2.6. Drying

The foam samples were oven-dried at 80 °C. The weight loss was monitored by periodically weighing the samples. The end time of drying was recorded as the point where less than 0.15% of the initial weight of the foam was lost over a 30 min drying interval. The initial and end times for drying were used to record the total drying time. Once drying was completed, the platen assemblies were dismantled, and the molded foam sample was removed and stored in a plastic bag at room temperature until further testing.

2.7. Mechanical Properties

The compressive and flexural properties of the samples were measured using a universal testing machine (Model ESM303, Mark-10, Copiague, NY, USA). The compressive properties of dry foams were measured on samples cut to dimensions approximately (5 cm × 5 cm) as per ASTM standard D-1621 [36]. Final dimensions were measured using calipers. The samples were conditioned for 48 h in a chamber with a small circulating fan. The relative humidity of the chamber was maintained near 50% using a saturated salt solution ($Mg(NO_3){\cdot}6H_2O$) as previously described [37]. Compression tests were performed using a deformation rate of 12.5 mm/min as per established methods (ASTM D 1621) [36]. Compressive strength was recorded as the stress at the yield point before 10% strain. The fiber foam without starch did not have a clear yield point, so the stress at 10% strain was recorded as the *CS* as per ASTM standard [36]. Samples were subjected to five load/unload cycles up to 50% strain using a deformation rate of 2.5 mm/min. The area under the loading curve was used to calculate toughness (Ω). A minimum of five replicates were made for each treatment.

2.8. Flexural Tests

Three-point flexural tests were performed using samples cut to dimensions approximately (20 cm × 5.0 cm × 2.6 cm). The final width and thickness measurements of samples were recorded using calipers. The flexural tests were performed using a deformation rate of 2.5 mm min^{-1}, a span distance of 152 mm, and a span/depth ratio of 5.85. Flexural stress (σ_f) and strain (ε_f) were calculated as per ASTM D790 [38].

2.9. Physical Properties

The dry bulk densities of the samples were determined from volume and weight measurements of oven-dried specimens [39]. Helium gas displacement pycnometry was used to determine the specific density (*dn*) of the foam solids. Porosity (*P*) was determined from the bulk density of the foams (*da*) and the specific density of the foam (*dn*) using Equation (2), which was obtained from the simple mixing rule with a negligible gas density [39]. The *dn* value of the foam solids from gas pyncnometry was 1.55 g/cm^3.

$$P(\%) = 100 \times (1 - da/dn) \quad (2)$$

2.10. Thermal Conductivity

Thermal conductivity was measured at a mean temperature of 22.7 °C on panel samples for each treatment according to standard methods (ASTM C-177-85) [40] using a thermal conductivity instrument (model GP-500, Sparrell Engineering, Damarascotta, ME, USA). Readings were taken at 1 h intervals as the instrument approached thermal equilibrium.

2.11. Respirometry

An automated respirometer system (Microoxymax, Columbus Instruments, Columbus, OH, USA) was used to monitor the mineralization of the fiber foams as per ASTM methods (D5338) with only minor modifications. Compost purchased locally was sieved (14 mesh) and stored overnight for moisture equilibration. Moisture content was determined gravimetrically by drying 10 g samples at 105 °C for 16 h. Fiber foam samples (with and without starch) were cut into small pieces (<5 mm) and weighed (~0.5 g) to the nearest 0.1 mg. The samples were added to a reaction jar along with compost (24.5 g), taking care to ensure uniform mixing. The moisture content was adjusted to 58% by adding water before beginning a run. Samples were kept for two days at 30 °C before raising the temperature to 58 °C. During the run, the CO_2 concentration was measured at 2 h intervals. Water (2 mL) was added daily to maintain the moisture content range between 50 and 60%. The carbon content of the samples was determined using a CHN Analyzer (Elementar Vario el Cube, Ronkonkoma, NY, USA). The theoretical percent biodegradation was calculated as the ratio of the moles of carbon in the sample versus the accumulated moles of CO_2 produced utilizing the ideal gas law as previously described [41].

2.12. Microscopy

Light micrographs were taken using a digital microscope (Dino-Lite model AM3113, Torrance, CA, USA) equipped with image capture software (Dinocapture 2.0). Cross-sectional slices (1 cm) of fiber foam samples were cut using a scroll saw. Backlighting was used to provide higher-contrast photomicrographs.

2.13. Thermogravimetric Analysis (TGA)

A Mettler Toledo TGA/DSC 3+ thermogravimetric analyzer (Greifensee, Switzerland) was used to determine the thermal stability of the foams. Each sample was first conditioned at 23 °C in a 50% relative humidity chamber for at least 48 h. The 8–11 mg sample was then heated from 30 °C to 650 °C in an alumina crucible at 10 °C/min. The sample chamber was purged with nitrogen gas at 40 cm^3/min.

2.14. Statistical Analysis

The data were analyzed by a one-way analysis of variance. A Tukey–Kramer Post Hoc test ($\alpha < 0.05$) was used to determine differences between treatment means. Significant differences were noted by the different letters following the mean values within rows in data tables.

3. Results and Discussion

Paperboard panels made with paperboard elements, including honeycomb, grid, and multi-cell lattice designs, can provide the mechanical strength and shock resistance needed in many internal packaging applications, but they lack the thermal insulation provided by plastic foams that may be needed for some packaging systems [27,29]. In the present study, cellulose fiber-based foam panels, both with and without a binder (starch), were produced that had good insulation properties and were attractive, flat (no warping, Figure 1), and had uniform thickness (Table 2). The foam comprised a core of entangled fibers with skin on the surface (see insert, Figure 1A). Three different paperboard reinforcing elements (angle, cylindrical, and interlocking grid) were embedded into the wet foam, which was then dried. The paperboard elements were tested to demonstrate that both interlocking and non-interlocking paperboard elements could be used with the foaming process (Figure 1). Non-interlocking elements, i.e., angle and cylindrical elements, were essentially anchored into position by the foam itself as it dried (Figure 1). This eliminated the need to glue face sheets onto the surfaces of the panels, as is common in paperboard honeycomb panels [29].

The foaming procedure provides the flexibility to test different permutations in the paperboard composite concept. For instance, different geometrical designs, fewer or more elements, varied spatial arrangement, and a variety of different paperboard thicknesses could be tested to help optimize the mechanical properties, density, and thermal conductivity of the fiber foam panels. The total weight of each set of angles, cylindrical, and grid elements used in making panels was 4.6, 6.2, and 22.4 g, respectively. Coincidentally, the D_d and thermal conductivity (TC) of the foam composites were positively correlated with the weight of the paperboard element sets used (Table 2).

All the composites tested had relatively low thermal conductivity and density but there were still significant differences among the samples themselves (Table 2). The EPS foam had the lowest density (D_d) and thermal conductivity (TC), while the grid composites had the highest. The volume (Va) of the wet foam and the pore volume of the dry foam ($\%P$) were inversely related to the wet (D_w) and dry (D_d) densities (Table 2). The addition of starch to the foam formulation (FF+S) decreased the Va and $\%P$, resulting in higher D_w and D_d than the FF-S sample. The decrease in Va and $\%P$ and the concomitant increase in D_d resulted in greater flexural and compressive strength (Tables 3 and 4).

Table 3. Flexural strength (σ_{fM}), strain (ε_{fM}), and modulus (E_f) of fiber foam (FF) with and without starch (+S and −S, respectively), fiber foam composites containing paperboard elements (angle, cylinder, grid), and EPS foam.

Sample	FF-S	FF+S	FF-S Angle	FF+S Angle	FF-S Cyl.	FF+S Cyl.	FF-S Grid	FF+S Grid	EPS Foam
σ_{fM} (kPa)	30.8 a*	94 bc	36 a	173 d	67 ab	224 e	191 de	460 f	133 cd
ε_{fM} (%)	5.0	3.57 ab	5.0	4.18 ab	5.0	4.40 ab	2.90 a	3.29 a	5.00 c
E_f (MPa)	1.12 a	4.32 ab	1.21 a	9.45 bc	2.57 a	11.9 c	10.6 c	22.9 d	4.29 a

* Mean values (n = 5) within rows followed by a different letter are significantly different ($p < 0.05$).

Table 4. Compressive strength (CS) strain (ε_c) and modulus (E_c) of fiber foam (FF) with and without starch (+S and −S, respectively) and fiber foam composites containing paperboard elements (angle, cylinder, and grid).

Sample	FF-S	FF+S	FF-S Angle	FF+S Angle	FF-S Cyl.	FF+S Cyl.	FF-S Grid	FF+S Grid	EPS Foam
CS (kPa)	1.6 ^{a*}	10 ^{ab}	30 ^{abc}	63 ^{c}	121 ^{d}	187 ^{e}	192 ^{e}	305 ^{f}	55 ^{bc}
ε_c (%)	10 ^{e}	9.22 ^{e}	6.0 ^{cd}	2.1 ^{a}	6.8 ^{d}	3.9 ^{abc}	4.5 ^{bc}	3.7 ^{ab}	4.0 ^{cde}
E_c (MPa)	0.016 ^{a}	0.16 ^{a}	0.70 ^{a}	3.4 ^{bc}	2.1 ^{ab}	5.0 ^{bc}	5.1 ^{bc}	8.7 ^{d}	1.7 ^{ab}
Ω_c (J) $\varepsilon_c = 10\%$	0.0056 ^{a}	0.041 ^{a}	0.13 ^{ab}	0.24 ^{b}	0.46 ^{c}	0.74 ^{d}	0.77 ^{d}	1.1 ^{e}	0.34 ^{c}
Ω_c (J) $\varepsilon_c = 50\%$	0.14 ^{a}	0.63 ^{a}	0.66 ^{a}	2.1 ^{b}	2.2 ^{b}	4.8 ^{d}	2.5 ^{bc}	6.2 ^{e}	3.1 ^{c}

* Mean values (n = 5) within rows followed by a different letter are significantly different ($\alpha < 0.05$).

The drying time (T_d) required at 80 °C was considerable (Table 2), but other approaches to drying the foam could be explored. For instance, sample thickness could be reduced, or other efficient drying technologies could be tested, including ambient air drying and solar-assisted, infrared-assisted, microwave-assisted, and similar hybrid drying technologies [42].

TGA analysis was performed to study the thermal decomposition properties of the individual foam components. These data were used to establish an upper temperature range for the drying oven. The derivative of the wt.% curve was used to determine the decomposition temperatures as shown in Figure 2. The TGA data indicate that the least thermally stable component was SDS which had a decomposition temperature of 237 °C. Decomposition temperatures for starch, PVA, and SWF were 301, 308, and 354 °C, respectively.

Based on the TGA results, oven drying temperatures in excess of 200 °C might be considered the upper temperature range for drying conditions. However, in preliminary drying tests, a strong odor was detected when samples were dried at only 120 °C. There was little or no odor detected when the samples were dried at 80 °C. The odor produced at 120 °C was attributed to the greater thermal instability of SDS in an aqueous environment. In the presence of water, SDS is reported to degrade into fatty alcohols and sodium sulfate after prolonged heating at relatively low temperatures [43]. Alternative foaming agents have been used for making cellulose foam and could be evaluated to determine whether they are more thermally stable and/or compatible with alternative drying methods [31].

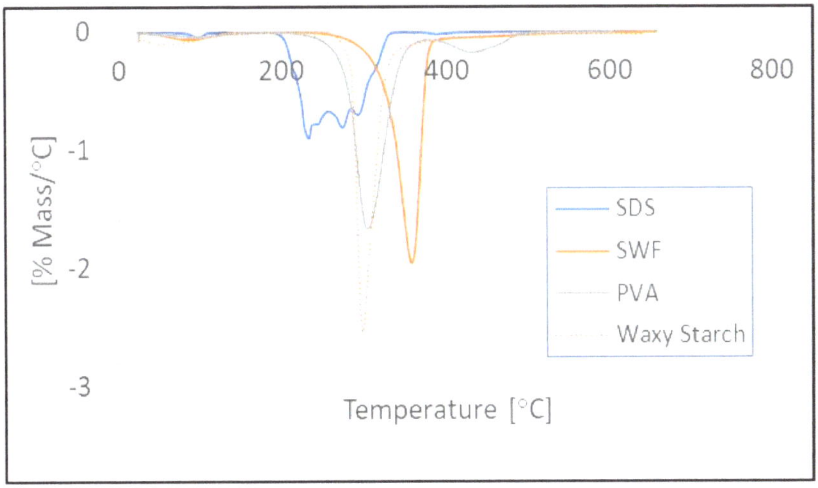

Figure 2. Graph of the DTG curves for Sodium dodecyl sulfate (SDS), polyvinyl alcohol (PVA), starch, and softwood fiber (SWF) from thermogravimetric analysis.

The drying times (T_d) for samples tested ranged from 333 to 553 min (Table 2). The addition of starch to the foam formulation (FF+S) significantly (paired t-test, t = 6.44×10^{-4}) increased the drying time (T_d) compared to samples without starch (FF-S, Table 2). The results highlight the effect a single ingredient can have on T_d and underscore the importance of assessing the impact different ingredients or even the ratio of ingredients may have on processing parameters.

As previously mentioned, legislative measures are being taken to phase out the use of EPS foam primarily in single-use packaging due to the resistance of EPS foam to biodegradation and its negative environmental impact [44]. In contrast to EPS foam, cellulose fiber-based products, such as paper and paperboard, are renewable, recyclable, compostable, and biodegradable [26]. Under favorable composting conditions, paper waste partially mineralizes to CO_2 while the residue forms humus, which is an excellent soil amendment [45]. Humus can slowly degrade further by fungi, bacteria, and soil organisms such as earthworms. The rate at which paper fiber biodegrades is dependent upon various factors, including how the fiber was originally processed, the lignan content, the type of paper additives used, and the environmental conditions [45]. Respirometry data from the present study showed that FF-S and FF+S samples mineralized in the range of 37–49% over a 46-day period (Figure 3). This is consistent with previous studies that report a 43–79% rate of mineralization for different papers over a 45-day period under composting conditions [46].

Cellulose fiber-based packaging materials are desirable partly because they can be reused and recycled. Paper products are reported to be recyclable up to seven times [26]. There is a well-established infrastructure for recycling paper and paperboard, especially in developed countries [26]. Fiber foam/paperboard composites are well suited for recycling using existing paper recycling streams partly because all the components used in making the fiber foam composites are already used in varying amounts in paper products.

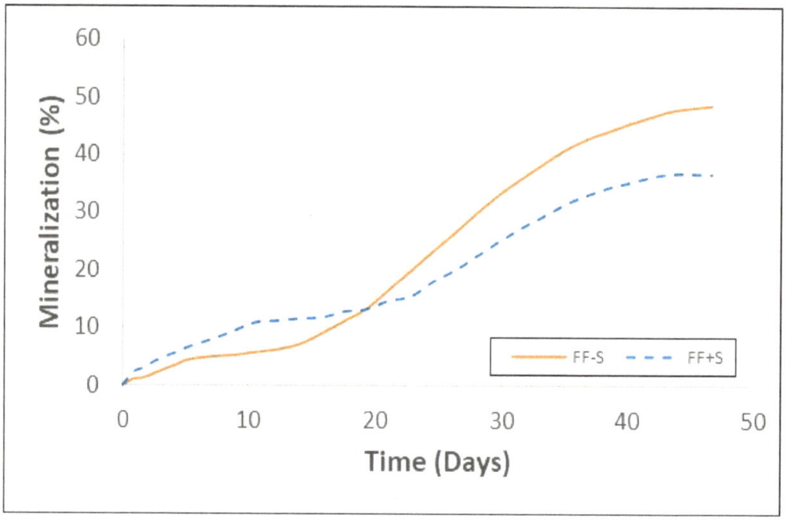

Figure 3. Mineralization rate of fiber foam samples both with and without starch.

The fiber foam and composites pose much less of an environmental concern compared to EPS foam, in part because all the ingredients, including the cellulose fiber, starch, SDS, and PVA, are biodegradable, at least to various extents. The chemical interaction of the ingredients is likely to be carried out mostly via hydrogen bonds between the hydroxyl groups on the fibers, starch, and PVA. These bonds can be easily disrupted by SDS under moist conditions encountered in composting environments, which then facilitates microbial access and biodegradation. It is well established that cellulose fiber and starch readily biodegrade in many environments. The fate of SDS in the environment has also been investigated [47]. Among its many applications, SDS is used as a fat emulsifier and as an ingredient in cosmetics, pharmaceuticals, and toothpaste [48]. It is also used extensively in deinking recycled paper and in various other processes of paper production [49]. In aerobic and anaerobic environments, SDS readily biodegrades into simple, nontoxic components and does not persist in the environment [47].

The most persistent component of fiber foams is PVA. PVA is a biocompatible polymer and can be manufactured economically from non-petroleum routes [50]. The environmental fate of PVA has been a subject of wide debate [37,51,52]. PVA is commonly used in the paper and textile industries as a sizing agent and is known to biodegrade in the presence of specific microorganisms [50,52]. The PVA degrading microorganisms are present in wastewater and compost environments [51–53]. Although PVA can degrade at rates similar to cellulose under optimal conditions [50], its biodegradation is much slower in many environments. For instance, PVA is known to accumulate as a pollutant in wastewater [52]. The PVA used in fiber foam formulations could be removed if necessary to further reduce the environmental footprint of the fiber foam. PVA-free fiber foam has been reported, but it requires a higher amount of water, which may lengthen the drying time [33]. Regardless of whether they contain PVA or not, the paperboard/fiber foam composites provide a more sustainable and environmentally benign option compared to EPS foam.

As previously mentioned, paper/paperboard grid/honeycomb packaging has outstanding toughness and provides excellent shock resistance in internal packaging applications [29]. The fiber foam composites, which were shown earlier to have good insulative properties, were tested under flexural and compressive strain as a means of assessing their mechanical strength and suitability as a replacement for EPS foam. In flexural tests, EPS foam samples failed abruptly in the range of 7–8% strain (ε_{fM}, Table 3, Figure 4). The fiber foam samples and composites, however, typically reached a peak force of resistance, and then yielded and eventually formed a bend but did not break except for the grid composite (Figures 4 and 5).

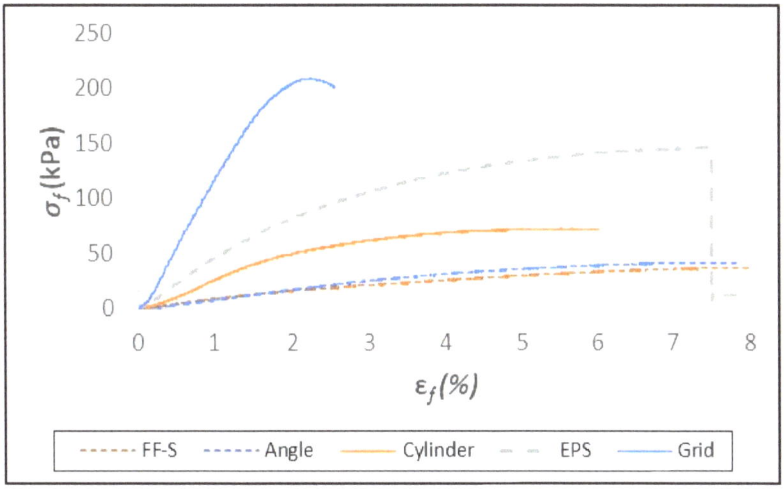

Figure 4. Typical flexural stress (σ_f)/strain (ε_f) curves for fiber foam without starch (FF-S), EPS foam, and FF-S composites containing paperboard elements (angle, cylinder, and grid).

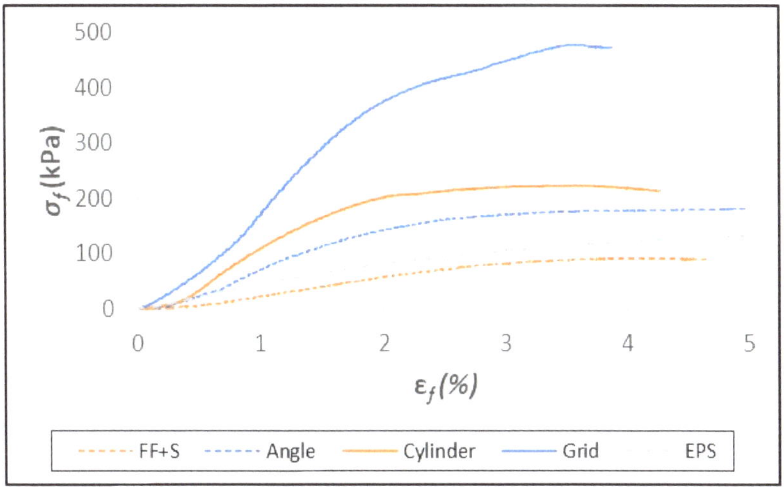

Figure 5. Typical flexural stress (σ_f)/strain (ε_f) curves for fiber foam with starch (FF+S), EPS foam, and FF+S composites containing paperboard elements (angle, cylinder, and grid).

The interlocking grid structure provided reinforcement against flexural strain, resulting in much higher strength (σ_{fM}) and modulus (E_f) values compared to the other composites (Table 3). The breakage of the grid structure sometimes occurred due to tearing that originated from the slots that were cut to form the interlocking grid. The FF-S foam and composites had lower flexural strength (σ_{fM}) than EPS foam except for the FF-S grid sample (Table 3, Figure 4).

In contrast to the FF-S samples, the FF+S foam composites all had flexural strength (σ_{fM}) and modulus (E_f) values in the same range or higher than the EPS foam (Table 3, Figure 5). The greater strength in samples containing starch is likely due to the ability of starch to act as an adhesive that binds fibers together and to the paperboard elements thus helping to reinforce the foam component and anchor the paperboard elements in place. The tearing and failure of the grid structure observed in the FF-S sample was less apparent in the FF+S grid sample (Figure 5). This may have been due to the higher interfacial forces exerted between cellulose fibers and the polymer matrix as well as the binding effect of starch that helped to strengthen the walls of the paperboard elements and better distribute the flexural stress throughout the structure [54].

The behavior of EPS foam under compressive strain (ε_c) was very different than its behavior under flexural strain (compare Figures 4 and 6). Compression stress/strain curves over an extremely large range (0–95%) in ε_c were obtained for EPS foam, FF-S, and FF+S fiber foam samples (Figure 6). The compression curves for the FF-S and FF+S samples revealed a compression behavior similar to the EPS foam and other elastomeric foams (Figure 6) [55,56]. A linear elastic region was observed at the beginning of the curves, followed by a plateau region where σ_c values increased at a relatively slow rate compared to the change in ε_c. Under very high ε_c values, the foams continued to densify and behave more like a solid than a foam (Figure 6). Despite densification, the foams remained intact and did not shatter or fracture. The results indicate that even under extreme levels of compaction/densification, the fiber foams behave as elastomeric foams similar to EPS foams and are able to withstand excessive ε_c without fracturing. The compression results also showed that foam formulations containing starch (FF+S) had higher σ_c values than FF-S, although neither of the fiber foam samples were in the range of the EPS foam sample (Figure 6).

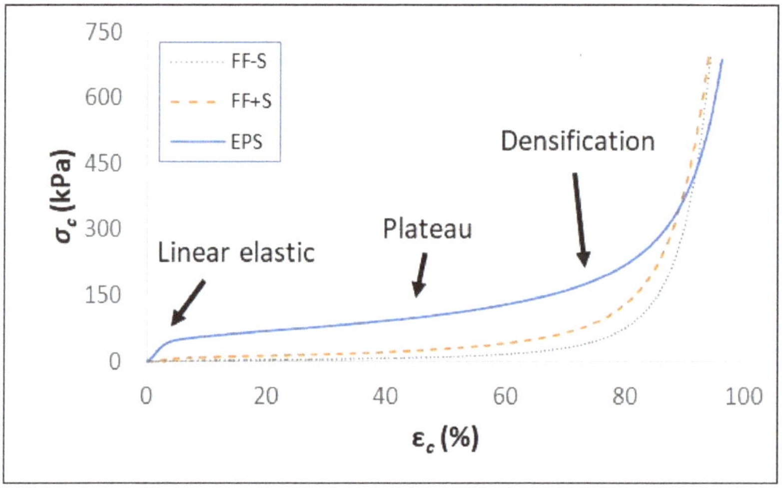

Figure 6. Stress–strain compression curves for EPS foam, fiber foam without starch (FF-S), and fiber foam with starch (FF+S).

Embedding paperboard elements (angle, cylinder, and grid) in the fiber foam (FF-S, FF+S) as a reinforcement had a significant effect on their compressive properties (Table 4). Compressive strength (CS) and toughness (Ω_c) have been used to assess the ability of a material to resist compressive strain (ε_c) and absorb shock during shipping [29,31]. The paperboard elements increased the *CS* and Ω_c values, which were measured at 10% and 50% ε_c. (Table 4). The angle paperboard composites had compressive strength (*CS*) and modulus (E_c) values in the range of EPS foam but had lower Ω_c values (Table 4). Meanwhile, the composites containing cylinder and grid elements had significantly higher *CS*, E_c, and Ω_c values than EPS foam, whether they contained starch or not (Table 4). The FF+S composites had higher mean values for *CS*, E_c, and Ω_c than the corresponding FF-S samples (Table 4).

Although the FF-S and FF+S samples behaved like typical elastomeric foams under a wide range of compressive strain (Figure 6), the paperboard composites did not

(Figures 7 and 8). Li et al. (2022) [29] reported that, initially, the compression curves for paperboard honeycomb panels increased linearly, peaked, and then decreased linearly before plateauing. A similar pattern was observed in the data for both the FF-S and FF+S paperboard composites (Figures 7 and 8). The rapid drop in the linear region of the compression curves was due to the failure/buckling of the walls of the paperboard elements under excessive compressive strain (ε_c) [29].

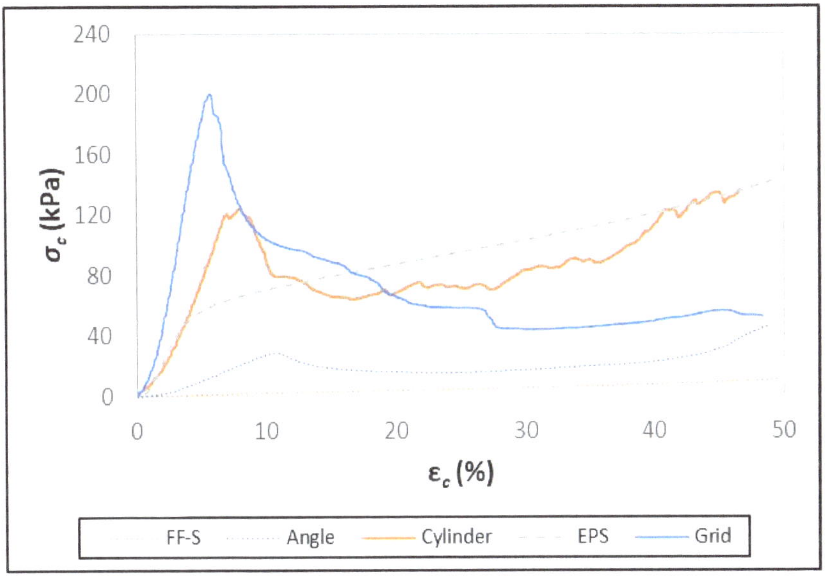

Figure 7. Typical compressive stress–strain curves for EPS foam, fiber foam without starch (FF-S), and FF-S composites containing angular, cylindrical, or grid paperboard elements.

The differences observed between the FF-S and FF+S samples were greater than might be expected from the sum of the fiber foam and the individual paperboard elements themselves. This was particularly evident for the FF-S and FF+S cylinder and grid samples, which had a difference of 66 and 108 kPa, respectively (Table 4).

It was somewhat surprising that the σ_c values for the FF-S grid sample decreased continuously until almost 30% ε_c before plateauing (Figure 7). Observation of the grid sample during compression tests revealed that the FF-S grid structure was not well anchored compared to the FF+S sample. As such, the walls of the FF-S grid were more easily able to dislodge and bend slightly at an angle thus preventing proper uniaxial loading. Furthermore, the slots cut for assembling and interconnecting the walls that formed the grid provided sites for tearing to occur as excessive compressive strain was applied.

In the FF+S compression curves, the σ_c values did not plateau horizontally as reported previously for paperboard structures [29] but rather continued to rise as ε_c increased (Figure 8). The higher σ_c values for the FF+S composites were likely due to the binding properties of starch that helped anchor the paperboard elements and allow better uniaxial loading. Additionally, when the paperboard elements were initially embedded in the wet foam, they became saturated and wet. The starch contained in the liquid phase likely made the paperboard elements stiffer and stronger upon drying.

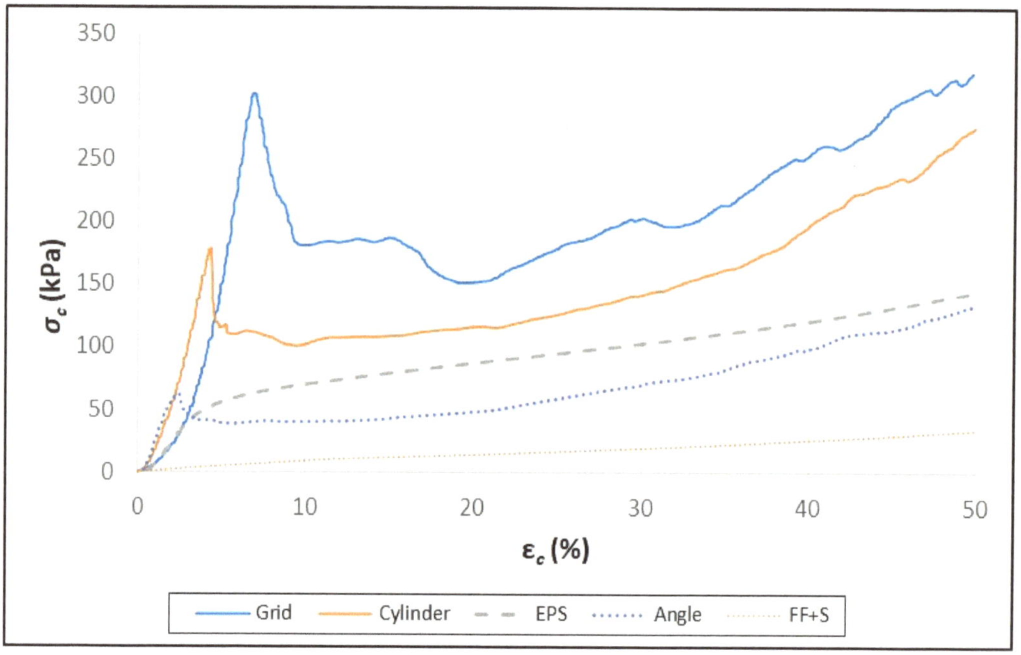

Figure 8. Stress–strain curves for starch containing fiber foam (1), starch/fiber/paperboard composites (angle = 2, cylinder = 3, grid = 4), and EPS foam (dashed line). These data are for the first (initial) stress/strain cycle.

4. Conclusions

Fiber foam and fiber foam/paperboard composites provide the thermal and mechanical properties needed in many packaging applications and that is currently provided by the EPS foam packaging slated to be phased out by legislative mandate. The composites do not require the use of face sheets as with honeycomb paperboard panels. The fiber foam process allows for the incorporation of starch as a binder and the use of both non-interlocking and interlocking reinforcing paperboard elements with different designs, spatial arrangement, and weight. Specific mechanical properties, densities, and thermal conductivity in the range of EPS foam can be targeted by varying the number and placement of paperboard elements and/or by incorporating a starch binder. Of the paperboard elements tested, composites made with cylindrical paperboard elements had the best overall performance in terms of flexural/compressive strength, modulus, and toughness while still maintaining a low density and thermal conductivity. The fiber foam/paperboard composites contained ingredients that are already used in the paper industry. This makes the foam composites more suitable for recycling in existing paper recycling streams. Further research is needed to identify a more thermally stable foaming agent and a more efficient drying technology that could reduce the time and energy needed for drying.

Author Contributions: Conceptualization, G.M.G.; Validation, G.H.D.T., L.E.S., D.W., C.L., W.H.-C. and W.O.; Formal analysis, A.P.K., B.-S.C. and Z.M.; Investigation, G.M.G.; Writing—original draft, G.M.G. All authors have read and agreed to the published version of the manuscript.

Funding: This research received no external funding.

Institutional Review Board Statement: Not applicable.

Data Availability Statement: Data are contained within the article.

Conflicts of Interest: The authors declare no conflict of interest.

References

1. Ge, C.; Huang, H. Corner foam versus flat foam: An experimental comparison on cushion performance. *Packag. Technol. Sci.* **2015**, *28*, 217–225. [CrossRef]
2. Marcondes, P.D.G. Minimum Sample Size Needed to Construct Cushion Curves Based on the Stress-Energy Method. Clemson University, All Theses. 135, 2007. Available online: https://tigerprints.clemson.edu/all_theses/135 (accessed on 15 January 2024).
3. Statistica. Global Parcel Shipping Volume between 2013 and 2027 (in Billion Parcels). Available online: https://www.statista.com/statistics/1139910/parcel-shipping-volume-worldwide/#:~:text=In%202022,%20global%20parcel%20shipping,growth%20rate%20of%2059%20percent (accessed on 15 January 2024).
4. Geyer, R.; Jambeck, J.R.; Law, K.L. Production, use, and fate of all plastics ever made. *Sci. Adv.* **2017**, *3*, e1700782. [CrossRef]
5. PEMRG. The Compelling Facts about Plastics. 2008. Available online: https://plasticseurope.org/wp-content/uploads/2021/10/2006-Compelling-facts.pdf (accessed on 9 January 2024).
6. Bökman, A.; Le, A. Driving Forces Behind Global Trade with Plastic Waste-Based on Reported Trade Statistics. Bachelor's Thesis, 2023. Available online: http://hdl.handle.net/20.500.12380/307287 (accessed on 25 March 2024).
7. Andena, L.; Caimmi, F.; Leonardi, L.; Nacucchi, M.; De Pascalis, F. Compression of polystyrene and polypropylene foams for energy absorption applications: A combined mechanical and microstructural study. *J. Cell. Plast.* **2019**, *55*, 49–72. [CrossRef]
8. Chen, W.; Hao, H.; Hughes, D.; Shi, Y.; Cui, J.; Li, Z.-X. Static and dynamic mechanical properties of expanded polystyrene. *Mater. Des.* **2015**, *69*, 170–180. [CrossRef]
9. Hidalgo-Crespo, J.; Moreira, C.; Jervis, F.; Soto, M.; Amaya, J.; Banguera, L. Circular economy of expanded polystyrene container production: Environmental benefits of household waste recycling considering renewable energies. *Energy Rep.* **2022**, *8*, 306–311. [CrossRef]
10. Ghoshal, T.; Parmar, P.R.; Bhuyan, T.; Bandyopadhyay, D. Polystyrene Foams: Materials, Technology, and Applications. In *Polymeric Foams: Fundamentals and Types of Foams*; ACS Publications: Washington, DC, USA, 2023; Volume 1, pp. 121–141.
11. Li, X.; Lin, Y.; Liu, M.; Meng, L.; Li, C. A review of research and application of polylactic acid composites. *J. Appl. Polym. Sci.* **2023**, *140*, e53477. [CrossRef]
12. Ho, B.T.; Roberts, T.K.; Lucas, S. An overview on biodegradation of polystyrene and modified polystyrene: The microbial approach. *Crit. Rev. Biotechnol.* **2018**, *38*, 308–320. [CrossRef]
13. iSustain. Can Polystyrene Be Cost-Effectively Recycled? 2024. Available online: https://isustainrecycling.com/can-polystyrene-cost-effectively-recycled/#:~:text=The%20good%20news%20is%20that,percent%20of%20post-industrial%20EPS (accessed on 9 January 2024).
14. Noguchi, T.; Miyashita, M.; Inagaki, Y.; Watanabe, H. A new recycling system for expanded polystyrene using a natural solvent. Part 1. A new recycling technique. *Packag. Technol. Sci. Int. J.* **1998**, *11*, 19–27. [CrossRef]
15. Shin, C. Filtration application from recycled expanded polystyrene. *J. Colloid Interface Sci.* **2006**, *302*, 267–271. [CrossRef]
16. Colorado General Assembly. HB21-1162. Management of Plastic Products. 2021. Available online: https://leg.colorado.gov/bills/hb21-1162 (accessed on 25 March 2024).
17. Emblidge, D. State Ban on Some Materials to Take Effect New Year's Day. 2024. Available online: https://13wham.com/news/local/state-ban-on-some-products-to-take-effect-new-years-day (accessed on 9 January 2024).
18. Murphy, P. Murphy Administration Marks First Anniversary of Law Limiting Single-Use Plastics in New Jersey. 2023. Available online: https://www.nj.gov/dep/newsrel/2023/23_0027.htm (accessed on 9 January 2024).
19. Walker, J.-C.; Pecht, A.M. *California Enacts EPR Law Aimed at Single-Use Plastic Packaging and Food Service Ware*; Keller & Heckman: Washington, DC, USA, 2022; Available online: https://www.khlaw.com/insights/california-enacts-epr-law-aimed-single-use-plastic-packaging-and-food-service-ware?language_content_entity=en (accessed on 9 January 2024).
20. House-of-Comons. Single Use Plastic: How Do Bans Differ across the UK and EU? UK Parliament, 2022. Available online: https://commonslibrary.parliament.uk/single-use-plastic-how-do-bans-differ-across-the-uk-and-eu/#:~:text=The%20UK%20Government%E2%80%99s%20proposed%20restrictions,the%20UK/EU%20Withdrawal%20Agreement (accessed on 9 January 2024).
21. Tapia-Blácido, D.R.; Aguilar, G.J.; de Andrade, M.T.; Rodrigues-Júnior, M.F.; Guareschi-Martins, F.C. Trends and challenges of starch-based foams for use as food packaging and food container. *Trends Food Sci. Technol.* **2022**, *119*, 257–271. [CrossRef]
22. Muniyasamy, S.; Ofosu, O.; John, M.J.; Anandjiwala, R.D. Mineralization of poly (lactic acid)(PLA), poly (3-hydroxybutyrate-co-valerate)(PHBV) and PLA/PHBV blend in compost and soil environments. *J. Renew. Mater.* **2016**, *4*, 133. [CrossRef]
23. Skye, C. Bioplastics, Biodegradable Plastics, & Compostable Plastics: What's the Difference? Earth911. 2021. Available online: https://earth911.com/business-policy/bioplastics-biodegradable-plastics-compostable-plastics/ (accessed on 9 January 2024).
24. Villamil Jiménez, J.A.; Le Moigne, N.; Bénézet, J.-C.; Sauceau, M.; Sescousse, R.; Fages, J. Foaming of PLA composites by supercritical fluid-assisted processes: A review. *Molecules* **2020**, *25*, 3408. [CrossRef] [PubMed]
25. EPA. National Overview: Facts and Figures on Materials, Wastes and Recycling. US EPA. 2023. Available online: https://www.epa.gov/facts-and-figures-about-materials-waste-and-recycling/national-overview-facts-and-figures-materials#Recycling/Composting (accessed on 9 January 2024).
26. Ozola, Z.U.; Vesere, R.; Kalnins, S.N.; Blumberga, D. Paper waste recycling. Circular economy aspects. *Environ. Clim. Technol.* **2019**, *23*, 260–273. [CrossRef]

27. Wang, D.; Yang, R. Study on damping characteristic of honeycomb paperboard and vibration reduction mechanism of packaging system. *J. Vib. Control* **2019**, *25*, 1536–1542. [CrossRef]
28. Xing, Y.; Sun, D.; Deng, Z. An Analysis of the Vibration Transmission Properties of Assemblies Using Honeycomb Paperboard and Expanded Polyethylene. *Materials* **2023**, *16*, 6554. [CrossRef] [PubMed]
29. Li, K.; Wang, J.; Lu, L.; Qin, Q.; Chen, J.; Shen, C.; Jiang, M. Mechanical properties and energy absorption capability of a new multi-cell lattice honeycomb paperboard under out-plane compression: Experimental and theoretical studies. *Packag. Technol. Sci.* **2022**, *35*, 273–290. [CrossRef]
30. Abd Kadir, N.; Aminanda, Y.; Ibrahim, M.; Mokhtar, H. Experimental study on energy absorption of foam filled kraft paper honeycomb subjected to quasi-static uniform compression loading. In *IOP Conference Series: Materials Science and Engineering*; IOP Publishing: Bristol, UK, 2016; Volume 152, p. 012048.
31. Nechita, P.; Năstac, S.M. Overview on foam forming cellulose materials for cushioning packaging applications. *Polymers* **2022**, *14*, 1963. [CrossRef] [PubMed]
32. Enso, S. Cellulose Foam Papira®. 2024. Available online: https://www.storaenso.com/en/products/bio-based-materials/cellulose-foam (accessed on 1 March 2024).
33. Glenn, G.; Orts, W.; Klamczynski, A.; Shogren, R.; Hart-Cooper, W.; Wood, D.; Lee, C.; Chiou, B.-S. Compression molded cellulose fiber foams. *Cellulose* **2023**, *30*, 3489–3503. [CrossRef]
34. Xiaolin, C.; Deng, X.J.; Nicolas, D.; Lebel, S.; Brunette, G.; Dorris, G.M.; Ben, Y.; Ricard, M.; Zhang, Y.Z.; Yang, D.-Q. Method of Producing Ultra-Low Density Fiber Composite Materials. US Patent No. 9,994,712, 12 June 2018.
35. Niu, Z.; Chen, F.; Zhang, H.; Liu, C. High Content of Thermoplastic Starch, Poly (butylenes adipate-co-terephthalate) and Poly (butylene succinate) Ternary Blends with a Good Balance in Strength and Toughness. *Polymers* **2023**, *15*, 2040. [CrossRef]
36. *ASTM D1621-16*; Standard Test Method for Compressive Properties of Rigid Cellular Plastics. ASTM: West Conshohocken, PA, USA, 2016. [CrossRef]
37. Kawai, F.; Hu, X. Biochemistry of microbial polyvinyl alcohol degradation. *Appl. Microbiol. Biotechnol.* **2009**, *84*, 227–237. [CrossRef]
38. *ASTM D790-17*; Standard Test Methods for Flexural Properties of Unreinforced and Reinforced Plastics and Electrical Insulating Materials. ASTM: West Conshohocken, PA, USA, 2017. [CrossRef]
39. Liu, Y.; Kong, S.; Xiao, H.; Bai, C.; Lu, P.; Wang, S. Comparative study of ultra-lightweight pulp foams obtained from various fibers and reinforced by MFC. *Carbohydr. Polym.* **2018**, *182*, 92–97. [CrossRef] [PubMed]
40. *ASTM C177-19*; Standard Test Method for Steady-State Heat Flux Measurements and Thermal Transmission Properties by Means of the Guarded-Hot-Plate Apparatus. ASTM: West Conshohocken, PA, USA, 2019. [CrossRef]
41. Imam, S.H.; Gordon, S.H. Biodegradation of coproducts from industrially processed corn in a compost environment. *J. Polym. Environ.* **2002**, *10*, 147–154. [CrossRef]
42. Hnin, K.K.; Zhang, M.; Mujumdar, A.S.; Zhu, Y. Emerging food drying technologies with energy-saving characteristics: A review. *Dry. Technol.* **2018**, *37*, 1465–1480. [CrossRef]
43. Sigma-Aldrich. Sodium Dodecyl Sulfate -Product Information. Available online: https://www.sigmaaldrich.com/deepweb/assets/sigmaaldrich/product/documents/263/218/l4522pis.pdf (accessed on 25 March 2024).
44. Turner, A. Foamed Polystyrene in the Marine Environment: Sources, Additives, Transport, Behavior, and Impacts. *Environ. Sci. Technol.* **2020**, *54*, 10411–10420. [CrossRef] [PubMed]
45. Venelampi, O.; Weber, A.; Rönkkö, T.; Itävaara, M. The biodegradation and disintegration of paper products in the composting environment. *Compost. Sci. Util.* **2003**, *11*, 200–209. [CrossRef]
46. Alvarez, J.V.L.; Larrucea, M.A.; Bermúdez, P.A.; Chicote, B.L. Biodegradation of paper waste under controlled composting conditions. *Waste Manag.* **2009**, *29*, 1514–1519. [CrossRef] [PubMed]
47. Bondi, C.A.; Marks, J.L.; Wroblewski, L.B.; Raatikainen, H.S.; Lenox, S.R.; Gebhardt, K.E. Human and environmental toxicity of sodium lauryl sulfate (sls): Evidence for safe use in household cleaning products. *Environ. Health Insights* **2020**, *9*. [CrossRef] [PubMed]
48. PubChem. *Sodium Dodecyl Sulfate*; National Institute of Health: Bethesda, MD, USA, 2024. Available online: https://pubchem.ncbi.nlm.nih.gov/compound/Sodium-dodecyl-sulfate (accessed on 10 January 2024).
49. Wang, T.; Chang, D.; Huang, D.; Liu, Z.; Wu, Y.; Liu, H.; Yuan, H.; Jiang, Y. Application of surfactants in papermaking industry and future development trend of green surfactants. *Appl. Microbiol. Biotechnol.* **2021**, *105*, 7619–7634. [CrossRef] [PubMed]
50. Liu, B.; Zhang, J.; Guo, H. Research progress of polyvinyl alcohol water-resistant film materials. *Membranes* **2022**, *12*, 347. [CrossRef]
51. Chiellini, E.; Corti, A.; D'Antone, S.; Solaro, R. Biodegradation of poly (vinyl alcohol) based materials. *Prog. Polym. Sci.* **2003**, *28*, 963–1014. [CrossRef]
52. Halima, N.B. Poly (vinyl alcohol): Review of its promising applications and insights into biodegradation. *RSC Adv.* **2016**, *6*, 39823–39832. [CrossRef]
53. Li, M.; Zhang, D.; Du, G.; Chen, J. Enhancement of PVA-degrading enzyme production by the application of pH control strategy. *J. Microbiol. Biotechnol.* **2012**, *22*, 220–225. [CrossRef] [PubMed]

54. Marcuello, C.; Chabbert, B.; Berzin, F.; Bercu, N.B.; Molinari, M.; Aguié-Béghin, V. Influence of surface chemistry of fiber and lignocellulosic materials on adhesion properties with polybutylene succinate at nanoscale. *Materials* **2023**, *16*, 2440. [CrossRef] [PubMed]
55. Mane, J.; Chandra, S.; Sharma, S.; Ali, H.; Chavan, V.; Manjunath, B.; Patel, R. Mechanical property evaluation of polyurethane foam under quasi-static and dynamic strain rates-an experimental study. *Procedia Eng.* **2017**, *173*, 726–731. [CrossRef]
56. Sadighi, M.; Salami, S. An investigation on low-velocity impact response of elastomeric & crushable foams. *Open Eng.* **2012**, *2*, 627–637.

polymers

MDPI

Article

Improvement in Crystallization, Thermal, and Mechanical Properties of Flexible Poly(L-lactide)-*b*-poly(ethylene glycol)-*b*-poly(L-lactide) Bioplastic with Zinc Phenylphosphate

Kansiri Pakkethati [1], Prasong Srihanam [1], Apirada Manphae [1,2], Wuttipong Rungseesantivanon [3], Natcha Prakymoramas [3], Pham Ngoc Lan [4] and Yodthong Baimark [1,*]

1 Biodegradable Polymers Research Unit, Department of Chemistry and Centre of Excellence for Innovation in Chemistry, Faculty of Science, Mahasarakham University, Mahasarakham 44150, Thailand; kansiri.p@msu.ac.th (K.P.); prasong.s@msu.ac.th (P.S.); apirada.m@msu.ac.th (A.M.)
2 Scientific Instrument Academic Service Unit, Faculty of Science, Mahasarakham University, Mahasarakham 44150, Thailand
3 National Metal and Materials Technology Centre (MTEC), 114 Thailand Science Park (TSP), Phahonyothin Road, Khlong Nueng, Khlong Luang, Pathum Thani 12120, Thailand; wuttir@mtec.or.th (W.R.); natchap@mtec.or.th (N.P.)
4 Faculty of Chemistry, University of Science, Vietnam National University-Hanoi, 19 Le Thanh Tong Street, Phan Chu Trinh Ward, Hoan Kiem District, Hanoi 10000, Vietnam; phamngoclan49@gmail.com
* Correspondence: yodthong.b@msu.ac.th

Abstract: Poly(L-lactide)-*b*-poly(ethylene glycol)-*b*-poly(L-lactide) (PLLA-PEG-PLLA) shows promise for use in bioplastic applications due to its greater flexibility over PLLA. However, further research is needed to improve PLLA-PEG-PLLA's properties with appropriate fillers. This study employed zinc phenylphosphate (PPZn) as a multi-functional filler for PLLA-PEG-PLLA. The effects of PPZn addition on PLLA-PEG-PLLA characteristics, such as crystallization and thermal and mechanical properties, were investigated. There was good phase compatibility between the PPZn and PLLA-PEG-PLLA. The addition of PPZn improved PLLA-PEG-PLLA's crystallization properties, as evidenced by the disappearance of the cold crystallization temperature, an increase in the crystallinity, an increase in the crystallization temperature, and a decrease in the crystallization half-time. The PLLA-PEG-PLLA's thermal stability and heat resistance were enhanced by the addition of PPZn. The PPZn addition also enhanced the mechanical properties of the PLLA-PEG-PLLA, as demonstrated by the rise in ultimate tensile stress and Young's modulus. We can conclude that the PPZn has potential for use as a multi-functional filler for the PLLA-PEG-PLLA composite due to its nucleating-enhancing, thermal-stabilizing, and reinforcing ability.

Keywords: poly(lactic acid); poly(ethylene glycol); block copolymer; nucleating agent; thermal stabilizer; reinforcing filler

Citation: Pakkethati, K.; Srihanam, P.; Manphae, A.; Rungseesantivanon, W.; Prakymoramas, N.; Lan, P.N.; Baimark, Y. Improvement in Crystallization, Thermal, and Mechanical Properties of Flexible Poly(L-lactide)-*b*-poly(ethylene glycol)-*b*-poly(L-lactide) Bioplastic with Zinc Phenylphosphate. *Polymers* **2024**, *16*, 975. https://doi.org/10.3390/polym16070975

Academic Editors: Masoud Ghaani and Stefano Farris

Received: 25 February 2024
Revised: 30 March 2024
Accepted: 31 March 2024
Published: 3 April 2024

1. Introduction

Biodegradable bioplastics have been widely investigated because they are considered environmentally friendly, sustainable, and renewable, and they also reduce pollution of wastes of conventional petroleum-based plastics. Poly(L-lactide) (PLLA) derived from renewable resources, such as starch-rich crops and sugarcane, is an important biodegradable bioplastic alternative to petroleum-based plastics because of its good biodegradability, biocompatibility, and processability, and also because it is the cheapest biodegradable bioplastic [1–3]. PLLA has been applied for use in biomedical applications [4–6], agriculture [7], sport [8], and packaging [9–12]. However, the low flexibility of PLLA has restricted its practical use and wider applications [3,13].

In order to increase PLLA's flexibility by increasing its chain mobility, an extensive range of plasticizers has been blended with PLLA [14–16]. The selection of a good plasticizer for PLLA has considered a number of factors, including plasticizer efficiency, non-toxicity, good miscibility, and durability. Among all of the plasticizers, poly(ethylene glycol) (PEG) was the most effective for PLLA because it is non-toxic, it has good miscibility, and it has the ability to accelerate PLLA's hydrolytic degradation [16]. Nevertheless, the primary drawback has been identified as phase separation and PEG migration from PLLA matrices [17–19], which has a direct impact on the stability and durability of the PLLA/PEG blends throughout storage and use [20,21].

High-molecular-weight PLLA-*b*-poly(ethylene glycol)-*b*-PLLA (PLLA-PEG-PLLA) triblock copolymers have more flexibility than PLLA [22,23]. This is due to the PEG middle-blocks acting as plasticizing sites to improve the chain mobility of PLLA end-blocks and decrease the glass transition temperature of PLLA end-blocks from about 60 °C to about 30 °C [23]. Moreover, the crystallization properties of the PLLA end-blocks were improved by the plasticizing effect of PEG middle-blocks. However, the melt strength of PLLA-PEG-PLLA is too low because of the plasticization effect of PEG middle-blocks. The melt strength of the PLLA-PEG-PLLA can be improved for conventional melt processing, such as injection molding, by reacting with a chain extender to form branching structures through post-treatment [23] and in situ copolymerization [24]. However, the obtained chain-extended PLLA-PEG-PLLA showed a slower crystallization rate and lower heat-resistant properties than the non-chain-extended PLLA-PEG-PLLA because the branching structures of chain-extended PLLA-PEG-PLLA inhibited the crystallization of PLLA end-blocks, thereby limiting its wider applications [25].

Depending on the thermal history of PLLA, it can be either amorphous or semicrystalline. PLLA will become highly amorphous upon quenching it from the melt phase (for example, during the extrusion and injection processes) [26]. Addition of a nucleating agent to the PLLA is an effective method for improving the crystallization of PLLA during its processing. Crystallization at high temperature can occur because the surface free energy barrier for nucleation is lowered upon cooling. Many nucleating agents have been used to enhance the crystallization properties and to increase the crystallinity of PLLA [16]. Examples of these nucleating agents are talcum [16,27,28], zinc phenylphosphate (PPZn) [16,27,29–31], stereocomplex PLA [16,27,32], and starch [16,33]. Highly crystalline PLLA exhibits good heat-resistance properties due to the improved stiffness [16,34]. It has been reported that PPZn is a more effective nucleating agent than both talcum powder and PLA stereocomplex [16].

For chain-extended PLLA-PEG-PLLA, various nucleating agents, such as talcum [35], calcium carbonate [24], and native starch [33], have been investigated to improve the crystallization properties of PLLA end-blocks of PLLA-PEG-PLLA. Among these nucleating agents, talcum was an effective nucleating agent, improving both the crystallization and heat-resistant properties of the chain-extended PLLA-PEG-PLLA. To the best of our knowledge, the nucleation effects of PPZn on crystallization and heat-resistant properties of chain-extended PLLA-PEG-PLLA have not been reported so far. Thus, the objective of this work is to study the effect of PPZn on non-isothermal and isothermal crystallization, heat resistance, mechanical properties, and PLLA thermal stability of the chain-extended PLLA-PEG-PLLA.

2. Materials and Methods

2.1. Materials

The chain-extended PLLA-PEG-PLLA was synthesized through ring-opening polymerization of L-lactide monomer, as described in our previous works [24,36]. PEG with molecular weight of 20,000 and stannous octoate were used as the initiating system. The number-averaged molecular weight (M_n) and dispersity ($Đ$) obtained from gel permeation chromatography (GPC) of the obtained PLLA-PEG-PLLA were 108,500 and 2.2, respectively. PPZn was synthesized from phenylphosphonic acid (98%, Acros Organics, Geel, Belgium) and zinc chloride ($ZnCl_2$, 98%, Acros Organics, Geel, Belgium) according to the literature [29]. Figure 1a shows a SEM image of the obtained PPZn powder. The average particle size of 100 PPZn particles determined from the SEM image using an ImageJ program was 4.12 ± 1.60 μm. The particle size distribution of PPZn powder is presented in Figure 1b. The thermal decomposition of PPZn powder determined from the thermogravimetric (TG) thermogram was in the range of 550–700 °C, as shown in Figure 1c. The residue weight at 800 °C of PPZn powder was about 73%. The XRD pattern of PPZn powder is illustrated in Figure 1d, with XRD peaks at 6.5°, 12.6°, and 18.5°.

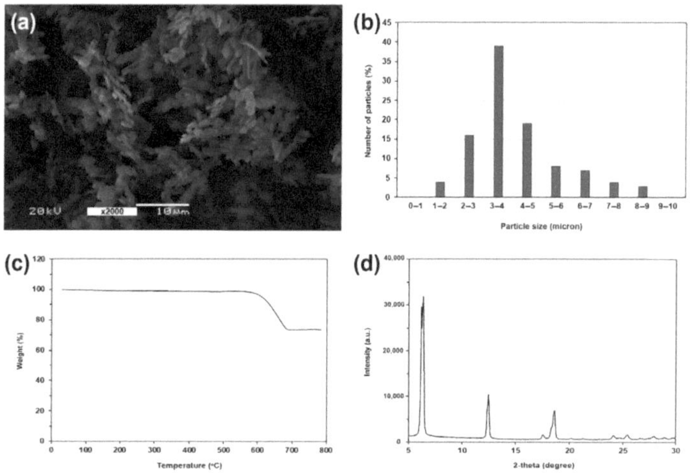

Figure 1. (**a**) SEM image, (**b**) particle size distribution, (**c**) TG thermogram, and (**d**) XRD pattern of PPZn powder.

2.2. Preparation of PLLA-PEG-PLLA/PPZn Composites

PLLA-PEG-PLLA and PPZn were dried in a vacuum oven at 50 °C overnight before melt mixing with a HAAKE internal mixer Polylab OS System (Waltham, MA, USA) at 190 °C for 6 min. A rotor speed of 100 rpm was used. The PLLA-PEG-PLLA composites with PPZn contents of 0.5, 1, 2, and 4 %wt were investigated. The ground particles of composites were dried in a vacuum oven at 50 °C overnight before compression molding with a Carver compression molding machine Auto CH Carver (Wabash, IN, USA) at 190 °C for 3 min without compression force, followed by a 5 MPa compression force for 2 min. The obtained film was immediately cooled for 3 min under 5 MPa compression force with water-cooled plates. The thicknesses of the obtained films were in the range of 0.2–0.3 mm.

2.3. Characterization of PLLA-PEG-PLLA/PPZn Composites

Differential scanning calorimetry (DSC) was conducted on a PerkinElmer DSC Pyris Diamond (Waltham, MA, USA) under a nitrogen atmosphere. For non-isothermal analysis, the samples were maintained at 200 °C for 3 min to erase the previous thermal history before quickly quenching to 0 °C with a cooling rate of 100 °C/min. The samples were then scanned from 0 °C to 200 °C with a heating rate of 10 °C/min for DSC heating scans. For the DSC cooling scan, the samples were heated at 200 °C for 3 min to erase the previous thermal history before scanning from 200 °C to 0 °C with a cooling rate of 10 °C/min. The degree of crystallinity of the samples, as determined by the DSC ($DSC\text{-}X_c$) of PLLA crystallites, was calculated using the following equation:

$$DSC\text{-}X_c\ (\%) = [(\Delta H_m - \Delta H_{cc})/(93.6 \times W_{PLLA})] \times 100 \quad (1)$$

where ΔH_m and ΔH_{cc} are the melting and cold-crystallization enthalpies, respectively. For 100%$DSC\text{-}X_c$ of PLLA, the ΔH_m value is 93.6 J/g [37]. W_{PLLA} is the weight fraction of PLLA.

For isothermal analysis, the samples were isothermally crystallized at 120 °C according to the literature [38], as follows. The samples were first heated at 200 °C for 3.0 min, cooled to 120 °C at a rate of 50 °C/min, and subsequently isothermal scanned at 120 °C until the crystallization process was completed. The time needed to achieve 50% of the final crystallinity is known as the crystallization half-time ($t_{1/2}$).

Wide-angle X-ray diffractometry (XRD) was performed on a Bruker Corporation XRD D8 Advance (Karlsruhe, Germany) to determine crystalline structures with CuKα radiation at 40 kV and 40 mA. The scan speed was 3°/min. The degree of crystallinity as determined according to the XRD ($XRD\text{-}X_c$) of PLLA crystallites was calculated using the following equation:

$$XRD\text{-}X_c\ (\%) = [(A_c)/(A_c + A_a)] \times 100 \quad (2)$$

where A_c is the peak area of PLLA crystallites and A_a is the halo area of the amorphous phase.

Thermogravimetric analysis (TGA) was performed on a TA Instruments TGA SDT Q600 (New Castle, DE, USA) from room temperature to 800 °C with a heating rate of 20 °C/min under nitrogen flow with a rate of 100 mL/min.

Dynamic mechanical analysis (DMA) was conducted on a TA Instrument DMA Q800 DMA (New Castle, DE, USA) with a tension mode from 30 °C to 150 °C at a heating rate of 2 °C/min. The scan amplitude was 10 µm, and the scanning frequency was 1 Hz.

The heat resistance of the composite films was investigated by testing the dimensional stability to heat at 80 °C under a 200 g load for 30 s [39,40]. The initial gauge length of the films was 20 mm. The dimensional stability to heat of the film samples was calculated using the following equation. The average value was obtained from five different determinations of each film sample.

$$\text{Dimensional stability to heat (\%)} = [\text{initial gauge length}/\text{final gauge length}] \times 100 \quad (3)$$

Cryo-fractured surfaces of the composite films were analyzed to observe the phase separation between the PLLA-PEG-PLLA matrix and the PPZn particles through scanning electron microscopy (SEM) on a JEOL JSM-6460LV SEM (Tokyo, Japan) at 15 kV. Before scanning, the film samples were gold sputter coated.

A tensile test was performed on a LY-1066B universal testing machine (Dongguan Liyi Environmental Technology Co., Ltd., Dongguan, China) according to the ASTM D882 at 25 °C with a crosshead speed of 50 mm/min. The initial gauge length was 50 mm, and the load cell was 100 kg. Each film sample was tested with a minimum of five determinations.

The film's opacity was determined using a Thermo Scientific Genesys 20 visible spectrophotometer (Loughborough, UK) and calculated with the following equation [41]:

$$\text{Opacity (mm}^{-1}) = A_{600}/X \quad (4)$$

where A_{600} is the absorbance of the film at 600 nm and X is the thickness of the film sample (mm).

3. Results

3.1. Thermal Transition Properties

The PLLA-PEG-PLLA composites with different PPZn contents were prepared to compare the nucleating effect of PPZn on the non-isothermal and isothermal DSC scans of the PLLA-PEG-PLLA matrix, as shown in Figures 2 and 3, respectively. In Figure 2a, the glass transition (T_g), cold crystallization (T_{cc}), and melting (T_m) temperatures are seen on the DSC heating thermograms. In Figure 2b, the crystallization temperature (T_c) is seen on the DSC cooling thermograms. The DSC results of both the DSC heating and cooling scans are summarized in Table 1. The T_g, T_{cc}, and T_m values of PLLA end-blocks of pure PLLA-PEG-PLLA were 31 °C, 81 °C, and 152 °C, respectively. The T_g and T_{cc} values of pure PLLA-PEG-PLLA disappeared when the PPZn was incorporated. This may be due to the glassy-to-rubbery transition in amorphous regions of PLLA end-blocks after quenching from 200 °C to 0 °C being difficult to detect by DSC for PLLA-PEG-PLLA composites because the free volume of the polymer chains was reduced for high-crystallinity polymers, which leads to restricting the motion of polymer chains in amorphous regions [42]. The disappearance of a T_{cc} peak of the PLLA-PEG-PLLA matrix indicates that the composites underwent complete crystallization during quenching from 200 °C to 0 °C [38,43,44]. It should be noted that lower-temperature shoulder T_m peaks were also detected when the PPZn was incorporated, which suggests that imperfect crystals of PLLA end-blocks were formed [29].

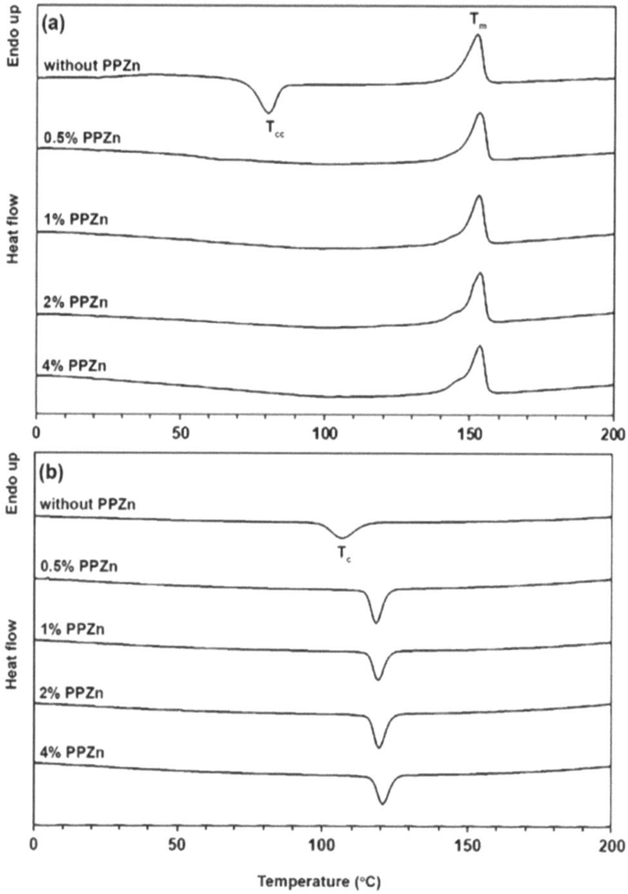

Figure 2. DSC thermograms of (**a**) heating scans and (**b**) cooling scans of PLLA-PEG-PLLA/PPZn composites with various PPZn contents.

Table 1. DSC results obtained from heating and cooling thermograms of PLLA-PEG-PLLA/PPZn composites.

PPZn Content (%wt)	W_{PLLA} [1]	T_g [2] (°C)	T_{cc} [2] (°C)	ΔH_{cc} [2] (J/g)	T_m [2] (°C)	ΔH_m [2] (J/g)	$DSC\text{-}X_c$ [2] (%)	T_c [3] (°C)
-	0.830	31	81	16.5	152	26.1	12.4	107
0.5	0.826	-	-	-	153	32.6	42.2	118
1	0.822	-	-	-	153	33.4	43.4	119
2	0.813	-	-	-	153	34.7	45.6	120
4	0.797	-	-	-	153	37.0	49.6	120

[1] Weight fraction of PLLA (W_{PLLA} of PLLA for PLLA-PEG-PLLA is 0.830 [24,36]). [2] Obtained from DSC heating thermograms. [3] Obtained from DSC cooling thermograms.

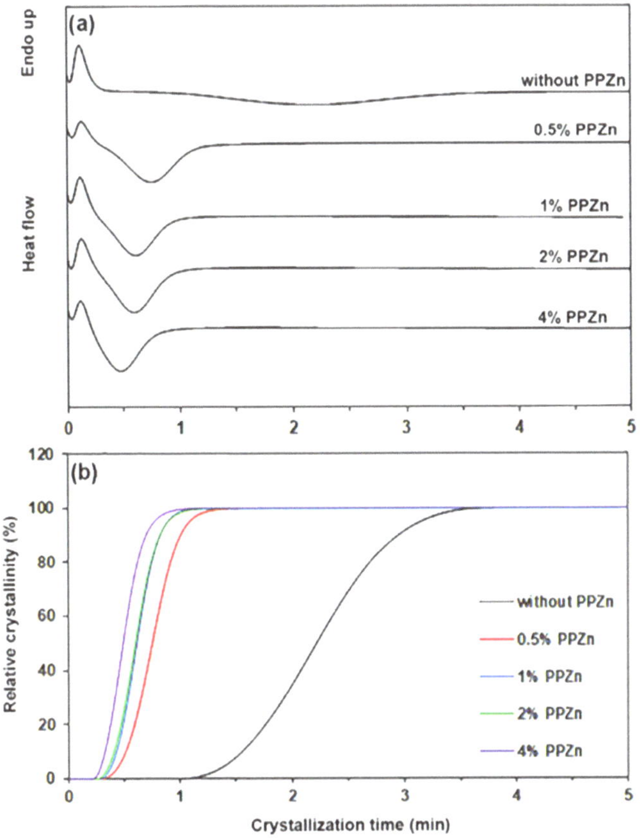

Figure 3. (**a**) Isothermal crystallization curves at 120 °C and (**b**) relative crystallinity–crystallization time curves of PLLA-PEG-PLLA/PPZn composites with various PPZn contents.

Degrees of crystallinity assessed from DSC ($DSC\text{-}X_c$) calculated from Equation (1) are also reported in Table 1. The $DSC\text{-}X_c$ value of pure PLLA-PEG-PLLA was 12.3% due to the crystallites of PLLA end-blocks. When the 0.5 %wt PPZn was incorporated, the $DSC\text{-}X_c$ value of the PLLA-PEG-PLLA composite was dramatically increased up to 42.2%. This suggests that the PPZn acted as an effective nucleating agent for PLLA-PEG-PLLA. The $DSC\text{-}X_c$ values slightly increased as the PPZn contents increased higher than 0.5 %wt. It should be noted that the $DSC\text{-}X_c$ value of PLLA-PEG-PLLA composite containing 4 %wt talcum was 48.5% in our previous work [35], whereas the PLLA-PEG-PLLA composite containing 4 %wt PPZn in this work was 49.6%. This suggests the PPZn had a better nucleating effect than the talcum when comparing data reported in the literature [16].

From DSC cooling thermograms, the T_c peaks of the PLLA-PEG-PLLA matrix dramatically shifted from 107 °C to 118 °C when the 0.5 %wt PPZn was incorporated, thus supporting the conclusion that added PPZn enhanced crystallization of PLLA-PEG-PLLA through a nucleating effect [45–47]. The T_c peaks slightly shifted to higher temperatures as the PPZn contents were higher than 0.5 %wt.

Furthermore, the crystallization properties of the composites were studied as derived from the crystallization half-time ($t_{1/2}$) that was obtained from isothermal crystallization curves at 120 °C, as shown in Figure 3a. The 50% relative crystallinity of polymer samples was obtained during isothermal scans at the time of $t_{1/2}$, as shown in Figure 3b. The resulting $t_{1/2}$ values are summarized in Table 2. The $t_{1/2}$ value of pure PLLA-PEG-PLLA was 2.18 min. The time was significantly decreased to 0.75 min when 0.5 %wt PPZn was added, thereby supporting the effective nucleating effect of added PPZn. The $t_{1/2}$ values steadily decreased as the PPZn contents increased higher than 0.5 %wt. The crystallization kinetics of the composites were investigated using the following Avrami equation [38,48]:

$$1 - X_t = \exp(-kt^n) \quad (5)$$

where X_t is the relative crystallinity as a function of time, t is the crystallization time, n is the Avrami exponent, and k is the crystallization rate constant.

Table 2. Crystallization half-time ($t_{1/2}$) and Avrami parameters (n and k) of PLLA-PEG-PLLA/PPZn composites obtained from isothermal curves at 120 °C.

PPZn Content (%wt)	$t_{1/2}$ (min)	n	k (min^{-k})	R^2
-	2.18	3.8569	0.0208	0.9986
0.5	0.75	3.2228	2.2596	0.9968
1	0.69	3.2092	4.2982	0.9983
2	0.58	3.2012	4.6353	0.9955
4	0.47	2.5142	5.9607	0.9970

The resulting Avrami parameters (n and k values) determined from $\log[-\ln(1 - X_t)]$ versus $\log(t)$ curves are also reported in Table 2. Graphs with good linear regression were obtained. All R^2 were higher than 0.99. Pure PLLA-PEG-PLLA appeared to have the slowest crystallization, as indicated by its highest n and lowest k values [48,49]. The n values of the composites steadily decreased and the k values significantly increased as the PPZn content increased, supporting a conclusion that the addition of PPZn improved the crystallization properties of PLLA-PEG-PLLA matrices [48,49]. All of the n values were higher than 2.0, suggesting a heterogeneous nucleation effect [50]. The DSC results from both non-isothermal and isothermal scans justified a conclusion that the PPZn acted as a good nucleating agent to increase the crystallinity of PLLA-PEG-PLLA.

3.2. Thermal Decomposition Behaviors

The thermal decomposition behaviors of the composites were determined from thermogravimetric (TG) and derivative TG (DTG) thermograms as well as their expanded thermograms, as shown in Figure 4. From Figure 4a, the pure PLLA-PEG-PLLA indicated by the black curve had two-step thermal decompositions of PLLA end-blocks (200–375 °C) and PEG middle-blocks (375–450 °C) [23,24]. All of the composite samples also had two-step thermal decompositions on TG thermograms similar to the pure PLLA-PEG-PLLA. As would be expected, the residue weight at 800 °C of the composites increased as the PPZn content increased (see Table 3). This is because the PPZn did not completely thermally decompose at 800 °C, as shown in Figure 1c.

Table 3. Thermal decomposition of PLLA-PEG-PLLA/PPZn composites.

PPZn Content (%wt)	Residue Weight at 800 °C [1] (%)	PLLA-$T_{d,max}$ [2] (°C)	PEG-$T_{d,max}$ [2] (°C)
-	0.42	323	416
0.5	0.97	328	417
1	1.22	328	418
2	2.15	328	416
4	3.66	329	415

[1] Obtained from TG thermograms. [2] Obtained from DTG thermograms.

The DTG thermograms in Figure 4b exhibited peaks of temperature at the maximum decomposition rate ($T_{d,max}$) for PLLA (*PLLA-$T_{d,max}$*) and for PEG (*PEG-$T_{d,max}$*). The results of the $T_{d,max}$ values are also summarized in Table 3. For pure PLLA-PEG-PLLA, the *PLLA-*$T_{d,max}$ peak was at 323 °C and the *PEG-*$T_{d,max}$ peak was at 416 °C. The *PLLA-*$T_{d,max}$ peaks of the composites were in the range of 328–329 °C, which were higher temperatures than that of the pure PLLA-PEG-PLLA (323 °C) when the PPZn was incorporated. The higher *PLLA-*$T_{d,max}$ values of the composites might indicate increased thermal stability of the composites compared to pure PLLA-PEG-PLLA. This may be due to a good heat transfer from the PLLA-PEG-PLLA matrices to the high-heat-stability PPZn that improved the thermal stability of the PLLA-PEG-PLLA, attributed to good phase adhesion between the matrix and filler [51]. However, the addition of PPZn did not significantly change the *PEG-*$T_{d,max}$ peaks. The *PEG-*$T_{d,max}$ peaks were in the range of 415–418 °C. This may be due to the phases of PLLA end-blocks being completely thermally decomposed at around 380 °C [see Figure 4b]. Then, almost all of the PLLA-PEG-PLLA matrices would have been destroyed, and there would be no more heat transfer from the PLLA-PEG-PLLA matrices to the PPZn. From the TGA results, a good heat transfer between the PLLA-PEG-PLLA matrix and PPZn also suggested good phase compatibility between them that SEM analysis would subsequently verify. In addition, Yang et al. have reported that the thermal stability of PLLA was improved with increasing crystallinity of PLLA [52].

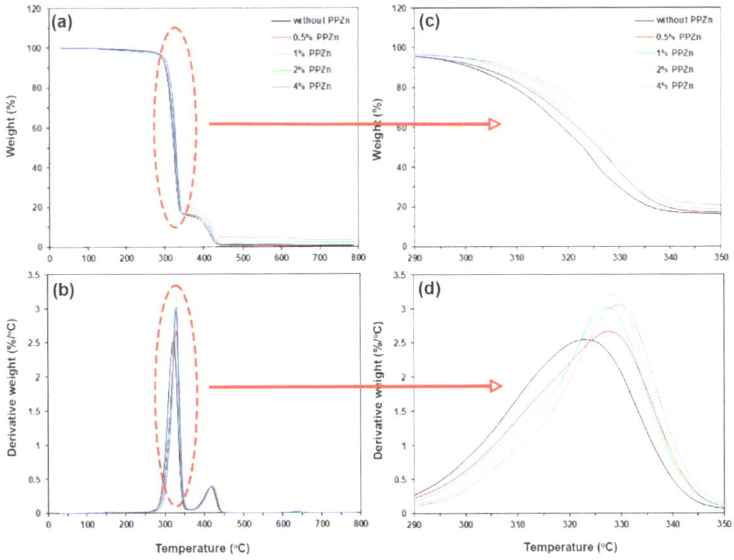

Figure 4. (**a**) TG and (**b**) DTG thermograms of PLLA-PEG-PLLA/PPZn composites with various PPZn contents as well as (**c**) expanded TG and (**d**) expanded DTG thermograms in decomposition region of PLLA end-blocks.

3.3. Crystalline Structures

The crystalline structures of film samples were investigated from XRD patterns, as shown in Figure 5. The pure PLLA-PEG-PLLA in Figure 5a had a broad XRD peak at 16.9° due to PLLA end-block crystallites [47,50]. For the composite films in Figure 5b–e, the XRD peaks of added PPZn were detected at 6.5°, 12.6°, and 18.5°. It was found that the intensities of XRD peaks at 16.9° of composite films steadily increased as the PPZn contents increased. In addition, the XRD peaks appeared at 14.8°, 19.1°, and 22.5° of the composite film containing 4 %wt PPZn due to PLLA crystallites [38,44]. Thus, the addition of PPZn did not change the crystalline structures of the PLLA end-blocks.

The *XRD-*X_c value of pure PLLA-PEG-PLLA film calculated from Equation (2) was 7.7%. The *XRD-*X_c values of film samples increased when the added PPZn content was increased. The *XRD-*X_c values were 9.4%, 10.4%, 15.6%, and 28.1% for the composite films containing PPZn contents of 0.5 %wt, 1 %wt, 2 %wt, and 4 %wt, respectively. The XRD results supported the conclusion that the PPZn acted as a nucleating agent according to the above DSC results. The difference between the *DSC-*X_c and *XRD-*X_c values could be due to the differences in crystallization conditions during the cooling process of the samples [27,53]. In addition, the film samples used in the XRD test were compressed and cooled to limit the mobility of the polymer chains for crystallization.

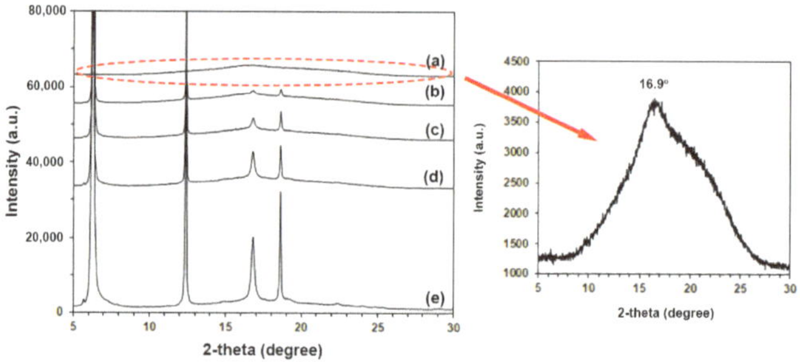

Figure 5. XRD patterns of PLLA-PEG-PLLA/PPZn composites (a) without PPZn and with PPZn contents of (b) 0.5 %wt, (c) 1 %wt, (d) 2 %wt, and (e) 4 %wt.

3.4. Thermo-Mechanical Properties

The thermo-mechanical properties of PLLA and PLLA-PEG-PLLA are related to their heat resistance properties, which have been widely determined through DMA analysis [27,40,43,45,54–59]. The variation in the storage modulus of PLLA-PEG-PLLA as a function of temperature is presented in Figure 6. Pure PLLA-PEG-PLLA, indicated by the black curve, first displayed a drop curve of storage modulus in the range of 30–70 °C, suggesting that it had low stiffness [40] because of the rubbery character and low crystallinity of PLLA-PEG-PLLA. At higher temperatures, this was followed by an increase in storage modulus again because of the cold crystallization of the PLLA end-blocks [27,40]. This indicates that pure PLLA-PEG-PLLA had poor heat resistance because the pure PLLA-PEG-PLLA film had low crystallinity, as described in the above XRD results (*XRD-*X_c of pure PLLA-PEG-PLLA is 7.7%). It has been reported that the phases of PLLA crystallites induce high storage modulus and increase heat-resistant properties [54–56,59].

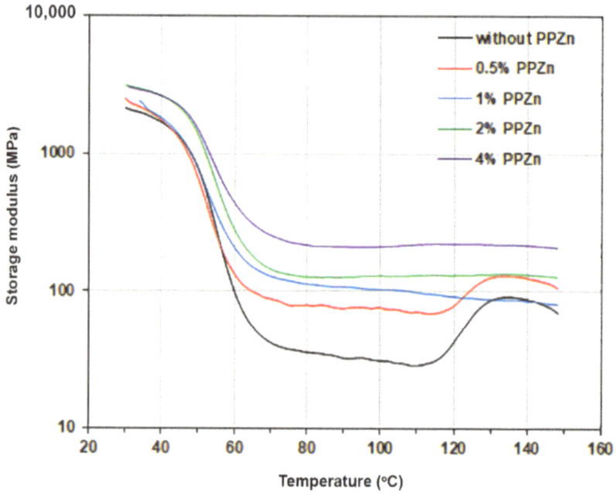

Figure 6. Storage modulus as a function of temperature of PLLA-PEG-PLLA/PPZn composite films with various PPZn contents.

The cold-crystallization effect of DMA curves from Figure 6 disappeared when the PPZn content was increased up to 1 %wt (see blue curve), indicating that the crystallinity of composite film was high enough to erase the cold-crystallization behavior [33]. The lowest storage modulus of pure PLLA-PEG-PLLA in the temperature range of 30–120 °C was only 29 MPa. The lowest storage moduli of the composites containing PPZn contents of 0.5, 1.0, 2.0, and 4.0 %wt were 70, 91, 125, and 209 MPa, respectively. The lowest storage moduli of the composites increased as the PPZn content increased. From the DMA results, it was concluded that the addition of PPZn increased the film stiffness of the PLLA-PEG-PLLA matrices during the DMA heating scan, enabling it to resist film deformation and improve film heat resistance. This may be explained by the fact that the addition of PPZn increased the crystallinity of PLLA-PEG-PLLA, thus maintaining the storage modulus of composite films during the DMA heating scan [19,40,43,45].

3.5. Dimensional Stability to Heat

The dimensional stability to heat of film samples at 80 °C under the load of 200 g for 30 s was also used to examine their heat-resistant properties. The film samples with higher heat resistance exhibited shorter final lengths and higher values of dimensional stability to heat [39,40]. Images of pure PLLA-PEG-PLLA and composite films before and after dimensional stability testing are displayed in Figure 7. It can be seen that the pure PLLA-PEG-PLLA showed the longest film extension, indicating that it had the lowest heat resistance. The film extension significantly decreased as the PPZn content increased.

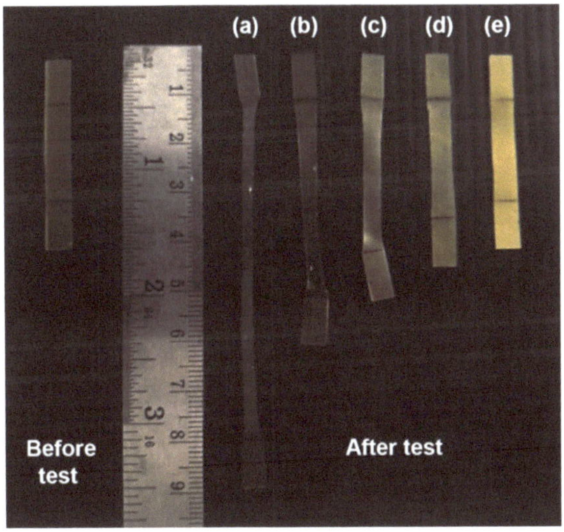

Figure 7. Photographs of PLLA-PEG-PLLA/PPZn composite films (**a**) without PPZn and with PPZn contents of (**b**) 0.5 %wt, (**c**) 1 %wt, (**d**) 2 %wt, and (**e**) 4 %wt after testing dimensional stability to heat at 80 °C under a 200 g load for 30 s.

The values of dimensional stability to heat were calculated from Equation (3) and compared in bar graphs, as shown in Figure 8. The pure PLLA-PEG-PLLA had the lowest dimensional stability to heat at about 28.3% because it had the lowest crystallinity. The values of dimensional stability to heat steadily increased as the PPZn content increased. These values were about 47.6%, 58.7%, 82.8%, and 96.1% for PPZn content of 0.5, 1, 2, and 4 %wt, respectively. This may be explained by the fact that the dimensional stability to heat of film samples strongly depended on the $XRD\text{-}X_c$ values of film samples [39,40,55]. The crystalline phases of PLLA end-blocks of PLLA-PEG-PLLA matrices improved film stiffness to resist film extension at 80 °C. The results supported the conclusion that the addition of PPZn improved the heat resistance of composited films according to the above DMA analysis.

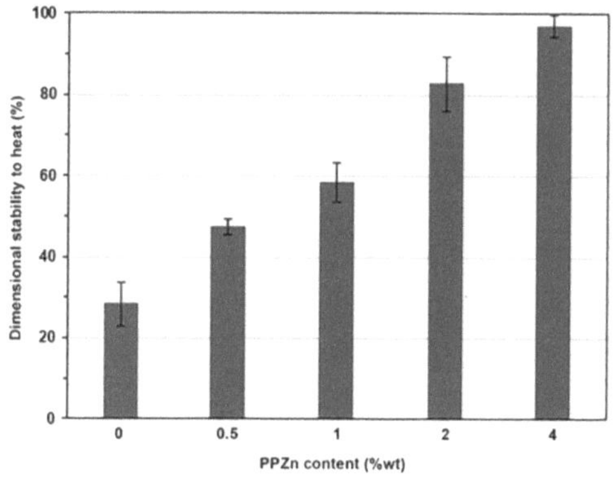

Figure 8. Dimensional stability to heat of PLLA-PEG-PLLA/PPZn composite films with various PPZn contents.

3.6. Phase Compatibility

The phase compatibility between the PLLA-PEG-PLLA matrix and PPZn of the composite films was investigated based on their cryo-fractured surfaces, as shown in Figure 9. The pure PLLA-PEG-PLLA film in Figure 9a showed rougher surfaces, suggesting that the film was flexible. Some small, white, needle-shaped streaks were also observed on the fractured surfaces. This could be because the PLLA-PEG-PLLA matrix was somewhat stretched before breaking in the cryo-fracture step. These tiny, white needles were also found on the fractured surfaces of all of the PLLA-PEG-PLLA/PPZn composite films, indicating that they were also flexible. Some PPZn particles of composite films in Figure 9b–e are indicated by white circles. It can be seen that the PPZn particles were well-distributed and dispersed on the PLLA-PEG-PLLA matrices, suggesting that the PPZn particles and PLLA-PEG-PLLA matrix had good phase compatibility.

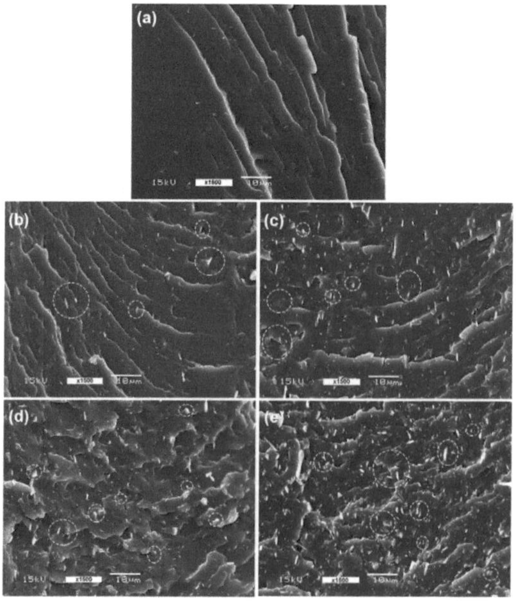

Figure 9. SEM images of cryo-fractured surfaces of PLLA-PEG-PLLA/PPZn composite films (**a**) without PPZn and with PPZn contents of (**b**) 0.5 %wt, (**c**) 1 %wt, (**d**) 2 %wt, and (**e**) 4 %wt (some PPZn particles were labeled by white circles; all bar scales = 10 μm).

3.7. Tensile Properties

Stress–strain curves from tensile testing were used to characterize the mechanical properties of the composite films, as shown in Figure 10. The tensile results are summarized in Table 4 and are clearly compared in bar graph types, as shown in Figure 11. The pure PLLA-PEG-PLLA film had an ultimate tensile stress of 13.5 MPa, a strain at break of 125%, and a Young's modulus of 190 MPa. All composite films, including the pure PLLA-PEG-PLLA film, exhibited a yield point. This result confirms the conclusion of the above SEM study that the composite films were flexible. With increased PPZn content, the ultimate tensile stress and Young's modulus of the composite films significantly increased, and the strain at break steadily decreased. The tensile results suggested that the PPZn acted as a reinforcing agent for PLLA-PEG-PLLA. This may be due to the crystallinity of composite films increasing as the PPZn was incorporated, as explained in the above XRD analysis. The crystalline phases could act as physical cross-linking sites to improve the tensile stress and Young's modulus of the flexible polymers [43,52]. In addition, the increase in tensile stress of PLLA-PEG-PLLA/PPZn films demonstrated the effective interaction between PLLA-PEG-PLLA and PPZn [60], and the reinforcing effect of PPZn subsequently restricted the chain mobility and deformation of PLLA-PEG-PLLA.

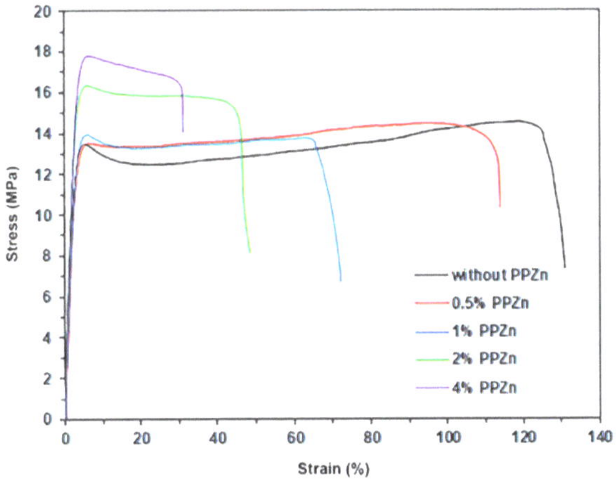

Figure 10. Tensile curves of PLLA-PEG-PLLA/PPZn composite films with various PPZn contents.

Table 4. Tensile properties and opacity of PLLA-PEG-PLLA/PPZn composite films.

PPZn Content (%wt)	Ultimate Tensile Stress (MPa)	Strain at Break (%)	Young's Modulus (Mpa)	Opacity (mm^{-1})
-	13.5 ± 3.1	125 ± 11	190 ± 21	0.402 ± 0.088
0.5	14.1 ± 2.5	112 ± 15	200 ± 25	0.754 ± 0.045
1	14.6 ± 2.7	76 ± 8	211 ± 22	0.936 ± 0.027
2	16.7 ± 3.4	51 ± 6	230 ± 24	1.960 ± 0.067
4	18.2 ± 2.8	29 ± 6	279 ± 31	3.829 ± 0.074

It was shown in our previous study [28] that the addition of 2 %wt talcum also improved the ultimate tensile stress of PLLA-PEG-PLLA. The talcum became aggregated when its content was higher than 2 %wt, reducing the tensile properties of the PLLA-PEG-PLLA composite films. This suggested poor phase compatibility between the PLLA-PEG-PLLA and talcum. However, this study showed that the ultimate tensile stress steadily increased with increasing the PPZn-content up to 4 %wt. This suggested a better phase compatibility of the PLLA-PEG-PLLA with PPZn than with talcum. Thus, PPZn can be considered a good reinforcement for PLLA-PEG-PLLA.

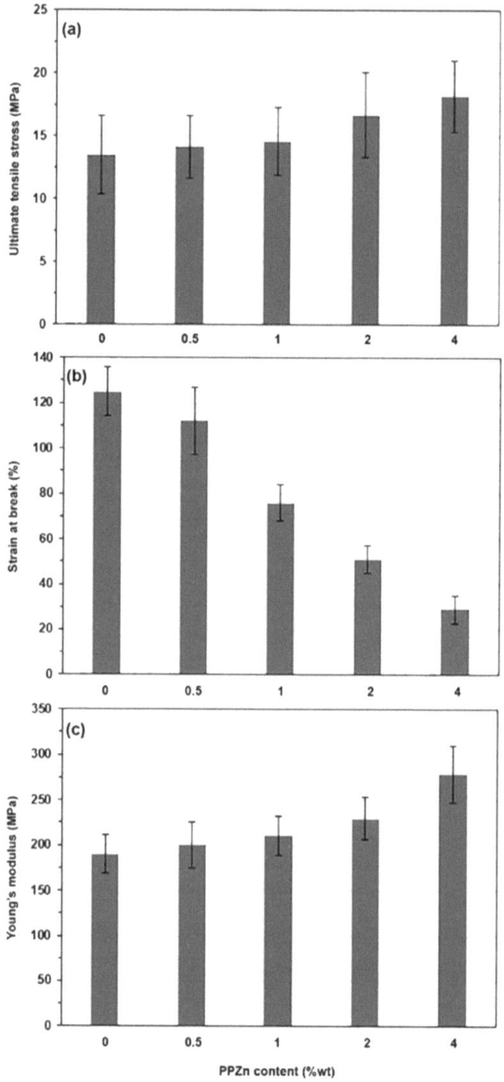

Figure 11. (**a**) Ultimate tensile stress, (**b**) strain at break, and (**c**) Young's modulus of PLLA-PEG-PLLA/PPZn composite films with various PPZn contents from Table 4.

3.8. Film's Opacity

The opacity of pure PLLA-PEG-PLLA film calculated from Equation (4) was 0.402 mm^{-1}, as also reported in Table 4. It was found that the opacity of film samples increased when the added PPZn content was increased. The composite films had higher opacity compared with the pure PLLA-PEG-PLLA film, as shown in Figure 12. However, the words covered by the composite films were still clearly visible, and they were legible. It has been reviewed that the polymer film's opacity increases as the crystallinity of polymers increases [61].

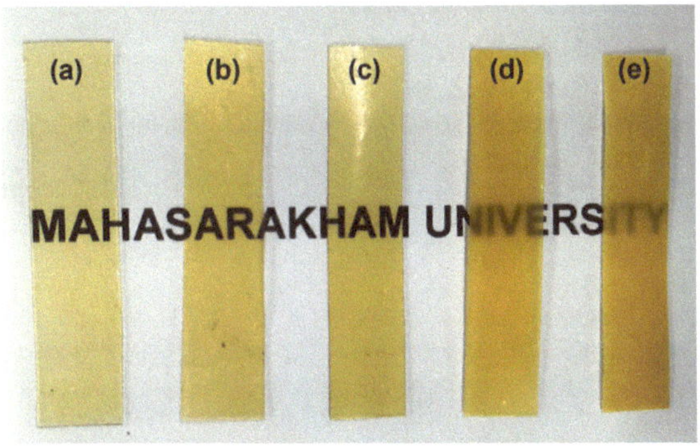

Figure 12. Photographs of PLLA-PEG-PLLA/PPZn composite films (**a**) without PPZn and with PPZn contents of (**b**) 0.5 %wt, (**c**) 1 %wt, (**d**) 2 %wt, and (**e**) 4 %wt.

4. Conclusions

This study aimed to understand the role and behavior of zinc phenyl phosphate (PPZn) when incorporated with the poly(L-lactide)-*b*-poly(ethylene glycol)-*b*-poly(L-lactide) (PLLA-PEG-PLLA) flexible bioplastic. The addition of the PPZn to PLLA-PEG-PLLA effectively improved the crystallization properties, thermal stability, heat resistance, and tensile properties of PLLA-PEG-PLLA matrices as follows. The T_{cc} peaks of the PLLA-PEG-PLLA/PPZn composites disappeared, the T_c peaks shifted to higher temperature, the crystallinity increased, and the $t_{1/2}$ decreased as the PPZn content increased, suggesting that the PPZn acted as a good nucleating agent. The PLLA-Td,max of the PLLA-PEG-PLLA composites shifted to a higher temperature with the PPZn contents, indicating that the PPZn acted as a thermal stabilizer. Increasing the stiffness and the dimensional stability to heat of PLLA-PEG-PLLA composites by adding PPZn is conductive to improving the heat resistance of PLLA-PEG-PLLA matrices because the crystallinity increases. The PLLA-PEG-PLLA/PPZn composites showed good phase compatibility, which led to enhanced crystallization properties and thermal stability of the PLLA-PEG-PLLA matrices. The ultimate tensile stress and Young's modulus of PLLA-PEG-PLLA matrices steadily increased as the PPZn content increased, suggesting that the PPZn acted as a reinforcing filler. The film's opacity increased with the PPZn content. Thus, PPZn can be used as an efficient multi-functional filler to develop the PLLA-PEG-PLLA's properties for widespread bioplastic applications.

Author Contributions: Conceptualization, K.P. and Y.B.; Methodology, K.P., P.S., A.M., W.R., N.P. and Y.B.; Investigation, K.P., P.S., A.M., W.R., N.P. and Y.B.; Resources, Y.B.; Visualization, K.P. and Y.B.; Writing—original draft, K.P., P.N.L. and Y.B.; Writing—review and editing, K.P., P.N.L. and Y.B. All authors have read and agreed to the published version of the manuscript.

Funding: This research project was financially supported by Mahasarakham University (Funding number 6717013/2567). Y.B. is also grateful for the partial support provided by the Centre of Excellence for Innovation in Chemistry (PERCH-CIC), Office of the Higher Education Commission, Ministry of Education, Thailand.

Institutional Review Board Statement: Not applicable.

Data Availability Statement: Data are contained within the article.

Conflicts of Interest: All authors declare that the research was conducted in the absence of any commercial or financial relationships that could be construed as potential conflicts of interest.

References

1. Castro-Aguirre, E.; Iñiguez-Franco, F.; Samsudin, H.; Fang, X.; Auras, R. Poly(lactic acid)–mass production, processing, industrial applications, and end of life. *Adv. Drug Deliv. Rev.* **2016**, *107*, 333–366. [CrossRef] [PubMed]
2. Hamad, K.; Kaseem, M.; Ayyoob, M.; Joo, J.; Deri, F. Polylactic acid blends: The future of green, light and tough. *Prog. Polym. Sci.* **2018**, *85*, 83–127. [CrossRef]
3. Mastalygina, E.E.; Aleksanyan, K.V. Recent approaches to the plasticization of poly(lactic acid) (PLA) (A review). *Polymers* **2024**, *16*, 87. [CrossRef] [PubMed]
4. de França, J.O.C.; Lima, Q.d.S.; Barbosa, M.M.d.M.; Fonseca, A.L.F.; Machado, G.d.F.; Dias, S.C.L.; Dias, J.A. Sonochemical synthesis of magnetite/poly(lactic acid) nanocomposites. *Polymers* **2023**, *15*, 4662. [CrossRef]
5. Guo, W.; Bu, W.; Mao, Y.; Wang, E.; Yang, Y.; Liu, C.; Guo, F.; Mai, H.; You, H.; Long, Y. Magnesium hydroxide as a versatile nanofiller for 3D-printed PLA bone scaffolds. *Polymers* **2024**, *16*, 198. [CrossRef] [PubMed]
6. Jamnongkan, T.; Sirichaicharoenkol, K.; Kongsomboon, V.; Srinuan, J.; Srisawat, N.; Pangon, A.; Mongkholrattanasit, R.; Tammasakchai, A.; Huang, C.-F. Innovative electrospun nanofiber mats based on polylactic acid composited with silver nanoparticles for medical applications. *Polymers* **2024**, *16*, 409. [CrossRef]
7. Durpekova, S.; Bergerova, E.D.; Hanusova, D.; Dusankova, M.; Sedlarik, V. Eco-friendly whey/polysaccharide-based hydrogel with poly(lactic acid) for improvement of agricultural soil quality and plant growth. *Int. J. Biol. Macromol.* **2022**, *212*, 85–96. [CrossRef]
8. Wu, Y.; Gao, X.; Wu, J.; Zhou, T.; Nguyen, T.T.; Wang, Y. Biodegradable polylactic acid and its composites: Characteristics, processing, and sustainable applications in sports. *Polymers* **2023**, *15*, 3096. [CrossRef] [PubMed]
9. Nasution, H.; Harahap, H.; Julianti, E.; Safitri, A.; Jaafar, M. Smart packaging based on polylactic acid: The effects of antibacterial and antioxidant agents from natural extracts on physical–mechanical properties, colony reduction, perishable food shelf life, and future prospective. *Polymers* **2023**, *15*, 4103. [CrossRef]
10. Swetha, T.A.; Bora, A.; Mohanrasu, K.; Balaji, P.; Raja, R.; Ponnuchamy, K.; Muthusamy, G.; Arun, A. A comprehensive review on polylactic acid (PLA)—Synthesis, processing and application in food packaging. *Int. J. Biol. Macromol.* **2023**, *234*, 123715. [CrossRef]
11. An, L.; Perkins, P.; Yi, R.; Ren, T. Development of polylactic acid based antimicrobial food packaging films with N-halamine modified microcrystalline cellulose. *Int. J. Biol. Macromol.* **2023**, *242 Pt 1*, 124685. [CrossRef]
12. Alexeeva, O.V.; Olkhov, A.A.; Konstantinova, M.L.; Podmasterev, V.V.; Petrova, T.V.; Martirosyan, L.Y.; Karyagina, O.K.; Kozlov, S.S.; Lomakin, S.M.; Tretyakov, I.V.; et al. A novel approach for glycero-(9,10-trioxolane)-trialeate incorporation into poly(lactic acid)/poly(ε-caprolactone) blends for biomedicine and packaging. *Polymers* **2024**, *16*, 128. [CrossRef] [PubMed]
13. Fortunati, E.; Puglia, D.; Iannoni, A.; Terenzi, A.; Kenny, J.M.; Torre, L. Processing conditions, thermal and mechanical responses of stretchable poly(lactic acid)/poly(butylene succinate) films. *Materials* **2017**, *10*, 809. [CrossRef]
14. Li, D.; Jiang, Y.; Lv, S.; Liu, X.; Gu, J.; Chen, Q.; Zhang, Y. Preparation of plasticized poly (lactic acid) and its influence on the properties of composite materials. *PLoS ONE* **2018**, *13*, e0193520. [CrossRef] [PubMed]
15. Greco, A.; Ferrari, F. Thermal behavior of PLA plasticized by commercial and cardanol-derived plasticizers and the effect on the mechanical properties. *J. Therm. Anal. Calorim.* **2021**, *146*, 131–141. [CrossRef]
16. Saeidlou, S.; Huneault, M.A.; Li, H.; Park, C.B. Poly(lactic acid) crystallization. *Prog. Polym. Sci.* **2012**, *37*, 1657–1677. [CrossRef]
17. Baiardo, M.; Frisoni, G.; Scandola, M.; Rimelen, M.; Lips, D.; Ruffieux, K.; Wintermantel, E. Thermal and mechanical properties of plasticized poly(L-lactic acid). *J. Appl. Polym. Sci.* **2003**, *90*, 1731–1738. [CrossRef]
18. Kulinski, Z.; Piorkowska, E. Crystallization, structure and properties of plasticized poly(l-lactide). *Polymer* **2005**, *46*, 10290–10300. [CrossRef]
19. Sungsanit, K.; Kao, N.; Bhattacharya, S.N. Properties of linear poly(lactic acid)/polyethylene glycol blends. *Polym. Eng. Sci.* **2012**, *52*, 108–116. [CrossRef]
20. Ma, P.; Shen, T.; Lin, L.; Dong, W.; Chen, M. Cellulose-g-poly(d-lactide) nanohybrids induced significant low melt viscosity and fast crystallization of fully bio-based nanocomposites. *Carbohydr. Polym.* **2017**, *155*, 498–506. [CrossRef]
21. Carbonell-Verdu, A.; Garcia-Garcia, D.; Dominici, F.; Torre, L.; Sanchez-Nacher, L.; Balart, R. PLA films with improved flexibility properties by using maleinized cotton-seed oil. *Eur. Polym. J.* **2017**, *91*, 248–259. [CrossRef]
22. Yun, X.; Li, X.; Jin, Y.; Sun, W.; Dong, T. Fast crystallization and toughening of poly(L-lactic acid) by incorporating with poly(ethylene glycol) as a middle block chain. *Polym. Sci. Ser. A* **2018**, *60*, 141–155. [CrossRef]
23. Baimark, Y.; Rungseesantivanon, W.; Prakymorama, N. Improvement in melt flow property and flexibility of poly(L-lactide)-*b*-poly(ethylene glycol)-*b*-poly(L-lactide) by chain extension reaction for potential use as flexible bioplastics. *Mater. Des.* **2018**, *154*, 73–80. [CrossRef]
24. Srihanam, P.; Thongsomboon, W.; Baimark, Y. Phase Morphology, mechanical, and thermal properties of calcium carbonate-reinforced poly(L-lactide)-b-poly(ethylene glycol)-b-poly(L-lactide) bioplastics. *Polymers* **2023**, *15*, 301. [CrossRef]
25. Cailloux, J.; Santona, O.O.; Franco-Urquiza, E.; Bou, J.J.; Carrasco, F.; Gamez-Perez, J.; Maspoch, M.L. Sheets of branched poly(lactic acid) obtained by one step reactive extrusion calendaring process: Melt rheology analysis. *Express Polym. Lett.* **2013**, *7*, 304–318. [CrossRef]
26. Lim, L.-T.; Auras, R.; Rubino, M. Processing technologies for poly(lactic acid). *Prog. Polym. Sci.* **2008**, *33*, 820–852. [CrossRef]

27. Zhang, X.; Meng, L.; Li, G.; Liang, N.; Zhang, J.; Zhu, Z.; Wang, R. Effect of nucleating agents on the crystallization behavior and heat resistance of poly(L-lactide). *J. Appl. Polym. Sci.* **2016**, *133*, 42999. [CrossRef]
28. Srisuwan, S.; Baimark, Y. Synergistic effects of PEG middle-blocks and talcum on crystallizability and thermomechanical properties of flexible PLLA-*b*-PEG-*b*-PLLA bioplastic. *e-Polymers* **2022**, *22*, 389–398. [CrossRef]
29. Wu, N.; Wang, H. Effect of zinc phenylphosphonate on the crystallization behavior of poly(L-lactide). *J. Appl. Polym. Sci.* **2013**, *130*, 2744–2752. [CrossRef]
30. Tabi, T.; Ageyeva, T.; Kovacs, J.G. The influence of nucleating agents, plasticizers, and molding conditions on the properties of injection molded PLA products. *Mater. Today Commun.* **2022**, *32*, 103936. [CrossRef]
31. Ageyeva, T.; Kovács, J.G.; Tabi, T. Comparison of the efficiency of the most effective heterogeneous nucleating agents for poly(lactic acid). *J. Therm. Anal. Calorim.* **2022**, *147*, 8199–8211. [CrossRef]
32. Tsuji, H.; Takai, H.; Saha, S.K. Isothermal and non-isothermal crystallization behavior of poly(L-lactic acid): Effects of stereocomplex as nucleating agent. *Polymer* **2006**, *47*, 3826–3837. [CrossRef]
33. Srisuwan, Y.; Baimark, Y. Improvement in thermal stability of flexible poly(L-lactide)-*b*-poly(ethylene glycol)-*b*-poly(L-lactide) bioplastic by blending with native cassava starch. *Polymers* **2022**, *14*, 3186. [CrossRef] [PubMed]
34. Ma, B.; Wang, X.; He, Y.; Dong, Z.; Zhang, X.; Chen, X.; Liu, T. Effect of poly(lactic acid) crystallization on its mechanical and heat resistance performances. *Polymer* **2021**, *212*, 123280. [CrossRef]
35. Thongsomboon, W.; Srihanam, P.; Baimark, Y. Preparation of flexible poly(L-lactide)-*b*-poly(ethylene glycol)-*b*-poly(L-lactide)/talcum/thermoplastic starch ternary composites for use as heat-resistant and single-use bioplastics. *Int. J. Biol. Macromol.* **2023**, *230*, 123172. [CrossRef] [PubMed]
36. Baimark, Y.; Rungseesantivanon, W.; Prakymoramas, N. Synthesis of flexible poly(L-lactide)-*b*-polyethylene glycol-*b*-poly(L-lactide) bioplastics by ring-opening polymerization in the presence of chain extender. *e-Polymers* **2020**, *20*, 423–429. [CrossRef]
37. Srihanam, P.; Srisuwan, Y.; Phromsopha, T.; Manphae, A.; Baimark, Y. Improvement in phase compatibility and mechanical properties of poly(L-lactide)-*b*-poly(ethylene glycol)-*b*-poly(L-lactide)/thermoplastic starch blends with citric acid. *Polymers* **2023**, *15*, 3966. [CrossRef] [PubMed]
38. Li, L.; Cao, Z.Q.; Bao, R.Y.; Xie, B.H.; Yang, M.B.; Yang, W. Poly(L-lactic acid)-polyethylene glycol-poly(L-lactic acid) triblock copolymer: A novel macromolecular plasticizer to enhance the crystallization of poly(L-lactic acid). *Eur. Polym. J.* **2017**, *97*, 272–281. [CrossRef]
39. Baimark, Y.; Kittipoom, S. Influence of chain-extension reaction on stereocomplexation, mechanical properties and heat resistance of compressed stereocomplex-polylactide bioplastic films. *Polymers* **2018**, *10*, 1218. [CrossRef]
40. Baimark, Y.; Pasee, S.; Rungseesantivanon, W.; Prakymoramas, N. Flexible and high heat-resistant stereocomplex PLLA-PEG-PLLA/PDLA blends prepared by melt process: Effect of chain extension. *J. Polym. Res.* **2019**, *26*, 218. [CrossRef]
41. Hasheminya, S.-M.; Mokarram, R.R.; Ghanbarzadeh, B.; Hamishekar, H.; Kafil, H.S.; Dehghannya, J. Development and characterization of biocomposite films made from kefiran, carboxymethyl cellulose and Satureja Khuzestanica essential oil. *Food Chem.* **2019**, *289*, 443–452. [CrossRef] [PubMed]
42. Seven, K.M.; Cogen, J.M.; Gilchrist, J.F. Nucleating agents for high-density polyethylene—A review. *Polym. Eng. Sci.* **2016**, *56*, 541–554. [CrossRef]
43. Shi, X.; Zhang, G.; Phuong, T.V.; Lazzeri, A. Synergistic effects of nucleating agents and plasticizers on the crystallization behavior of poly(lactic acid). *Molecules* **2015**, *20*, 1579–1593. [CrossRef] [PubMed]
44. Kong, D.; Zhang, D.; Guo, H.; Zhao, J.; Wang, Z.; Hu, H.; Xu, J.; Fu, C. Functionalized boron nitride nanosheets/poly(L-lactide) nanocomposites and their crystallization behavior. *Polymers* **2019**, *11*, 440. [CrossRef] [PubMed]
45. Li, H.; Huneault, M.A. Effect of nucleation and plasticization on the crystallization of poly(lactic acid). *Polymer* **2007**, *48*, 6855–6866. [CrossRef]
46. Wang, L.; Wang, Y.N.; Huang, Z.G.; Weng, Y.X. Heat resistance, crystallization behavior, and mechanical properties of polylactide/nucleating agent composites. *Mater. Des.* **2015**, *66*, 7–15. [CrossRef]
47. Chen, P.; Yu, K.; Wang, Y.; Wang, W.; Zhou, H.; Li, H.; Mi, J.; Wang, X. The effect of composite nucleating agent on the crystallization behavior of branched poly(lactic acid). *J. Polym. Environ.* **2018**, *26*, 3718–3730. [CrossRef]
48. Li, Y.; Han, C. Isothermal and nonisothermal cold crystallization behaviors of asymmetric poly(L-lactide)/poly(D-lactide) blends. *Ind. Eng. Chem. Res.* **2012**, *51*, 15927–15935. [CrossRef]
49. Jalali, A.; Huneault, M.A.; Elkoun, S. Effect of thermal history on nucleation and crystallization of poly(lactic acid). *J. Mater. Sci.* **2016**, *51*, 7768–7779. [CrossRef]
50. Gao, P.; Alanazi, S.; Masato, D. Crystallization of polylactic acid with organic nucleating agents under quiescent conditions. *Polymers* **2024**, *16*, 320. [CrossRef]
51. Vidovic, E.; Faraguna, F.; Jukic, A. Influence of inorganic fillers on PLA crystallinity and thermal properties. *J. Therm. Anal. Calorim.* **2017**, *127*, 371–380. [CrossRef]
52. Yang, J.-Y.; Kim, D.-K.; Han, W.; Park, J.-Y.; Kim, K.-W.; Kim, B.-J. Effect of nucleating agents addition on thermal and mechanical properties of natural fiber-reinforced polylactic acid composites. *Polymers* **2022**, *14*, 4263. [CrossRef]
53. Srithep, Y.; Nealey, P.; Turng, L.S. Effects of annealing time and temperature on the crystallinity and heat resistance behavior of injection-molded poly(lactic acid). *Polym. Mater. Sci. Eng.* **2013**, *53*, 580–588. [CrossRef]

54. Chauliac, D.; Pullammanappallil, P.C.; Ingram, L.O.; Shanmugam, K.T. A combined thermochemical and microbial process for recycling polylactic acid polymer to optically pure L-lactic acid for reuse. *J. Polym. Environ.* **2020**, *28*, 1503–1512. [CrossRef]
55. Yin, H.-Y.; Wei, X.-F.; Bao, R.-Y.; Dong, Q.-X.; Liu, Z.-Y.; Yang, W.; Xie, B.-H.; Yang, M.-B. Enhancing thermomechanical properties and heat distortion resistance of poly(L-lactide) with high crystallinity under high cooling rate. *ACS Sustain. Chem. Eng.* **2015**, *3*, 654–661. [CrossRef]
56. Vadori, R.; Mohanty, A.K.; Misra, M. The effect of mold temperature on the performance of injection molded poly(lactic acid)-based bioplastic. *Macromol. Mater. Eng.* **2013**, *298*, 981–990. [CrossRef]
57. Si, W.-J.; An, X.-P.; Zeng, J.-B.; Chen, Y.-K.; Wang, Y.-Z. Fully biobased, highly toughened and heat-resistant poly(L-lactide) ternary blends via dynamic vulcanization with poly(D-lactide) and unsaturated bioelastomer. *Sci. China Mater.* **2017**, *60*, 1008–1022. [CrossRef]
58. Masutani, K.; Kobayashi, K.; Kimura, Y.; Lee, C.W. Properties of stereo multi-block polylactides obtained by chain-extension of stereo tri-block polylactides consisting of poly(L-lactide) and poly(D-lactide). *J. Polym. Res.* **2018**, *25*, 74. [CrossRef]
59. Tabi, T.; Ageyeva, T.; Kovacs, J.G. Improving the ductility and heat deflection temperature of injection molded poly(lactic acid) products: A comprehensive review. *Polym. Test.* **2021**, *101*, 107282. [CrossRef]
60. Bindhu, B.; Renisha, R.; Roberts, L.; Varghese, T.O. Boron Nitride reinforced polylactic acid composites film for packaging: Preparation and properties. *Polym. Test.* **2018**, *66*, 172–177. [CrossRef]
61. Lin, Y.; Bilotti, E.; Bastiaansen, C.W.M.; Peijs, T. Transparent semi-crystalline polymeric materials and their nanocomposites: A review. *Polym. Eng. Sci.* **2020**, *60*, 2351–2376. [CrossRef]

polymers

Article

Physicochemical Characterization and In Vitro Activity of Poly(ε-Caprolactone)/Mycophenolic Acid Amorphous Solid Dispersions

Oroitz Sánchez-Aguinagalde [1], Eva Sanchez-Rexach [1], Yurena Polo [2], Aitor Larrañaga [1], Ainhoa Lejardi [1,*], Emilio Meaurio [1] and Jose-Ramon Sarasua [1]

[1] Department of Mining-Metallurgy Engineering and Materials Science, POLYMAT, Bilbao School of Engineering, University of the Basque Country (UPV/EHU), Plaza Ingeniero Torres Quevedo 1, 48013 Bilbao, Spain; oroitz.sanchez@ehu.eus (O.S.-A.); evagloria.sanchez@ehu.eus (E.S.-R.); aitor.larranagae@ehu.es (A.L.); emiliano.meaurio@ehu.eus (E.M.); jr.sarasua@ehu.es (J.-R.S.)
[2] Polimerbio SL, Paseo Miramon 170, 20014 Donostia-San Sebastian, Spain; ypolo@polimerbio.com
* Correspondence: ainhoa.lejardi@ehu.es

Abstract: The obtention of amorphous solid dispersions (ASDs) of mycophenolic acid (MPA) in poly(ε-caprolactone) (PCL) is reported in this paper. An improvement in the bioavailability of the drug is possible thanks to the favorable specific interactions occurring in this system. Differential scanning calorimetry (DSC) was used to investigate the miscibility of PCL/MPA blends, measuring glass transition temperature (T_g) and analyzing melting point depression to obtain a negative interaction parameter, which indicates the development of favorable inter-association interactions. Fourier transform infrared spectroscopy (FTIR) was used to analyze the specific interaction occurring in the blends. Drug release measurements showed that at least 70% of the drug was released by the third day in vitro in all compositions. Finally, preliminary in vitro cell culture experiments showed a decreased number of cancerous cells over the scaffolds containing MPA, presumably arising from the anti-cancer activity attributable to MPA.

Keywords: poly(ε-caprolactone) (PCL); mycophenolic acid (MPA); amorphous solid dispersions (ASDs); miscibility; interactions; drug release; cancer treatment

Citation: Sánchez-Aguinagalde, O.; Sanchez-Rexach, E.; Polo, Y.; Larrañaga, A.; Lejardi, A.; Meaurio, E.; Sarasua, J.-R. Physicochemical Characterization and In Vitro Activity of Poly(ε-Caprolactone)/Mycophenolic Acid Amorphous Solid Dispersions. *Polymers* **2024**, *16*, 1088. https://doi.org/10.3390/polym16081088

Academic Editors: Stefano Farris and Masoud Ghaani

Received: 10 November 2023
Revised: 27 March 2024
Accepted: 3 April 2024
Published: 13 April 2024

1. Introduction

As new treatments and drugs appear for all kinds of diseases, we are also faced with great challenges to achieve a satisfactory application of these remedies. Although they may be effective in theory, most of the drugs that are being approved are not feasible in terms of their biopharmacological properties. The main causes are low permeability, poor solubility, or rapid elimination from the body. In fact, 90% of the drugs being developed are molecules with low water solubility, in addition to almost 40% of the drugs already approved [1–3]. The dimensions of this problem can be seen, for example, in the case of the oral administration of doses. In order to reach systemic circulation, the drug must be dissolved in the intestinal fluids of the gastrointestinal tract, which is difficult in the case of low solubility [4]. The cause of this low bioavailability is the different molecular arrangements, where the crystalline compounds are the ones that present the greatest problem [5]. In order to solve this problem, one of the established strategies is amorphization, which transforms low-energy crystalline substances into high-energy amorphous compounds, giving them greater solubility and bioavailability [6]. However, these amorphous solids are not thermodynamically stable because of their excess enthalpy, entropy, and free energies, which cause them to tend to form crystals [7]. For this reason, achieving the stability of these compounds is a great challenge.

One of the strategies used for this purpose is developing amorphous solid dispersions (ASDs). In the 1970s, Chiou and Riegelman defined the term solid dispersions as the dispersion of an active pharmaceutical ingredient (API) in an amorphous carrier in a solid state prepared by solvent, melting, or solvent-melting methods [7]. In these systems, there is a mixture at the molecular level between a polymer and the drug in an amorphous state, increasing its bioavailability [8–12]. It is known that the low thermodynamic stability due to the high energy of the amorphous state causes relaxation, nucleation, and recrystallization under different variables [13–16]. Thus, the role of the polymeric matrix is to inhibit this process and maintain the mixture in a single homogeneous phase [17,18]. To avoid this crystallization and maintain the mixture in the metastable region of the binary phase diagram, miscibility between the API and polymer is essential [19–21]. The kinetic stability provided by storage below the glass transition temperature (T_g) must also be taken into account. In fact, according to Hancock et al., the stability of the mixture could be ensured for years by storing it at least 50 K below T_g [22,23]. One significant challenge in this system is the unpredictable nature of polymer–drug interactions [24].

One interesting drug to test this system is mycophenolic acid (MPA—$C_{17}H_{20}O_6$, 320 g/mol; aqueous solubility: 35.5 mg/L). Mycophenolic acid (Scheme 1) is an antibiotic produced by the *Penicillium* family and is best known for its use as an immunosuppressive agent to prevent rejection in organ transplants [25,26]. In addition, this drug has more biological properties, such as antifungal or antiviral properties [27]. It also has the potential to prevent and perhaps treat chronic allograft vasculopathy, as it can inhibit the proliferation of vascular smooth muscle cells (VSMCs), mesangial cells, and myofibroblasts [28]. However, one of the most striking properties is its ability to act against tumor cells of various types such as leukemia or lymphoma, among others [29]. This is because MPA is an inhibitor of inosine monophosphate dehydrogenase (IMPDH), which leads to the reduction of xanthine monophosphate (XMP), guanosine-5′-triphosphate (GTP), and deoxyguanosine triphosphate (dGTP), thus inhibiting the proliferation of lympholeukocytes and cancer cells [26,30]. Despite having so many favorable properties, the bioavailability of MPA in vivo is relatively poor due to the high clearance inside a living organism, which limits its possibility of clinical application [25]. This, in addition to its low aqueous solubility, makes it a perfect candidate for forming amorphous solid dispersions.

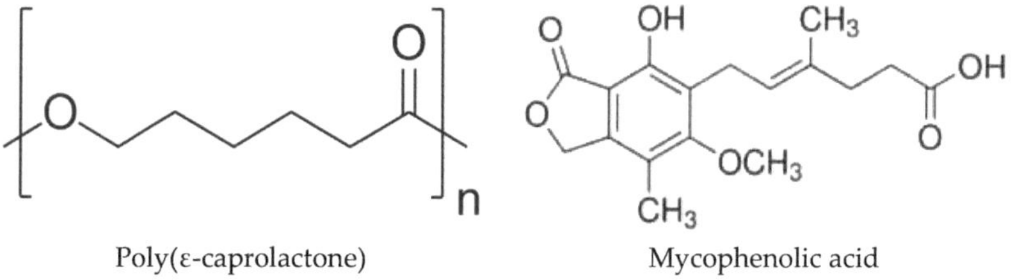

Scheme 1. Chemical structures of PCL and MPA.

In this work, the polymer selected as the matrix to disperse MPA in amorphous form is poly(ε-caprolactone) (PCL), a biodegradable semicrystalline polyester. Its glass transition temperature is around −60 °C and its melting point at around 60 °C. The biodegradation of this polymer under physiological conditions has been reported to last several months to years [31,32], making it suitable for long-term biomedical applications. In this work, miscibility and interactions between PCL and MPA are studied to verify the suitability of this mixture for the formation of an amorphous solid dispersion. In addition, we separately tested the interaction of the blends containing increasing concentrations of MPA with both a non-cancerous fibroblast cell line (MRC5), approved by ISO 10993 for cytotoxicity studies [33], and a widely used immortalized HeLa cell line derived from cervical cancer [34].

2. Experimental Section

2.1. Starting Materials

Poly(ε-caprolactone) (PURASORB® PC12 trade name) with an average molecular weight (M_w) of 1.3×10^5 g/mol and M_w/M_n = 1.76 was purchased from Purac Biochem (Gorinchem, The Netherlands). Mycophenolic acid ($C_{17}H_{20}O_6$, M = 320.34 g/mol) was obtained from Fluorochem Ltd. (Gossop, UK), and dichloromethane (DCM) was supplied by Labkem (Dublin, Ireland).

2.2. Blend Preparation

Films were prepared by solvent casting from dichloromethane (DCM) solutions containing 2.5 wt% of PCL/MPA blend at room temperature.

2.3. Differential Scanning Calorimetry (DSC)

A Modulated DSC Q200 from TA Instruments was used for thermal analyses. All the scans were performed in hermetic aluminum pans under nitrogen atmosphere with sample weights between 5 and 10 mg. Two scans from −80 °C to 160 °C with a scan rate of 20 °C/min were performed in order to measure glass transition temperatures (T_g) in the second one.

2.4. Melting Point Depression Analysis

The melting point depression of MPA was observed in MPA-rich blends containing 0–20 wt% PCL. To obtain the melting temperature of MPA crystals, samples were heated in the DSC with a scan rate of 1 °C/min.

The samples were weighed again after the DSC scans, and no weight loss was observed during the thermal treatments.

2.5. Fourier Transform Infrared Spectroscopy (FTIR)

A Nicolet AVATAR 370 Fourier transform infrared spectrophotometer was used to record FTIR spectra of the blends, with a resolution of 2 cm^{-1} and averaged over 64 scans in the range of 400–4000 cm^{-1}. Dichloromethane solutions containing 2 wt% of blends were cast on KBr pellets by evaporation of the solvent at room temperature. The absorbance of the samples was within the range where the Lambert–Beer law is obeyed.

2.6. In Vitro Drug Release

In vitro drug release experiments were performed for the PCL/MPA 99.95/0.05, 99.9/0.1, 99.8/0.2, 99.5/0.5, 99/1, and 98/2 blends. Round samples of PCL/MPA of Ø10 mm obtained by solvent casting were immersed in 1 mL of 0.1 M PBS buffer (pH 7.4) at 37 °C. At fixed intervals, samples of 200 μL were taken and replaced with fresh PBS at 37 °C. The drug concentration in solution was determined using a BioTech Sinergy H1M MicroPlate Reader (Minneapolis, MN, USA) using a calibration curve that was previously obtained measuring the absorbance at a wavelength of 305 nm for solutions of MPA in 0.1 M PBS.

The release kinetics of mycophenolic acid were examined by considering four mathematical models as follows:

$$\text{Zero} - \text{order}: \; C_t/C_\infty = k_0 t \tag{1}$$

$$\text{First} - \text{order}: \; \ln(1 - C_t/C_\infty) = -k_1 t \tag{2}$$

$$\text{Higuchi}: \; C_t/C_\infty = k_h t^{\frac{1}{2}} \tag{3}$$

$$\text{Korsmeyer–Peppas}: \; C_t/C_\infty = k t^n \tag{4}$$

where C_t is the cumulative amount of the drug released at time t, C_∞ is the starting amount of the drug, n is the release exponent, and k_0, k_1, k_h, and k are the kinetic constants. Zero-order kinetics (Equation (1)) represents a release process that is controlled by the relaxation of polymeric chains, independent of its concentration and with a constant release rate. The first-order kinetics (Equation (2)) model represents a drug release rate that depends on its concentration [32]. Higuchi (Equation (3)) describes drug release as a diffusion process based on Fick's law, square root time-dependent. If the release mechanism is not well known or when more than one type of release phenomena could be involved, the Korsmeyer–Peppas (Equation (4)) model is applied. It is possible to define whether the release happens by Fickian diffusion, anomalous transport, Case-II transport, or Super Case-II transport depending on the values obtained for the release exponent, n [35,36].

2.7. In Vitro Cell Culture Experiments

In vitro cell culture experiments were performed on the PCL/MPA 99.5/0.5, 99/1, and 98/2 blends. Circular samples of PCL/MPA of Ø6 mm were obtained, and each side was sterilized for 30 min under UV light. Either the immortalized HeLa cell line (ATCC, Manassas, VA, USA) derived from cervical cancer or the non-cancerous fibroblasts MRC5 (CCL-171, ATCC, Manassas, VA, USA) derived from lung tissue were drop-seeded over the materials at a concentration of 25,000 cells per scaffold. After 1 h, 480 mL of prewarmed DMEM (Fisher Scientific, Madrid, Spain) at 37 °C supplemented with 10% fetal bovine serum (FBS) (Fisher Scientific, Madrid, Spain), 1% L-glutamine (Fisher Scientific, Madrid, Spain), and penicillin/streptomycin (Fisher Scientific, Madrid, Spain) were added. PCL films were used as the negative control, and for the positive control, MPA in dissolution at a concentration of 300 ppm was dissolved on the culture media and filtrated (0.2 μm). Cells were incubated at 37 °C and 5% CO_2 in a standard cell culture incubator.

2.8. Immunostaining

After 1 or 3 days in vitro (DIV), samples were fixed with 4% paraformaldehyde (PFA) (Fisher Scientific, Spain) and permeabilized with 0.3% triton-X100 (Fisher Scientific, Spain) in PBS (Fisher Scientific, Spain) containing 1% Bovine Serum Albumin (BSA) (Sigma Aldrich, Spain). For the staining, rhodamine/phalloidin (Fisher Scientific, Madrid, Spain) and DAPI, 4′,6-diamidino-2-phenylindole dihydrochloride (Fisher Scientific, Madrid, Spain) were diluted in 1% PBS BSA and incubated for 1.5 h. After washing each sample 2 times in PBS containing 0.1% Tween-20 (Fisher Scientific, Madrid, Spain) and 1 time in PBS, the samples were mounted using mounting medium (Abcam, Waltham, MA, USA). The samples were analyzed in an inverted fluorescence microscope (Nikon Eclipse Ts2). For cell quantification studies, 5 different points were taken.

2.9. Cell Count and Statistical Analysis

For cell counts, five aleatory images of 0.1 mm^2 were taken for each of the triplicates in each condition, and nuclear DAPI labeling was used to calculate the total number of cells. The data were subjected to one-way analysis of variance (ANOVA) using Kruskal–Wallis followed by Dunn's post hoc test. The level of significance was set at $p < 0.05$. The results were presented as mean ± SD or SEM.

3. Results and Discussion

3.1. Miscibility Analysis by Differential Scanning Calorimetry (DSC)

When two components are miscible, a single glass transition temperature (T_g) between the T_g of each material, which changes progressively with the composition, is expected [37,38]. On the contrary, the detection of more than one single value would indicate a separation into individual amorphous phases within the system. Different methods have been employed to predict the glass transition temperature of amorphous binary systems, such as the Gordon–Taylor (GT), Couchman–Karasz (CK), and Fox equations (Equation (5)). Considering that the Fox equation was developed to analyze systems formed by components of equal densities, it is appropriate to use it to estimate this intermediate T_g, as the densities of PCL and MPA are 1.14 g/cm^3 and 1.3 g/cm^3, respectively [39]:

$$\frac{1}{T_{gb}} = \frac{w_1}{T_{g1}} + \frac{w_2}{T_{g2}} \tag{5}$$

where w_1 and w_2 are the weight fractions of components 1 and 2, respectively, T_{g1} and T_{g2} are the glass transition temperatures of the pure components, and T_{gb} is the glass transition temperature of the blend.

Figure 1 shows the first scan DSC traces obtained for the pure components and for different PCL/MPA blends. As can be seen, pure PCL is a semicrystalline polymer displaying a glass transition temperature located at about −60 °C and a melting endotherm at about 60 °C. On the other hand, MPA is a crystalline compound melting at 145 °C, which can be also supercooled to undergo a glass transition at 11 °C after reheating the quenched melt (see Figure 2).

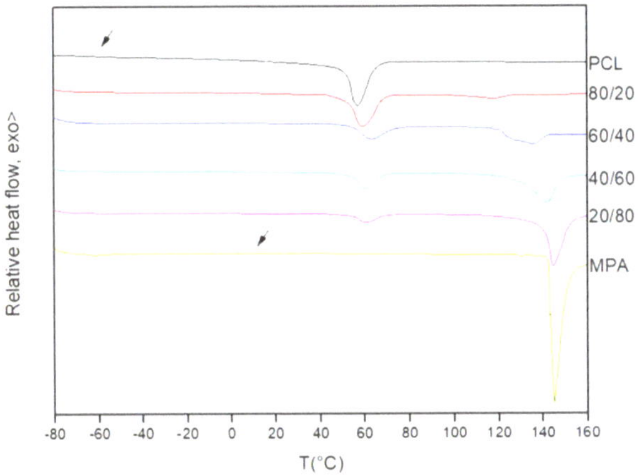

Figure 1. First scan DSC traces for PCL, MPA, and PCL/MPA blends.

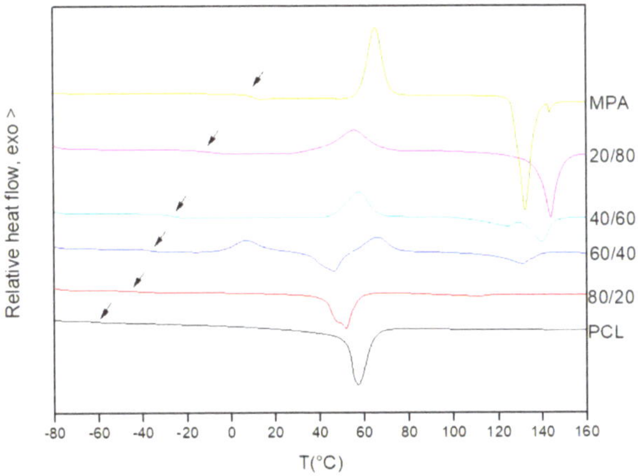

Figure 2. Second scan DSC traces for PCL, MPA, and PCL/MPA blends.

As can be seen in Figure 2, the PCL/MPA blends show composition-dependent single glass transitions located close to the values predicted using the Fox equation (see Table 1 and Figure 3). Consequently, it can be concluded that the two components are completely miscible in the amorphous phase. Furthermore, the melting temperature of PCL decreases as the content of MPA increases. Furthermore, the crystallization of PCL is totally suppressed when the drug composition exceeds 50 wt%.

Table 1. Thermal properties of PCL/MPA blends.

PCL/MPA	T_g Experimental (°C)	T_g Theoretical (Fox) (°C)	T_m PCL (°C)	ΔH_f PCL (J/g)
PCL	−60.0	-	57.2	66.4
80/20	−44.1	−48.8	51.7	49.8
60/40	−36.3	−36.3	46.4	25.9
40/60	−25.7	−22.4	-	-
20/80	−10.2	−6.8	-	-
MPA	11.1	-	-	-

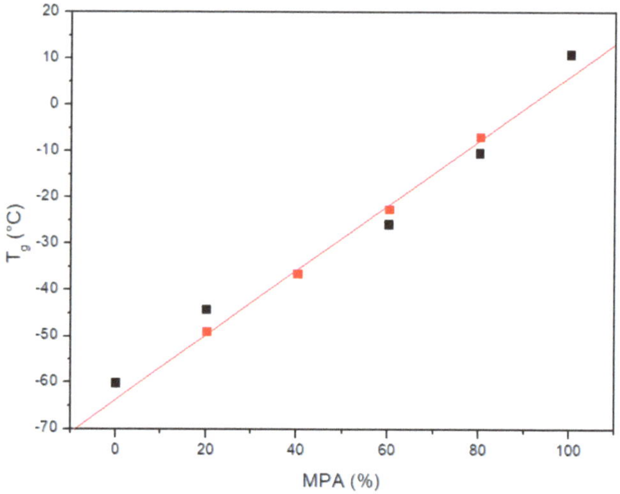

Figure 3. Glass transition temperature versus composition for the PCL/MPA system: (■) experimental values and (■) Fox equation.

3.2. Melting Point Depression Analysis

If the free energy of the mixing of the two components (ΔG_{mix}) is negative, a system can be considered thermodynamically miscible.

$$\Delta G_{mix} = \Delta H_{mix} - T\Delta S_{mix} \quad (6)$$

where ΔH_{mix} and ΔS_{mix} are the enthalpy and entropy of mixing, respectively. $T\Delta S_{mix}$ is always positive since the entropy of mixing is added to the entropy of melting, making the entropy change in a miscible blend larger than in the pure component. Consequently, the sign of ΔG_{mix} depends on the value of ΔH_{mix}. In order to avoid phase separation, the cohesive interactions need to be lower than the sum of adhesive interactions, generating a favorable enthalpy of mixing. The miscibility between two components in terms of the change in the Gibbs free energy can be described using the melting point depression method, based on Flory–Huggins theory. According to this method, the melting point temperature of the drug will decrease as the polymer content in the mixture increases if the cohesive forces in the pure components are weaker than the adhesive forces between the drug and the polymer [32,40]. Flory's relationship can be used to analyze the depression of the equilibrium melting point:

$$\frac{1}{T_m} - \frac{1}{T_m^0} = \frac{-R}{\Delta H_{2u}} \frac{V_{2u}}{V_{1u}} \left(\frac{\ln \phi_2}{m_2} + \left(\frac{1}{m_2} - \frac{1}{m_1} \right) \phi_1 + \chi_{12} \phi_1^2 \right) \quad (7)$$

where T_m^0 is the equilibrium melting point of the pure crystallizable component and T_m is the equilibrium melting point of its blends; the subscripts 1 and 2 refer to the amorphous and crystallizable components, respectively. R is the universal gas constant, while ΔH_{2u} is the heat of fusion per mole of crystalline repeat units. V_u is the molar volume of the repeating unit, m is the degree of polymerization, ϕ is the volumen fraction, and χ_{12} is the interaction parameter.

In order to apply Equation (7), the molar volume of MPA ($V_2 = 246.3$ cm^3/mol) can be considered as the molar volume of the lattice sites, resulting in $m_2 = 1$. The same volume can be taken as the molar volume of the polymeric repeat unit $V_2 = V_{1u}$. Since $m_1 = V_{pol}/V_{1u}$ is large, $1/m_1 \approx 0$. As a result, Equation (7) simplifies to:

$$\frac{1}{T_m} - \frac{1}{T_m^0} = \frac{-R}{\Delta H_2} \left(\ln \phi_2 + \phi_1 + \chi \phi_1^2 \right) \quad (8)$$

The melting points of pure components and different PCL/MPA blends were measured at a low heating rate (1 °C/min). The average melting point of pure MPA is $T_m^0 = 140.3$ °C, and this temperature is decreased by nearly 5 °C when 20 wt% PCL is added to the blend. The data obtained for each blend can be seen in Table 2. These results, with the average melting enthalpy of pure MPA ($\Delta H_{MPA} = 114.7$ J/g) were used to plot Equation (8) as a function of the square of the volume fraction of the polymer, ϕ_1^2. The slope of this plot, which can be seen in Figure 4, gives an approximation of the interaction parameter of $\chi = -1.18$. The negative values for the interaction parameter indicate an exothermic reaction, confirming a thermodynamically miscible blend. It is also possible to calculate the interaction energy density, B, at the melting temperature of MPA according to Equation (9):

$$\chi = \frac{BV_r}{RT} \quad (9)$$

where V_r is a reference volumen ($V_r = V_2 = 246.3$ cm^3/mol), yielding $B = -16.5$ J/cm^3.

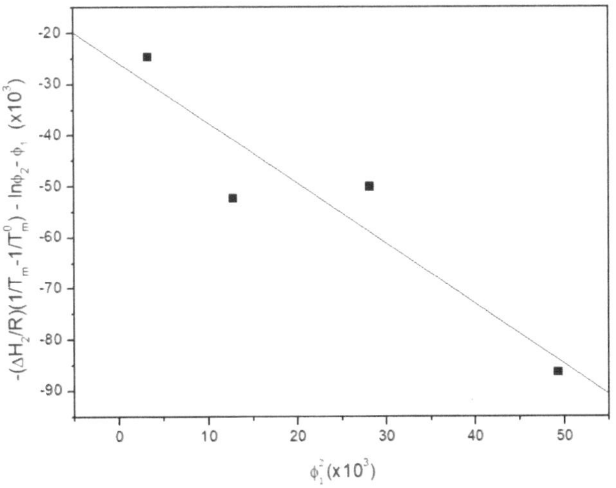

Figure 4. Analysis of the melting temperature of MPA according to Equation (8) for the PCL/MPA system. The slope of the plot gives the interaction parameter $\chi = -1.18$.

Table 2. Melting temperatures of MPA obtained from 1 °C min^{-1} scan rates.

MPA wt%	T_m (°C)		
	Sample 1	Sample 2	Sample 3
100	139.2	140.5	140.9
95	139.6	139.9	138.3
90	138.4	138.1	137.4
85	137.4	138.9	136.8
80	135.8	135.7	135.9

3.3. Fourier Transform Infrared Spectroscopy (FTIR)

The analysis of the changes observed in the infrared spectrum upon blending provides information about the changes in specific interactions and can eventually aid in explaining the energetic contributions driving the miscibility of the system. In the PCL/MPA system, both the carbonyl and the hydroxyl stretching regions are of main interest because hydrogen bonding interactions can be expected for those groups. Figure 5 shows the carbonyl stretching region for PCL, MPA, and their blends. The spectrum of pure PCL shows a peak at 1725 cm^{-1} attributable to crystalline PCL and a shoulder at 1735 cm^{-1} arising from the amorphous phase [32,40]. On the other hand, pure MPA shows two different peaks located at 1744 and 1708 cm^{-1} attributable, respectively, to the lactone carbonyl and the carboxylic acid carbonyl. Both locations are at the lower end of the spectral ranges corresponding to those functional groups [41] because of the hydrogen bonding interactions occurring in pure MPA. Figure 6 sketches these interactions as derived from XRD studies [42–44]. As it can be seen, in pure MPA, the molecules are joined in the crystal by carboxylic acid groups forming dimers, along with bifurcated hydrogen bonds between the hydroxyl group and the carboxylic acid carbonyl (absorption band at 1708 cm^{-1}). In addition, an intramolecular bifurcated hydrogen bond red shifts the absorption of the lactone carbonyl to the reported wavenumber (1744 cm^{-1}).

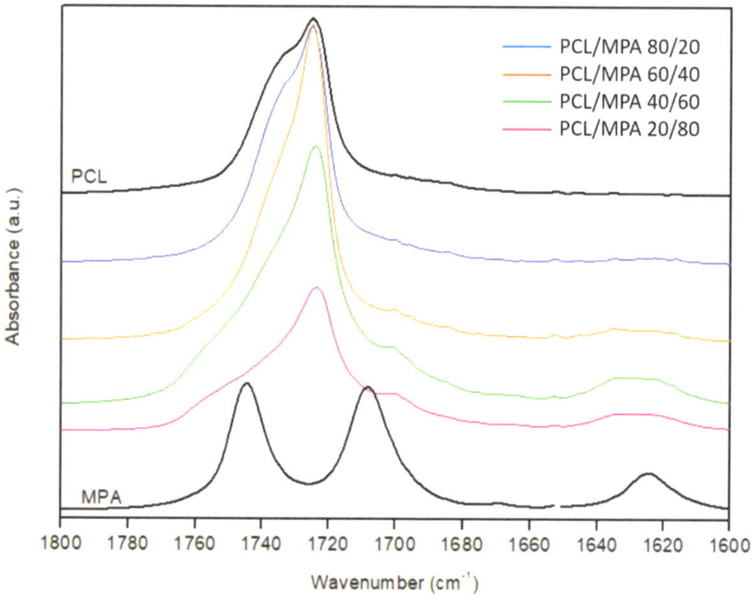

Figure 5. Carbonyl stretching region for pure PCL and MPA and PCL/MPA blends of different compositions.

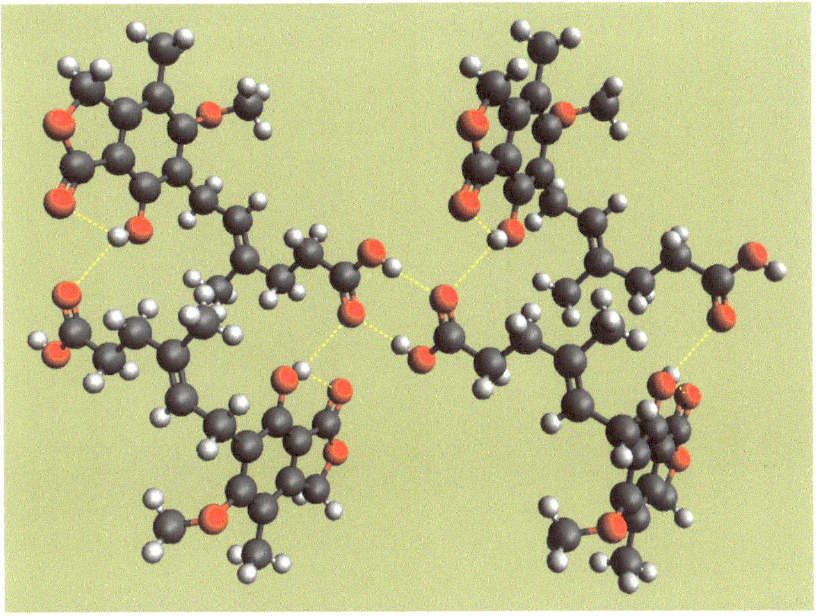

Figure 6. Hydrogen bonding in crystalline MPA (see text).

The PCL/MPA 20/80 and 40/60 blends show a major peak located at about 1724 cm^{-1}, accompanied by two shoulders at higher wavenumbers located at about 1735 cm^{-1} and 1750 cm^{-1}. At these compositions, PCL is almost in amorphous form according to the DSC results (hence, the contribution corresponding to crystalline PCL should be negligible), and the absorption bands corresponding to MPA are expected to prevail over those of PCL; hence, the band at 1724 cm^{-1} is most likely attributable carboxylic acid carbonyls forming dimers in the amorphous phase. This band is probably strongly overlapped with PCL carbonyls hydrogen bonded with hydroxyl groups present in MPA, but unfortunately, these two components are not distinguishable. The shoulder at about 1735 cm^{-1} can be attributed to free C=O groups in PCL and the one at about 1750 cm^{-1} to lactone carbonyls in the amorphous phase.

Finally, Figure 7 shows the hydroxyl stretching region for MPA and its blends with PCL. As can be seen, the OH stretching band in pure MPA is located at about 3416 cm^{-1}, and blending broadens the band and shifts it to higher wavenumbers. Band broadening is a consequence of the presence of amorphous MPA, while shifting to higher wavenumbers can be attributed to weaker hydrogen bonding interactions in the blends compared with pure MPA. Despite the weaker nature of the interactions, the energetic balance will still render favorable to miscibility as long as the blend achieves a larger number of interactions, arising from the introduction of additional interacting groups (the PCL carbonyls).

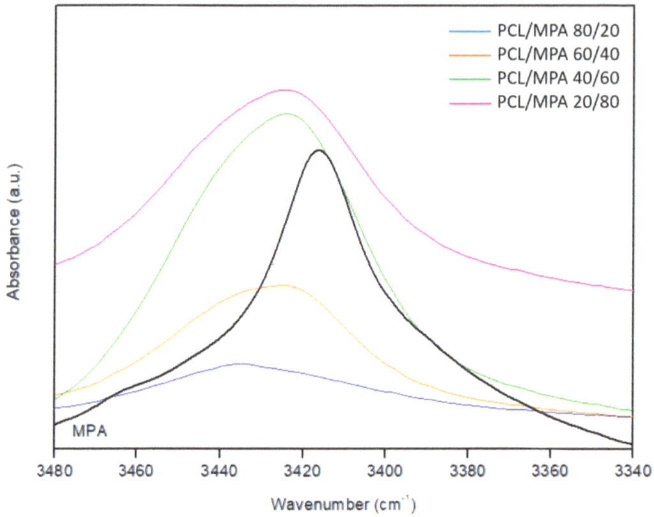

Figure 7. Hydroxyl stretching region for pure MPA and PCL/MPA blends of different compositions.

3.4. *In Vitro Drug Release*

Figure 8 shows the in vitro release profiles of MPA from different PCL/MPA blend (0.05, 0.1, 0.2, 0.5, 1, and 2 wt% MPA) samples for 3 days (72 h). From the first moment, the release of MPA starts, which is faster in the samples with the lowest drug content. By the third day, the samples of 0.05 and 0.1 wt% MPA had released all the drugs, while the rest of the samples had released between 69 and 79 wt% of the total amount. Figure 9 shows that the release rate slowed down after the third day following an asymptotic tendency; thus, it is thought that the remaining drug will be fully released when bulk erosion begins.

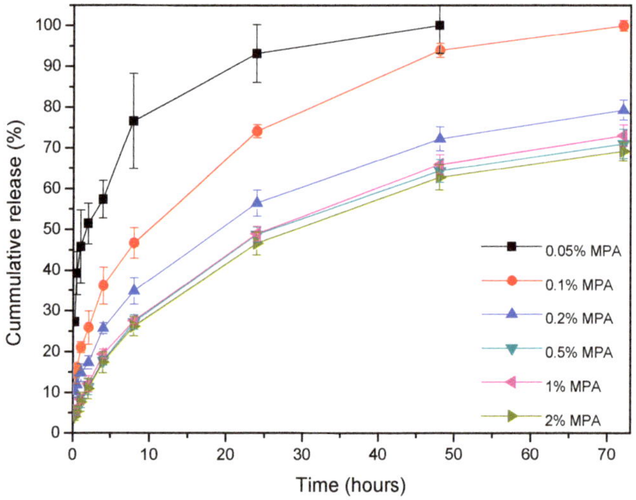

Figure 8. Drug release profiles of PCL/MPA films containing different drug concentrations immersed in 0.1 PBS buffer at 37 °C, shown in %.

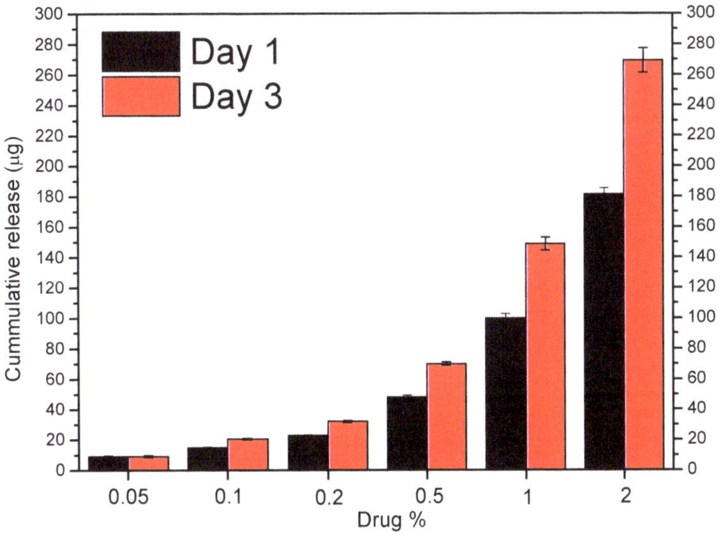

Figure 9. Drug release profiles of PCL/MPA films containing different drug concentrations immersed in 0.1 PBS buffer at 37 °C, shown in µg.

The amorphous MPA dissolved in the matrix can easily travel across the polymer matrix reaching the outer solution as the solution temperature (37 °C) is higher than the glass transition temperature of PCL (−60 °C), allowing the chains to have enough mobility.

According to the results observed in Table 3, the release mechanisms vary depending on the concentration of MPA. For those with the lowest concentration (0.05 and 0.1 wt% MPA), the model that fits the results is first-order [45], while for the rest of the concentrations, a trend toward the Higuchi model is seen [46]. Therefore, at very low concentrations, the release kinetics depends on the concentration, while as the amount of MPA increases, the drug is released by a diffusion process based on Fick's law, square root time-dependent.

Table 3. Fitting of the release data to the mathematical models for drug release kinetics. R^2 is the correlation coefficient and n is the release exponent.

MPA %	Zero-Order	First-Order	Higuchi	Korsmeyer–Peppas	
0.05	$R^2 = 0.64$	$R^2 = 0.94$	$R^2 = 0.87$	$R^2 = 0.81$	n = 0.58
0.1	$R^2 = 0.86$	$R^2 = 0.99$	$R^2 = 0.98$	$R^2 = 0.92$	n = 0.5
0.2	$R^2 = 0.74$	$R^2 = 0.88$	$R^2 = 0.94$	$R^2 = 0.95$	n = 0.46
0.5	$R^2 = 0.75$	$R^2 = 0.85$	$R^2 = 0.94$	$R^2 = 0.99$	n = 0.5
1	$R^2 = 0.77$	$R^2 = 0.89$	$R^2 = 0.95$	$R^2 = 0.99$	n = 0.5
2	$R^2 = 0.77$	$R^2 = 0.87$	$R^2 = 0.95$	$R^2 = 0.99$	n = 0.5

3.5. Cell Viability with HeLa Immortalized Cancer Cells

In the last decade, some groups have reported the suppression of proliferation and the enhancement of apoptosis mediated by MPA in cancerous cells [26,47,48]. These properties are attributed to several interactions of the MPA with molecules involved in the cell cycle, cell death, cell proliferation, and movement [49]. Here, the combination of PCL with MPA as a possible anti-cancer therapy was studied. For this purpose, HeLa cells, a widely used immortalized cell line derived from cervical cancer were chosen. First, the attachment and proliferation capabilities of HeLa cells were analyzed on PCL scaffolds containing increasing concentrations of MPA after 1 and 3 days post-seeding (DIV1 and DIV3 respectively) by immunofluorescence assays against DAPI and rhodamine/Phalloidin (Rh/Ph). Results at DIV1 suggested that the PCL/MPA substrates affected the number of attached HeLa cells in a dose-dependent manner, achieving even the same impairment on cell viability as the MPA in solution (Dis) (PCL/MPA 0.5 71.5 ± 6.1%; PCL/MPA 1 64.6 ± 5.3%; PCL/MPA 2 40.1 ± 7.6%; Dis 53.6 ± 5%; compared with the control of PCL 100 ± 9.2%; $p < 0.0001$, one-way ANOVA). The results were further demonstrated at DIV3, when again, the number of the HeLa cells over the scaffolds containing MPA was reduced with respect to the PCL control (PCL/MPA 0.5 32.6 ± 3.1%; PCL/MPA 1 39.1 ± 5.3%; PCL/MPA 2 28.2 ± 2.4%; Dis 30.0 ± 1.4%; PCL 100 ± 7.6%; $p < 0.0001$, one-way ANOVA) (Figure 10). The results are in accordance with other studies where MPA caused the impairment of HeLa cell viability [49,50]. But the clearance effect of the tissues in vivo [51] must also be taken into consideration, where the slow release of MPA by the scaffolds may be an advantage compared with the direct drug administration [52]. Moreover, in future experiments, these scaffolds could be further engineered to adjust the release of the MPA to the kinetics of the drug by just modifying the degradation ratio of the scaffolds or the anchoring mechanism of the drug [52], which is a promising tool for anti-cancer drug delivery.

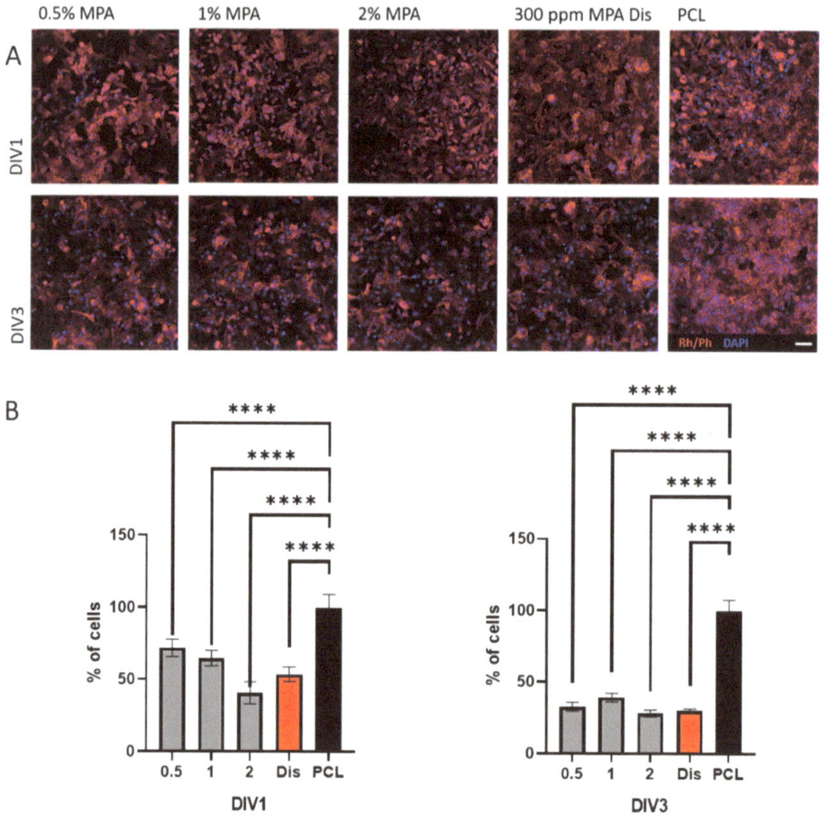

Figure 10. (**A**) Immunofluorescence assays against rhodamine/phalloidin (Rh/Ph) in red sowing the cytoskeleton and DAPI in blue showing the nuclei of HeLa cells cultured over PCL/MPA scaffolds containing increasing amounts of MPA. As a negative control, HeLa cells were cultured over PCL scaffolds, and as a positive control, HeLa cells cultured over PCL scaffolds were incubated with 300 ppm MPA in solution. (**B**) Quantification of the percentage of cells over the scaffolds at DIV1 and DIV3. (**** $p < 0.0001$ compared to the PCL scaffold at the same time points. Dunn's or Holm–Sidak method one-way ANOVA analysis of variance on ranks.) Scale bar 50 µm.

3.6. Cell Viability with Fibroblasts

One big hallmark of anti-cancer therapies is the possibility of treating cancer cells without affecting non-cancerous cells [53]. In this regard, the adhesion and proliferation of the non-cancerous fibroblast cell line MRC5 or CCL-171 were tested. This cell line is not immortalized, and according to ISO 10993-5:2009, it is considered a good control for cytotoxic experiments [33]. The results suggested that the number of MRC5 cells able to attach and proliferate on the PCL scaffolds containing increasing amounts of MPA at DIV1 (PCL/MPA 0.5 113.7 ± 17.2%; PCL/MPA 1 128.7 ± 13.7%; PCL/MPA 2 110.7 ± 11.6%; PCL 100 ± 7.1%) and DIV3 (PCL/MPA 0.5 109.1 ± 4.8%; PCL/MPA 1 113.3 ± 8.4%; PCL/MPA 2 91.6 ± 7.3%; PCL 100 ± 3. %) were similar to the pristine PCL scaffolds (Figure 11). Surprisingly, both at DIV1 (Dis 32.2 ± 3.5%; $p < 0.05$, one-way ANOVA) and DIV3 (Dis 69.4 ± 4.9%; $p < 0.05$, one-way ANOVA), the MRC5 cells seeded on the PCL scaffolds together with MPA in solution appeared to have an impaired proliferation. In this regard, several tumorigenic and non-tumorigenic cell lines showed resistance to MPA treatment by converting MPA into its inactive form 7-*O*-glucoronide [54–56], which might explain the results obtained with the scaffolds. However, further research is needed to study the anti-cancer capabilities of the MPA blended in PCL films with other cancerous and non-cancerous cell lines and the advantages and disadvantages compared with MPA in solution.

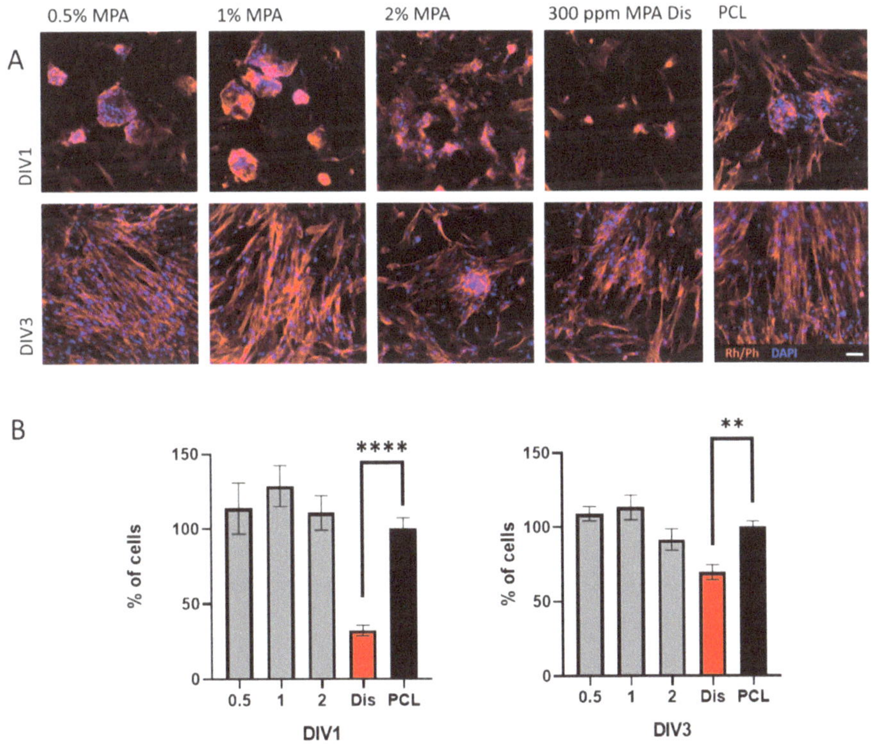

Figure 11. (**A**) Immunofluorescence assays against rhodamine/phalloidin (Rh/Ph) in red showing the cytoskeleton and DAPI in blue showing the nuclei of MRC5 cells cultured over PCL/MPA scaffolds containing increasing amounts of MPA. As a negative control, MRC5 cells were cultured over the PCL scaffolds, and as a positive control, MRC5 cells cultured over PCL scaffolds were incubated with 300 ppm MPA in dissolution. (**B**) Quantification of the percentage of cells over the scaffolds at DIV1 and DIV3. (** $p < 0.05$ and **** $p < 0.001$ compared to the PCL scaffold at the same time points. Dunn's or Holm–Sidak method one-way ANOVA analysis of variance on ranks.) Scale bar 50 μm.

4. Conclusions

In the present work, the possibility of forming an amorphous solid dispersion of mycophenolic acid using poly(ε-caprolactone) as a matrix was confirmed. Miscibility between the two compounds was observed in thermal properties. On the one hand, the intermediate glass transition temperature criterion was confirmed for all the compositions. On the other hand, the analysis of the melting point depression of the MPA crystals resulted in a negative interaction parameter, $\chi = -1.18$, indicating favorable interactions between the polymer and the bioactive molecule.

The analysis by FTIR spectroscopy for the PCL/MPA blends does not allow us to clearly confirm the occurrence of hydrogen bonding interactions between the hydroxyl groups present in MPA and the C=O groups of PCL because of the complex nature of the C=O stretching region. Nevertheless, the overall results observed in both the C=O and O-H stretching regions follow the typical trends observed in other polymer blends with miscibility driven by hydrogen bonding interactions.

The potential application of this PCL/MPA amorphous solid dispersion as drug delivery matrices was proven, as at least 70% of the drug was delivered by the third day in vitro in all compositions. In addition, in vitro cell culture experiments with HeLa and MRC5 cells showed that it is possible to maintain the active form of MPA for cancer treatment, as there is a decreased viability of cancerous cells when cultured over PCL materials containing MPA. Here, the potential beneficial outcome of dissolving MPA in PCL matrices for anti-cancer treatment was observed, although further research is needed to study the anti-cancer capabilities with other cell lines and the mechanism, advantages, and disadvantages these amorphous solid dispersions offer compared with crystalline MPA.

Author Contributions: Methodology, E.S.-R., A.L. (Aitor Larrañaga) and E.M.; Investigation, O.S.-A. and Y.P.; Writing—original draft, O.S.-A. and Y.P.; Writing—review & editing, A.L. (Aitor Larrañaga), A.L. (Ainhoa Lejardi), E.M. and J.-R.S.; Supervision, A.L. (Ainhoa Lejardi); Funding acquisition, J.-R.S. All authors have read and agreed to the published version of the manuscript.

Funding: This research was funded by Spanish Ministry of Science and Innovation MICINN (PID2019-106236GB-I00) and the Basque Government Department of Education, Culture and Language Policy (IT1766-22).

Institutional Review Board Statement: Not applicable.

Data Availability Statement: The data presented in this study are available on request from the corresponding author.

Acknowledgments: Financial support from the Spanish Ministry of Science and Innovation MICINN (PID2019-106236GB-I00) and the Basque Government Department of Education, Culture and Language Policy (IT1766-22) is gratefully acknowledged. A Bikaintekgrant (20-AF-W2-2018-00001) from the Basque Government is also gratefully acknowledged. The authors are thankful for the technical and human support provided by SGIker of UPV/EHU and European funding (ERDF and ESF).

Conflicts of Interest: There are no conflicts of interest to declare.

References

1. Kalepu, S.; Nekkanti, V. Insoluble drug delivery strategies: Review of recent advances and business prospects. *Acta Pharm. Sin. B* **2015**, *5*, 442–453. [CrossRef] [PubMed]
2. Loftsson, T.; Brewster, M.E. Pharmaceutical applications of cyclodextrins: Basic science and product development. *J. Pharm. Pharmacol.* **2010**, *62*, 1607–1621. [CrossRef]
3. Hodgson, J. ADMET—Turning chemicals into drugs. *Nat. Biotechnol.* **2001**, *19*, 722–726. [CrossRef] [PubMed]
4. Mudie, D.M.; Amidon, G.L.; Amidon, G.E. Physiological Parameters for Oral Delivery and in vitro Testing. *Mol. Pharm.* **2010**, *7*, 1388–1405. [CrossRef] [PubMed]
5. Shi, Q.; Li, F.; Yeh, S.; Wang, Y.; Xin, J. Physical stability of amorphous pharmaceutical solids: Nucleation, crystal growth, phase separation and effects of the polymers. *Int. J. Pharm.* **2020**, *590*, 119925. [CrossRef] [PubMed]
6. Shi, Q.; Wang, Y.; Moinuddin, S.M.; Feng, X.; Ahsan, F. Co-amorphous Drug Delivery Systems: A Review of Physical Stability, in vitro and in vivo Performance. *AAPS PharmSciTech* **2022**, *23*, 259. [CrossRef] [PubMed]

7. Chiou, W.L.; Riegelman, S. Pharmaceutical Applications of Solid Dispersion Systems. *J. Pharm. Sci.* **1971**, *60*, 1281–1302. [CrossRef] [PubMed]
8. Parulski, C.; Gresse, E.; Jennotte, O.; Felten, A.; Ziemons, E.; Lechanteur, A.; Evrard, B. Fused deposition modeling 3D printing of solid oral dosage forms containing amorphous solid dispersions: How to elucidate drug dissolution mechanisms through surface spectral analysis techniques? *Int. J. Pharm.* **2022**, *626*, 122157. [CrossRef] [PubMed]
9. Bhanushali, J.S.; Dhiman, S.; Nandi, U.; Bharate, S.S. Molecular interactions of niclosamide with hydroxyethyl cellulose in binary and ternary amorphous solid dispersions for synergistic enhancement of water solubility and oral pharmacokinetics in rats. *Int. J. Pharm.* **2022**, *626*, 122144. [CrossRef] [PubMed]
10. Becelaere, J.; Van Den Broeck, E.; Schoolaert, E.; Vanhoorne, V.; Van Guyse, J.F.; Vergaelen, M.; Borgmans, S.; Creemers, K.; Van Speybroeck, V.; Vervaet, C.; et al. Stable amorphous solid dispersion of flubendazole with high loading via electrospinning. *J. Control. Release* **2022**, *351*, 123–136. [CrossRef] [PubMed]
11. Nambiar, A.G.; Singh, M.; Mali, A.R.; Serrano, D.R.; Kumar, R.; Healy, A.M.; Agrawal, A.K.; Kumar, D. Continuous Manufacturing and Molecular Modeling of Pharmaceutical Amorphous Solid Dispersions. *AAPS PharmSciTech* **2022**, *23*, 249. [CrossRef] [PubMed]
12. Vasconcelos, T.; Sarmento, B.; Costa, P. Solid dispersions as strategy to improve oral bioavailability of poor water soluble drugs. *Drug Discov. Today* **2007**, *12*, 1068–1075. [CrossRef] [PubMed]
13. Holm, T.P.; Knopp, M.M.; Berthelsen, R.; Löbmann, K. Supersaturated amorphous solid dispersions of celecoxib prepared in situ by microwave irradiation. *Int. J. Pharm.* **2022**, *626*, 122115. [CrossRef] [PubMed]
14. Rumondor, A.C.F.; Dhareshwar, S.S.; Kesisoglou, F. Amorphous Solid Dispersions or Prodrugs: Complementary Strategies to Increase Drug Absorption. *J. Pharm. Sci.* **2016**, *105*, 2498–2508. [CrossRef] [PubMed]
15. Wu, J.; Van den Mooter, G. The influence of hydrogen bonding between different crystallization tendency drugs and PVPVA on the stability of amorphous solid dispersions. *Int. J. Pharm.* **2023**, *646*, 123440. [CrossRef] [PubMed]
16. Zhang, J.; Shi, X.; Tao, W. Curcumin amorphous solid dispersions benefit from hydroxypropyl methylcellulose E50 to perform enhanced anti-inflammatory effects. *Int. J. Biol. Macromol.* **2023**, *252*, 126507. [CrossRef] [PubMed]
17. Van Den Mooter, G. The use of amorphous solid dispersions: A formulation strategy to overcome poor solubility and dissolution rate. *Drug Discov. Today Technol.* **2012**, *9*, e79–e85. [CrossRef] [PubMed]
18. Pandi, P.; Bulusu, R.; Kommineni, N.; Khan, W.; Singh, M. Amorphous solid dispersions: An update for preparation, characterization, mechanism on bioavailability, stability, regulatory considerations and marketed products. *Int. J. Pharm.* **2020**, *586*, 119560. [CrossRef] [PubMed]
19. Qian, P.-Y.; Xu, Y.; Fusetani, N. Natural products as antifouling compounds: Recent progress and future perspectives. *Biofouling* **2009**, *26*, 223–234. [CrossRef] [PubMed]
20. Knopp, M.M.; Olesen, N.E.; Huang, Y.; Holm, R.; Rades, T. Statistical Analysis of a Method to Predict Drug-Polymer Miscibility. *J. Pharm. Sci.* **2016**, *105*, 362–367. [CrossRef] [PubMed]
21. Klueppelberg, J.; Handge, U.A.; Thommes, M.; Winck, J. Composition Dependency of the Flory–Huggins Interaction Parameter in Drug–Polymer Phase Behavior. *Pharmaceutics* **2023**, *15*, 2650. [CrossRef] [PubMed]
22. Hancock, B.C.; Shamblin, S.L.; Zografi, G. Molecular Mobility of Amorphous Pharmaceutical Solids Below Their Glass Transition Temperatures. *Pharm. Res.* **1995**, *12*, 799–806. [CrossRef] [PubMed]
23. Newman, A.; Knipp, G.; Zografi, G. Assessing the performance of amorphous solid dispersions. *J. Pharm. Sci.* **2012**, *101*, 1355–1377. [CrossRef] [PubMed]
24. DeBoyace, K.; Wildfong, P.L.D. The Application of Modeling and Prediction to the Formation and Stability of Amorphous Solid Dispersions. *J. Pharm. Sci.* **2018**, *107*, 57–74. [CrossRef] [PubMed]
25. Han, D.; Sasaki, M.; Yoshino, H.; Kofuji, S.; Sasaki, A.T.; Steckl, A.J. In-vitro evaluation of MPA-loaded electrospun coaxial fiber membranes for local treatment of glioblastoma tumor cells. *J. Drug Deliv. Sci. Technol.* **2017**, *40*, 45–50. [CrossRef]
26. Zheng, Z.H.; Yang, Y.; Lu, X.H.; Zhang, H.; Shui, X.X.; Liu, C.; He, X.B.; Jiang, Q.; Zhao, B.H.; Si, S.Y. Mycophenolic acid induces adipocyte-like differentiation and reversal of malignancy of breast cancer cells partly through PPARγ. *Eur. J. Pharmacol.* **2011**, *658*, 68. [CrossRef] [PubMed]
27. Bentley, R. Mycophenolic Acid: A One Hundred Year Odyssey from Antibiotic to Immunosuppressant. *Chem. Rev.* **2000**, *100*, 3801–3826. [CrossRef] [PubMed]
28. Park, J.; Ha, H.; Seo, J.; Kim, M.S.; Kim, H.J.; Huh, K.H.; Park, K.; Kim, Y.S. Mycophenolic Acid Inhibits Platelet-Derived Growth Factor-Induced Reactive Oxygen Species and Mitogen-Activated Protein Kinase Activation in Rat Vascular Smooth Muscle Cells. *Am. J. Transplant.* **2004**, *4*, 1982–1990. [CrossRef] [PubMed]
29. Floryk, D.; Huberman, E. Mycophenolic acid-induced replication arrest, differentiation markers and cell death of androgen-independent prostate cancer cells DU145. *Cancer Lett.* **2006**, *231*, 20–29. [CrossRef] [PubMed]
30. Hackl, A.; Ehren, R.; Weber, L.T. Effect of mycophenolic acid in experimental, nontransplant glomerular diseases: New mechanisms beyond immune cells. *Pediatr. Nephrol.* **2017**, *32*, 1315–1322. [CrossRef] [PubMed]
31. Sanchez-Rexach, E.; Meaurio, E.; Iturri, J.; Toca-Herrera, J.L.; Nir, S.; Reches, M.; Sarasua, J.R. Miscibility, interactions and antimicrobial activity of poly(ε-caprolactone)/chloramphenicol blends. *Eur. Polym. J.* **2018**, *102*, 30–37. [CrossRef]
32. Sanchez-Rexach, E.; de Arenaza, I.M.; Sarasua, J.R.; Meaurio, E. Antimicrobial poly(ε-caprolactone)/thymol blends: Phase behavior, interactions and drug release kinetics. *Eur. Polym. J.* **2016**, *83*, 288–299. [CrossRef]

33. *ISO 10993-5:2009*; Biological Evaluation of Medical Devices—Part 5: Tests for in vitro Cytotoxicity. ISO Standards: Geneva, Switzerland, 2009.
34. Rashid, F.; Saeed, A.; Iqbal, J. In Vitro Anticancer Effects of Stilbene Derivatives: Mechanistic Studies on HeLa and MCF-7 Cells. *Anti-Cancer Agents Med. Chem.* **2021**, *21*, 793–802. [CrossRef] [PubMed]
35. Costa, P.; Lobo, J.M.S. Modeling and comparison of dissolution profiles. *Eur. J. Pharm. Sci.* **2001**, *13*, 123–133. [CrossRef] [PubMed]
36. Sánchez-Aguinagalde, O.; Lejardi, A.; Meaurio, E.; Hernández, R.; Mijangos, C.; Sarasua, J.-R. Novel Hydrogels of Chitosan and Poly(vinyl alcohol) Reinforced with Inorganic Particles of Bioactive Glass. *Polymers* **2021**, *13*, 691. [CrossRef] [PubMed]
37. Pezzoli, R.; Lyons, J.G.; Gately, N.; Higginbotham, C.L. Investigation of miscibility estimation methods between indomethacin and poly(vinylpyrrolidone-co-vinyl acetate). *Int. J. Pharm.* **2018**, *549*, 50–57. [CrossRef]
38. Hernandez-Montero, N.; Ugartemendia, J.M.; Amestoy, H.; Sarasua, J.R. Complex phase behavior and state of miscibility in Poly(ethylene glycol)/Poly(l-lactide-co-ε-caprolactone) Blends. *J. Polym. Sci. B Polym. Phys.* **2014**, *52*, 111–121. [CrossRef]
39. Baird, J.A.; Taylor, L.S. Evaluation of amorphous solid dispersion properties using thermal analysis techniques. *Adv. Drug Deliv. Rev.* **2012**, *64*, 396–421. [CrossRef]
40. Sánchez-Aguinagalde, O.; Meaurio, E.; Lejardi, A.; Sarasua, J.-R. Amorphous solid dispersions in poly(ε-caprolactone)/xanthohumol bioactive blends: Physicochemical and mechanical characterization. *J. Mater. Chem. B* **2021**, *9*, 4219–4229. [CrossRef] [PubMed]
41. Colthup, N.B.; Daly, L.H.; Wiberley, S.E. *Chapter 9—Carbonyl Compounds*, 3rd ed.; Wiberley, E., Ed.; Academic Press: San Diego, CA, USA, 1990; pp. 289–325. [CrossRef]
42. Harrison, W.; Shearer, H.M.M.; Trotter, J. Crystal structure of mycophenolic acid. Journal of the Chemical Society. *Perkin Trans.* **1972**, *2*, 1542–1544. [CrossRef]
43. Covarrubias, A.; Zúñiga-Villarreal, N.; González-Lucas, A.; Díaz-Domínguez, J.; Espinosa-Pérez, G. Crystal Structure of Mycophenolic Acid: 6-(4-Hydroxy-6-methoxy-7-methyl-3-oxo-1,3-dihydroisobenzofuran-5-yl)-4-methyl-hex-4-enoic Acid. *Anal. Sci.* **2000**, *16*, 783–784. [CrossRef]
44. Zeng, Q.Z.; Ouyang, J.; Zhang, S.; Zhang, L. Structural characterization and dissolution profile of mycophenolic acid cocrystals. *Eur. J. Pharm. Sci.* **2017**, *102*, 140–146. [CrossRef]
45. Li, J.; Mooney, D.J. Designing hydrogels for controlled drug delivery. *Nat. Rev. Mater.* **2016**, *1*, 16071. [CrossRef] [PubMed]
46. Baishya, H. Application of Mathematical Models in Drug Release Kinetics of Carbidopa and Levodopa ER Tablets. *J. Dev. Drugs* **2017**, *6*, 1000171. [CrossRef]
47. Klangjorhor, J.; Chaiyawat, P.; Teeyakasem, P.; Sirikaew, N.; Phanphaisarn, A.; Settakorn, J.; Lirdprapamongkol, K.; Yama, S.; Svasti, J.; Pruksakorn, D. Mycophenolic acid is a drug with the potential to be repurposed for suppressing tumor growth and metastasis in osteosarcoma treatment. *Int. J. Cancer* **2020**, *146*, 3397–3409. [CrossRef] [PubMed]
48. Dun, B.; Sharma, A.; Teng, Y.; Liu, H.; Purohit, S.; Xu, H.; Zeng, L.; She, J.X. Mycophenolic acid inhibits migration and invasion of gastric cancer cells via multiple molecular pathways. *PLoS ONE* **2013**, *8*, e81702. [CrossRef] [PubMed]
49. Dun, B.; Sharma, A.; Xu, H.; Liu, H.; Bai, S.; Zeng, L.; She, J.X. Transcriptomic changes induced by mycophenolic acid in gastric cancer cells. *Am. J. Transl. Res.* **2014**, *6*, 28–42.
50. Dun, B.; Xu, H.; Sharma, A.; Liu, H.; Yu, H.; Yi, B.; Liu, X.; He, M.; Zeng, L.; She, J.X. Delineation of biological and molecular mechanisms underlying the diverse anticancer activities of mycophenolic acid. *Int. J. Clin. Exp. Pathol.* **2013**, *6*, 2880–2886. [PubMed]
51. Howgate, E.M.; Yeo, K.R.; Proctor, N.J.; Tucker, G.T.; Rostami-Hodjegan, A. Prediction of in vivo drug clearance from in vitro data. I: Impact of inter-individual variability. *Xenobiotica* **2006**, *36*, 473–497. [CrossRef] [PubMed]
52. Rambhia, K.J.; Ma, P.X. Controlled drug release for tissue engineering. *J. Control. Release* **2015**, *219*, 119–128. [CrossRef] [PubMed]
53. Hanahan, D.; Weinberg, R.A. The Hallmarks of Cancer. *Cell* **2000**, *100*, 57–70. [CrossRef] [PubMed]
54. Morath, C.; Reuter, H.; Simon, V.; Krautkramer, E.; Muranyi, W.; Schwenger, V.; Goulimari, P.; Grosse, R.; Hahn, M.; Lichter, P.; et al. Effects of mycophenolic acid on human fibroblast proliferation, migration and adhesion in vitro and in vivo. *Am. J. Transplant.* **2008**, *8*, 1786–1797. [CrossRef] [PubMed]
55. Chen, K.; Cao, W.; Li, J.; Sprengers, D.; Hernanda, P.Y.; Kong, X.; van der Laan, L.J.; Man, K.; Kwekkeboom, J.; Metselaar, H.J.; et al. Differential sensitivities of fast-and slow-cycling cancer cells to inosine monophosphate dehydrogenase 2 inhibition by mycophenolic acid. *Mol. Med.* **2015**, *21*, 792–802. [CrossRef] [PubMed]
56. Franklin, T.J.; Jacobs, V.; Bruneau, P.; Ple, P. Glucuronidation by human colorectal adenocarcinoma cells as a mechanism of resistance to mycophenolic acid. *Adv. Enzym. Regul.* **1995**, *35*, 91–100. [CrossRef] [PubMed]

polymers

MDPI

Article

Optimised Degradation of Lignocelluloses by Edible Filamentous Fungi for the Efficient Biorefinery of Sugar Beet Pulp

Zydrune Gaizauskaite [1,2,*], Renata Zvirdauskiene [1], Mantas Svazas [3], Loreta Basinskiene [1] and Daiva Zadeike [1,*]

[1] Department of Food Science and Technology, Faculty of Chemical Technology, Kaunas University of Technology, 50254 Kaunas, Lithuania; renata.zvirdauskiene@ktu.lt (R.Z.); loreta.basinskiene@ktu.lt (L.B.)
[2] Food Institute, Kaunas University of Technology, 50254 Kaunas, Lithuania
[3] Department of Applied Economics, Finance and Accounting, Agriculture Academy of Vytautas Magnus University, 53361 Kaunas, Lithuania; mantas@svazas.lt
* Correspondence: zydrune.gaizauskaite@ktu.lt (Z.G.); daiva.zadeike@ktu.lt (D.Z.)

Abstract: The degradation of the complex structure of lignocellulosic biomass is important for its further biorefinery to value-added bioproducts. The use of effective fungal species for the optimised degradation of biomass can promote the effectiveness of the biorefinery of such raw material. In this study, the optimisation of processing parameters (temperature, time, and *s/w* ratio) for cellulase activity and reducing sugar (RS) production through the hydrolysis of sugar beet pulp (SBP) by edible filamentous fungi of *Aspergillus*, *Fusarium*, *Botrytis*, *Penicillium*, *Rhizopus*, and *Verticillium* spp. was performed. The production of RS was analysed at various solid/water (*s/w*) ratios (1:10–1:20), different incubation temperatures (20–35 °C), and processing times (60–168 h). The *Aspergillus niger* CCF 3264 and *Penicillium oxalicum* CCF 3438 strains showed the most effective carboxymethyl cellulose (CMC) degrading activity and also sugar recovery (15.9–44.8%) from SBP biomass in the one-factor experiments. Mathematical data evaluation indicated that the highest RS concentration (39.15 g/100 g d.w.) and cellulolytic activity (6.67 U/g d.w.) could be achieved using *A. niger* CCF 3264 for the degradation of SBP at 26 °C temperature with 136 h of processing time and a 1:15 solid/water ratio. This study demonstrates the potential of fungal degradation to be used for SBP biorefining.

Keywords: sugar beet pulp; filamentous fungi; degradation; cellulase activity; process optimisation; biorefinery

Citation: Gaizauskaite, Z.; Zvirdauskiene, R.; Svazas, M.; Basinskiene, L.; Zadeike, D. Optimised Degradation of Lignocelluloses by Edible Filamentous Fungi for the Efficient Biorefinery of Sugar Beet Pulp. *Polymers* **2024**, *16*, 1178. https://doi.org/10.3390/polym16091178

Academic Editors: Stefano Farris, Bruno Medronho, Bin Li and Masoud Ghaani

Received: 10 January 2024
Revised: 17 April 2024
Accepted: 18 April 2024
Published: 23 April 2024

1. Introduction

The growing global demand for sustainable and green technologies is promoting the development of a circular bioeconomy, where industrial waste can be transformed into higher-value products. Lignocellulosic by-products of the agricultural industry are a large resource of irrationally used biomass. The bioconversion of secondary raw materials, generated at different stages of the food production chain into bioproducts can provide a sustainable solution to renew decreasing energy resources and a rational strategy to reduce the global growth of agro-industrial waste. For the industry, it is important to apply efficient and economically useful processes for the conversion of agrobiomass into fermentable sugars, which can further be upgraded by fermentative and biocatalytic routes to value-added bioproducts [1].

Sugar beet (*Beta vulgaris* var. Saccharifera) is one of the main crops for sugar production in Europe. Approximately 280 million metric tonnes of sugar beet were produced globally in the 2019–2020 period, and the EU is the largest producer of sugar from sugar beet (approximately 46% of the world's production) [2]. The most competitive producers in Europe are France, Germany, The Netherlands, Belgium, and Poland [3]. Lithuania is one of the largest sectors, and the sugar industry produces the most residual streams (estimated at 930,000 tons in 2023). The processing of sugar beet produces about 500,000 tons of spent sugar beet pulp (SBP) (moisture content 70–75%). Such huge quantities of biomass require reasonable implementation of a sugar beet processing by-product management system [3].

SBP, a lignocellulosic by-product of the sugar industry, was mainly used for animal feeding or as a raw material for the production of biogas [4]. Recent developments in the biotechnological valorisation of this raw material can increase its value as a raw material for the production of bioplastics, biofuels, chemicals, such as organic acids, microbial enzymes, feed proteins, and also pectic oligosaccharides [5,6].

On a dry-weight basis, SBP consists of polymeric carbohydrates (75–85% *w*/*w*), including 20–25% cellulose, 25–36% hemicelluloses, and 20–25% pectin with a low lignin content (1–3%) [7,8], and contains approximately 8% of the protein [9]. The low lignin content of SBP indicates that for the depolymerisation of this fraction, high-cost treatments are not required. The processing of biomass can be carried out by separate enzyme hydrolysis and fermentation or by simultaneous saccharification and fermentation [10]. The disadvantage of this processing might be the retardation of enzymatic hydrolysis by reaction products, which can reduce the yield of fermentable sugars released from polysaccharides. On the one hand, produced glucose and cellobiose can inhibit the activity of cellulases; however, on the other hand, the released saccharides can be consumed by microorganisms during fermentation [11]. Furthermore, cellulase production expenses in the industry are high due to the substantial enzyme loss, high energy consumption, and long processing time experienced during microbial fermentation [12]. The insolubility of cellulose leads to complex fermentation operations, making the process time and energy-consuming [13]. Even more, enzyme recovery can be largely impaired not only by the possible enzyme inactivation but also by the nonspecific and irreversible adsorption of the enzymes on the substrate, especially lignin [14].

Over the last decade, plant cell wall-degrading filamentous fungi have been analysed for the production of various enzymes with different catalytic activities that can be applied to the hydrolysis of renewable lignocellulosic feedstocks. Filamentous fungi are a large and diverse taxonomically group of microorganisms. They involve genera, such as *Aspergillus*, *Penicillium*, *Fusarium*, *Cladosporium*, *Emericella*, *Eurotium*, *Paecilomyces*, *Curvularia*, etc. [15]. Fungi have a unique extracellular enzyme system, including hydrolytic enzymes responsible for the degradation of various kinds of biomass polysaccharides, as well as oxidative enzymes capable of degrading lignin [16]. White-rot fungi can degrade cellulose, hemicelluloses, and lignin, whereas brown-rot fungi efficiently metabolise cellulose and hemicelluloses but can only slightly modify lignin [17]. The combination of fungal hydrolysis with other physical or chemical pretreatment methods could lead to a reduced biomass conversion time, as well as lower costs [18].

To our knowledge, there are not many reports on the optimisation of the fungal degradation-assisted production of reducing sugars in SBP. The optimisation of fermentation conditions for the production of pectinase and cellulase by *A. niger* NCIM 548 was performed on different substrates, such as wheat bran, corn bran, and kinnow peel [19]. The production of a hydrolysate of exhausted sugar beet pulp pellets as a generic microbial culture medium was carried out by Marzo et al. [20].

To develop a biotechnological pretreatment method for sugar beet pulp and its further utilisation as a medium for feed protein or bioethanol production, this study was conducted for the selection of the most active fungal strain as well as the optimisation of the processes' parameters (temperature, time, and *s/w* ratio) for cellulase activity and reducing sugar production through the hydrolysis of SBP by edible filamentous fungi of *Aspergillus*, *Fusarium*, *Botrytis*, *Penicillium*, *Rhizopus*, and *Verticillium* spp. In this way, more valuable bioproducts can be obtained, especially in terms of sustainable biorefinery and the rational use of bioresources.

2. Materials and Methods

2.1. Sugar Beet Pulp

Sugar beet pulp (SBP) was obtained from the company SC "Nordic Sugar Kėdainiai" (Kėdainiai, Lithuania) after saccharose extraction. The SBP material (moisture~72%) (Table 1) was frozen and stored at a −20 °C temperature. Part of the raw material was dried in a convection oven at 40 °C to a constant weight and ground using a laboratory mill (A10, IKA-Werke, Staufen, Germany) to powder (315–500 μm particles). The dried raw material was packed in tightly closed plastic bags and stored at 4 °C during the experiment.

Table 1. The chemical composition (g/100 g d.w.) and total count of microorganisms of SBP.

Components	SBP
Protein	5.37 ± 0.54
Fat	0.74 ± 0.01
Sugars	6.98 ± 0.31
SDF	18.64 ± 1.22
IDF	58.65 ± 1.14
Lignin *	3.61 ± 0.35
Ash	6.01 ± 0.12
TCM	1.08×10^6

* acid-insoluble lignin; SDF—soluble dietary fibre; IDF—insoluble dietary fibre; TCM—total count of microorganisms.

2.2. Fungal Strains

Ten fungal strains of *Aspergillus, Fusarium, Botrytis, Penicillium, Rhizopus*, and *Verticillium* spp., which were obtained from the Culture Collection of Fungi (CCF) of the Department of Botany of Charles University (Prague, Czech Republic) were used for the degradation of SBP. The fungi were preserved on a slanted potato dextrose agar (PDA, Liofilchem, Roseto degli Abruzzi, Italy) at 4 °C temperature. For the experiment, fungi were sub-cultured on the PDA slants and incubated at 25 °C, periodically sub-culturing on PDA before use.

2.3. Fungal Inoculum Preparation

Fungal spore suspensions were prepared from freshly sporulated (from 3- to 5-day-old) cultures according to the description by Aberkane et al. [21]. The fungal colonies were covered with 2–3 mL of a 1% Tween20 solution prepared in sterile-distilled water, and then the conidia were agitated carefully with a sterile spatula and transferred to a sterile tube. The resulting suspensions were mixed for 20 s with a vortex mixer at 2200× *g* (VWR Reax top, VWR International, Orange, CA, USA). The number of fungal spores of the inoculant was adjusted to 10^6 CFU/mL by counting with a hematocytometer (BLAUBRAND® Neubauer chamber; Merck, Madrid, Spain).

2.4. Hydrolytic Activity Evaluation

The cellulose-degrading potential of ten fungal strains was determined on carboxymethyl cellulose (CMC) agar. The agar base was prepared according to Maki et al. [22], using 1% (*w/w*) of CMC. The plates were incubated at 25 °C until visible colonies were formed. Then, colonies were stained with 0.1% Congo red dye and, after 10 min, were washed off with a 1 mol/L sodium chloride solution, followed by washing with deionised water. The fungal colonies with clear zones (mm) around them were considered cellulase-active fungi. The measurements of growth colonies and hydrolysis zones were used to evaluate the hydrolytic activity expressed as a relative activity index (RAI) [23].

2.5. Shake-Flask Experiments

For the shake-flask experiment, fungal cultures were kept in 250 mL flasks with 150 mL of a specific nutrition medium containing 3 g/L of $(NH_4)_2HPO_4$, 2 g/L of KH_2PO_4, 0.5 g/L of $MgSO_4$, 0.5 g/L of $Ca_3(PO_4)_2$, and 30 g/L of SBP (initial pH 5.0). Before the experiment, the medium was sterilised for 15 min at 121 °C. The tested samples were prepared by adding 1 mL of the suspension containing 10^6 CFU/mL of the fungal spores. The samples were incubated for 7 days at 25 °C. After the first 6 and 12 h, and further after every 12 h during the 36–168 h incubation period, the production of reducing sugars was analysed.

For the one-factor experiments, the most active fungal strains (*Aspergillus niger* CCF 3264, *Fusarium solani* CCF 2967, *Botrytis cinerea* CCF 2361, *Penicillium oxalicum* CCF 3438) were selected. The test samples were prepared by mixing SBP powder (50 g) with distilled water in the glass vessels at different solid/water (*s/w*) ratios (1:10; 1:12.5; 1:15; 1:17.5; 1:20). The samples were sterilised (121 °C; 15 min) and, after cooling to 20 °C in temperature, 5 mL of the fungal spore suspension was added to each vessel. After careful mixing, the samples were incubated for 7 days at 25 °C. The second experiment was carried out at a constant *s/w* ratio (1:15) and different incubation temperatures (20, 25, 30, and 35 °C). The reducing sugar analysis was performed every 12 h during a 60–168 h incubation period. The sugar recovery was expressed as a percentage of total biomass solids, excluding protein, ash, and lignin contents.

2.6. *Enzyme Activity Determination*

The sample extract for the enzyme activity assays was prepared according to Gasparotto et al. [24]. A homogenous SBP sample (5 ± 0.01 g) after 120 h of hydrolysis by fungi was taken and mixed with 20 mL of the 0.05 mol/L citrate buffer (pH 5). After orbital shaking (BIOSAN OS-20, SIA Biosan, Riga, Latvia) for 1 h at room temperature, the samples were centrifuged (4000× *g*; 10 min), and the obtained supernatants were collected and used as the enzyme extracts for the activity assays.

2.6.1. Cellulase Activity Determination

The cellulase activity was determined according to Ghose [25], using cellulosic filter paper as the substrate. The reaction mixture (total volume 1.5 mL), containing 50 mg of filter paper strip, 0.5 mL of the sample extract, and 1 mL of sodium citrate buffer (0.05 mol/L; pH 5), was incubated for 1 h in a 50 °C water bath with shaking (120 rpm). The reducing sugars produced in the medium were determined according to Miller [26]. Control samples, consisting only of the sample extract in the buffer and the substrate in the buffer, were prepared and used to correct absorbance values. The unit of cellulase activity was defined as the amount of enzyme that released 1 µmole of glucose from 50 mg of filter paper per gram of the dry solids of the substrate per minute under assay conditions.

2.6.2. Endoglucanase Activity Determination

The endoglucanase (endo-β-1,4-glucanase) activity was determined according to Ghose [25]. For the analysis, the 0.5 mL of 1% CMC solution in 0.05 mol/L sodium acetate buffer (pH 5) was mixed with 0.5 mL of an appropriately diluted enzyme extract and incubated for 30 min at 50 °C in a water bath with shaking (120 rpm). Released reducing sugars were determined as described above. The unit of endoglucanase activity was defined as the amount of enzyme used to release 1 µmole of glucose per min from the CMC substrate under the described conditions and was expressed as U/g d.w. of SBP.

2.6.3. *β*-Glucosidase Activity Determination

β-Glucosidase activity was determined according to the description by Verchot and Borelli [27]. The reaction mixture was prepared by mixing 0.2 mL of *p*-nitrophenyl-*β*-d-glucopyranoside (0.01 mol/L prepared in 0.05 mol/L sodium citrate buffer, pH 5) and 0.2 mL of the sample extract, which after, was incubated for 30 min in a 50 °C water bath with shaking (120 rpm). The reaction was then stopped by adding 4 mL of the 0.05 mol/L sodium hydroxide–glycine buffer (pH 10.6). The activity of *β*-glucosidase was determined by measuring the release of *p*-nitrophenol using a UV-1800 spectrophotometer (Shimadzu Corp., Kyoto, Japan) at 420 nm. The unit of *β*-glucosidase activity was defined as the amount of enzyme required to form 1 µmole of *p*-nitrophenol per minute under the assay conditions, using *p*-nitrophenyl-*β*-d-glucopyranoside as the substrate, which was expressed as U/g d.w. of SBP.

2.7. Chromatographic Quantification of Monosaccharides

The fermented sample (0.5 ± 0.01 g) was mixed with 5 mL of deionised water, which was ultrasonicated for 10 min (ArgoLab DU100; Chromservis s.r.o; Prague, Czech Republic), then diluted to 10 mL, cooled to room temperature, and centrifuged for 10 min at 4000× *g*. The liquid was filtered through a 0.22 μm membrane filter and used for the analysis. A High-Performance Liquid Chromatography (HPLC) system (Shimadzu Corp., Kyoto, Japan) with the ELSD detector and thermostatic column was used for arabinose, galactose, glucose, fructose, mannose, and xylose analysis. The column temperature was kept at 28 °C. The normal phase column was YMC-Pack Polyamine II (250 × 4.6 mm, I.D, 12 nm, s-5 μm) with the column guard YMC-Pack Polyamine II. The mobile phase (eluent) was a water/acetonitrile (25/75) isocratic system with a flow rate of 1 mL/min. The method's limit of detection (LOD) was 0.5 g/L, and the limit of quantification (LOQ) was 2 g/L. The quantitative analysis was performed using calibration curves of external standards with $R^2 > 0.99$.

2.8. Experimental Design and Statistical Analysis

The response surface method (RSM) and central composite design (CCD) were employed for the optimisation of processing conditions for the effective production of RS and cellulase activities in the SBP substrate, evaluating the effect of selected independent variables, such as temperature (*T*, °C), processing time (*t*, min), and solid/water ration (*s/w*) at different levels. Experiments were carried out according to the three-factorial Box–Behnken design, consisting of 17 experiments and including three central points for each factor group, which were performed in random order at three replications (Table 2). Based on the preliminary investigations and data from the literature, the following conditions were adopted for the investigated process parameters that were assigned as the independent variables: *s/w* ratio (X1), *T* (X2), and *t* (X3).

Table 2. Processing variables for the three-factorial Box–Behnken design.

Variable	Symbol	Coding Level		
		−1	0	+1
Solid/water ratio (*s*/*w*)	X1	12.5	15	17.5
Temperature (*T*, °C)	X2	20	25	30
Time (*t*, h)	X3	120	132	144

The processing conditions were optimised by developing the simplest possible mathematical models with a determination coefficient higher than 80%. The responses were the RS content (Y1, g/100 g d.w.) and cellulase activity (Y2, U/g d.w.) exclusively obtained under the influence of degradation by *A. niger* CCF 3264. Equation (1) was used for the estimation of the surface response area.

$$Y = \beta_0 + \beta_1 X1 + \beta_2 X2 + \beta_3 X3 + \beta_4 X1^2 + \beta_5 X2^2 + \beta_6 X3^2 + \beta_7 X1X2 + \beta_8 X1X3 + \beta_9 X2X3, \quad (1)$$

where βi—coefficient in the quadratic equation.

All analyses were performed at least in triplicate. The results were analysed using Minitab version 21.4.2 software (Minitab LLC, State College, PA, USA). The analysis of variance (ANOVA) was conducted for the assessment of the suitability of the mathematical models using the coefficient of '*lack of fit*' and the Fisher value (*F*). Statistical analysis of the data was performed using SPSS software (ver. 27.0, IBM, Armonk, NY, USA). The significant differences between means were evaluated by the one-way ANOVA at a significance level of 0.05.

3. Results and Discussion

3.1. Selection of Potential Fungal Strains for Sugar Beet Pulp Degradation

In this study, ten strains of filamentous fungi were tested for their cellulose-degrading potential. The evaluation was based on the analysis of the ability of fungi to grow on the CMC-based agar (Figure S1) and also by comparing the relative hydrolytic activity index (RAI) values. Enzyme activities, such as cellulase, endoglucanase, and β-glucosidase, and the production of reducing sugars (RSs) were evaluated in the shake-flask experiments.

The *Aspergillus niger*, *Penicillium oxalicum*, *Botrytis cinerea*, and *Fusarium solani* fungi were observed to have CMC hydrolytic activity with 42.8, 29.5, 61.8, and 43.2 mm, respectively, clear cellulolytic zones around their colony when plated on the CMC-based agar (Table 3).

Table 3. Hydrolytic activity of tested fungal strains on CMC-based agar.

Fungal Strain	Clear Zone Diameter, mm	Colony Diameter, mm	RHA
Botrytis cinerea CCF 2361	61.84 ± 2.61	53.71 ± 1.89	1.15
Aspergillus nidulants CCF 2912	31.71 ± 0.93	26.34 ± 0.89	1.20
Aspergillus niger CCF 3264	42.78 ± 1.07	33.17 ± 1.07	1.29
Fusarium avenaceum CCF 3306	71.68 ± 1.84	62.81 ± 0.83	1.14
Fusarium solani CCF 2967	43.20 ± 5.23	37.43 ± 4.16	1.16
Fusarium oxysporum CCF 1389	44.88 ± 2.63	39.13 ± 1.94	1.15
Fusarium graminearum CCF 1626	54.33 ± 3.06	47.33 ± 1.08	1.15
Penicillium oxalicum CCF 3438	29.47 ± 1.53	22.67 ± 0.79	1.30
Rhizoctonia solani CCF 1360	43.17 ± 2.06	41.24 ± 1.79	1.05
Verticillium spp. CCF 1896	44.21 ± 3.17	41.07 ± 2.83	1.07

According to the David and Stout classification, the clear zones > 20 mm show high cellulose-degrading activity [28]. In the case of cellulose hydrolytic activity, the maximum RAI values were observed for *Penicillium oxalicum* (1.30), *Aspergillus* spp. (1.20–1.29), *Fusarium* spp. (1.14–1.16), and *Botrytis cinerea* (1.15), indicating their potential as promising cellulase producers [29]. Minimum RAI values were recorded for *Rhizoctonia solani* (1.05) and *Verticillium* spp. (1.07).

The obtained results are consistent with Namnuch et al. [23], who demonstrated *A. flavus* KUB2 showing maximum values of the hydrolytic activity index between 1.10 and 1.12 at a 30 °C temperature, depending on the pH of the medium.

The four most active strains were tested for cellulase, endoglucanase, and β-glucosidase enzyme activity production in the SBP substrate at selected conditions (25 °C; 120 h). The results of the analysis are shown in Table 4. The fungi-secreted enzyme activities in the SBP substrate varied significantly between fungal strains.

Table 4. Cellulolytic enzymes activities (U/g d.w.) in shake-flask experiments.

Fungi	Cellulase	Endoglucanase	β-Glucosidase
A. niger CCF 3264	7.35 ± 0.56 ^{a}	1.72 ± 0.31 ^{a}	0.77 ± 1.36 ^{a}
B. cinerea CCF 2361	5.33 ± 0.41 ^{b}	1.39 ± 0.14 ^{b}	0.62 ± 1.76 ^{b}
F. solani CCF 2967	3.17 ± 0.27 ^{c}	0.94 ± 0.12 ^{c}	0.31 ± 1.95 ^{d}
P. oxalicum CCF 3438	5.31 ± 0.49 ^{b}	0.83 ± 0.08 ^{d}	0.46 ± 0.22 ^{c}

Values are mean ± SD (n = 3). Different superscript letters in the same column represent significant differences at $p < 0.05$.

A. niger CCF 3264 exhibited the highest cellulase activity (7.35 U/g d.w.), while *P. oxalicum* CCF 3438 and *B. cinerea* CCF 2361 showed similar cellulase production (5.31–5.32 U/g d.w.), while the lowest activity was determined for *F. solani* CCF 2967 (3.17 U/g d.w.). In the case of other enzyme activities, *F. solani* CCF 2967 and *P. oxalicum* CCF 3438 displayed on average 36.3–51.4% lower endoglucanase and 37.9–50.1% lower β-glucosidase activities compared to other strains (Table 4). Our study showed that *A. niger*

CCF 3264, *P. oxalicum* CCF 3438, and *B. cinerea* CCF 2361 have the potential to produce cellulose and hemicellulose degrading enzymes.

These observations are in agreement with several reports showing the production of appropriate cellulolytic enzyme activity by several *ascomycetes* on different substrates. In the research of Vaithanomsat et al. [30], the fungus *A. niger* SOI017 was shown to be effectively producing enzyme β-glucosidase compared to other tested fungal strains. *A. niger* produced relatively high cellulase activity (8.89 U/g d.w.) after 72 h of incubation on Coir waste as the substrate [31], and also, the production of cellulase activity of 6.23 U/g d.w. was reported on brewery-spent grain [32]. In the study of Kumar et al., *A. niger* showed the highest cellulase activity of 10.81 U/g on the wheat bran substrate at solid-state fermentation conditions [19]. *Penicillium* sp. AKB-24, among other enzymes, produced β-glucosidase (6 IU/g d.s.) activity in the wheat bran substrate over 7 days of incubation at 30 °C [33]. de Oliveira Júnior et al. reused guarana processing by-products for endoglucanase (0.84 U/g) and xylanase (1.0 U/g) production by the fungus *Myceliophthora heterothallic* [34].

The adaptation of enzymatic hydrolysis for the cellulosic raw material conversion into glucose involves the synergistic hydrolysis effect on at least three different enzymes: endoglucanase, exoglucanase, and β-glucosidase [35]. β-glucosidases are the key enzymes in cellulose hydrolysis, as rate-limiting enzymes that are regulated by the feedback inhibition of the formed product of glucose, contributing to the efficiency of this process. Therefore, these enzymes are of considerable interest as constituents of cellulose-degrading systems applied to biomass conversion [36]. According to the literature, several fungal species, such as *Aspergillus*, *Fusarium*, and *Trichoderma*, can produce glucose-tolerant β-glucosidases [37]. Glucose-tolerant β-glucosidases, prevalent naturally in filamentous fungi, have a significant effect on eliminating the inhibitory effect of hexoses on alcoholic fermentation [37].

The differences in the activity of the enzymes for degrading the SBP polysaccharides were evaluated by comparing concentrations of the total RS, glucose, and arabinose that were released from SBP over 120 h of hydrolysis at 25 °C. The chromatographic analysis of saccharides showed the presence of glucose, arabinose, xylose, mannose, and fructose (Table 5).

Table 5. The total reducing sugars (RSs) and monosaccharides (g/100 g d.w.) produced during the 120 h shake-flask cultivation of fungi on SBP substrate.

Fungi	Total RS	Fructose	Glucose	Arabinose	Xylose	Mannose
A. niger CCF 3264	25.13 ± 1.34	0.927	5.704	6.572	4.806	3.897
B. cinerea CCF 2361	24.86 ± 0.87	1.906	4.633	5.326	5.649	4.385
F. solani CCF 2967	21.06 ± 0.57	1.995	4.272	4.501	3.109	3.141
P. oxalicum CCF 3438	24.17 ± 0.36	1.404	4.726	5.608	5.109	4.549

Values are mean ± SD (n = 3).

The highest concentrations of sugars in the SBP hydrolysates were determined for arabinose, glucose, and xylose; mannose was found at slightly lower concentrations, and the lowest content was determined for fructose. The highest amounts of glucose and arabinose produced were *A. niger* CCF 3264 (5.70 and 6.57 g/100 g d.w., respectively), following *P. oxalicum* CCF 3438 (4.73 and 5.61 g/100 g d.w., respectively) and *B. cinerea* CCF 2361 (4.63–5.33 g/100 g d.w.) (Table 5).

According to the literature, relatively high glucose, arabinose, and galacturonic acid concentrations (from 12.5 to 17.0 g/L) were produced by *A. niger* AACC 11414 in the SBP hydrolysate after 166 h of incubation at a 50 °C temperature [38]. In addition, an appropriate concentration (54.8 g/L) of other saccharides, such as xylose, mannose, and galactose, was obtained [38]. In our study, the total concentration of RS produced by *A. niger* CCF 3264 in the SBP medium after 120 h of hydrolysis was 25.13 g/100 g d.w. and the total concentration of glucose and arabinose was 12.28 g/100 g d.w., corresponding to 7.90 g/L and 3.86 g/L, respectively. The lower concentration of sugars was possibly obtained due to the specificity of the strain, shorter hydrolysis time, and lower temperature. From the consistently high concentration of released arabinose, it can be concluded that almost all analysed cultures produced pectinolytic activity, even the *F. solani* culture, which had significantly lower activity. The highest glucose concentration confirmed by the highest cellulase activity (Table 4) was found after SBP hydrolysis with the *A. niger* CCF 3264 culture.

The difference in the sugar profile might be due to the different adaptations of the strains in the SBP medium and their enzyme systems. According to the literature [10], after the enzymatic hydrolysis of SBP, raffinose, arabinose, galactose, glucose, fructose, xylose, mannose, and galacturonic acid can be detected. While glucose is mainly obtained during the degradation of cellulose, pectin and hemicelluloses are providers for pentoses, such as arabinose and galactose. Hemicelluloses are also a source of xylose and mannose [39]. As the main effect on galactose release was pectinolytic activity, and it was not detected in the SBP hydrolysates, it can be assumed that the tested fungi might have had low pectinolytic activity or the process conditions were not suitable for the production of pectinases [40]. In addition, microorganisms can produce ribose when consuming xylose after the depletion of glucose. In the study of Park et al. [41], *E. coli* SGK013 were shown to produce 0.75 g/L of ribose during the batch fermentation of the medium containing glucose and xylose.

The production of different enzyme activities by fungi is highly dependent on the composition of the substrate and feedback inhibition by the hexoses. In this case, it is possible that the hydrolysis of polysaccharides and the consumption of carbon sources by fungi were more or less efficient. Moreover, the maximum catalytic activity of fungal cellulolytic enzymes, in many cases, is observed at pH 5 [42].

The fungal strains *A. niger* CCF 3264 and *P. oxalicum* CCF 3438, showing the highest SBP-degrading potential, except *F. solani* CCF 2967, which showed the lowest hydrolytic activity, were selected for further one-factor experiments.

3.2. One-Factor Experiments

The one-factor experiments were carried out in order to ascertain the most active fungal strain and determine the ranges of SBP hydrolysis parameters for further process optimisation (Figures 1 and 2). Factors affecting the RS production due to the fungal degradation of the SBP lignocellulose matrix were studied, measuring the amount of RS produced at different fermentation points of time and temperature (the initial RS concentration in raw material was 2.29 g/100 g d.w.).

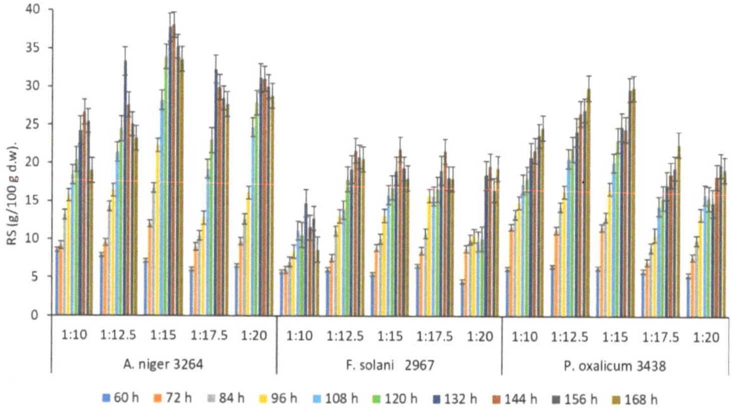

Figure 1. The effect of solid/water ratio (1:10–1:20) on production of reducing sugars (RSs) by selected fungal strains during 60–168 h of (25 °C, pH 5) incubation on the SBP substrate.

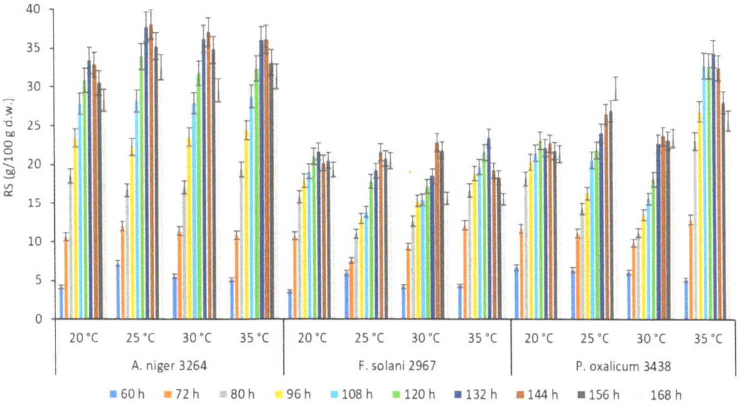

Figure 2. The effect of temperature on the production of reducing sugars (RSs) by selected fungal strains during 60–168 h (25 °C, pH 5) of incubation on the SBP substrate.

According to the literature, temperatures between 24 and 37 °C, pH in a range of 4.0–6.5, and water activity >0.95 are optimal conditions for the growth of the most fungal strains [43,44], whereas yeast fermentation temperatures near 30 °C and pHs of 5.0–5.5 are the most suitable [45]. Thus, for the experiment, the initial pH of the medium was adjusted to 5.0 and kept constant in the case of further biomass degradation with fungi.

As the degradation of lignocellulosic biomass is affected by moisture content, experiments on different initial solid/water ratios (*s/w*) were carried out (Figure 1). It can be seen that the RS concentration initially increased and then decreased with increasing water content. The RS concentration reached the maximum when the s/*w* ratio was varied between 1:12.5 and 1:17.5, depending on the fungal strain. Each strain showed a specific moisture optimum for maximum RS production. In the case of *F. solani* CCF 2967, the selected moisture range did not initiate significant differences in the RS contents, which depended on the duration of the process. In the case of *P. oxalicum* CCF 3438 and *A. niger* CCF 3264, the RS concentrations increased considerably, with an increasing *s/w* ratio from 1:10 to 1:15 within the 60–168 h and 60–144 h processing periods, respectively (Figure 1).

The effect of time and temperature on the RS production during the incubation of fungal strains on the SBP substrate is demonstrated in Figure 2. The *s/w* ratio was selected for each fungal strain based on the previous experiment results (Figure 1).

The highest sugar recovery rate (15.9–44.8%) was obtained at the late period of incubation (108–168 h). In most cases, processing times up to 60–80 h gave relatively low sugar recovery (3.9–27.6%). A significant increase in the RS production was observed from 96 h of processing, e.g., the highest sugar recovery was fixed for *A. niger* CCF 3264 (26.2–44.8%), following *P. oxalicum* CCF 3438 (15.9–38.2%) and *F. solani* CCF 2967 (13.3–27.6%).

In the case of processing temperature, a noticeable increase in the sugar recovery was achieved at 25–35 °C temperatures (5.0–44.8%), compared to 20 °C (4.3–39.3%) for all tested fungi, depending on the strain (Figure 2).

The results obtained are in agreement with the reports, showing that filamentous fungi, especially basidiomycetes and ascomycetes, efficiently degrade plant lignocelluloses [46,47]. The fungi *Aspergillus niger*, *A. oryzae*, and *Trichoderma reesei* were shown to possess a variety of enzymes that could degrade the complex structure of lignocellulosic substrates [48]. Fungi, such as *A. niger*, were reported to have a flexible regulatory network adapted to the utilisation of hemicellulose and cellulose when pectin degradation was impaired [49]. In the literature, it was reported that the cultivation pH and temperature are significant factors affecting fungal hydrolytic activity [19,21]. For example, sugarcane bagasse waste samples treated by *A. flavus* KUB2 did not show cellulase activity at an incubation temperature lower than 30 °C due to the slower movement of the substrate across the fungal cells at lower temperatures, resulting in the lower yield of the products [21]. In the case of moisture content, solid-state fermentation was reported to be more effective for pectinase and cellulase production by *A. niger* compared to submerged fermentation [19].

The optimisation of fermentation conditions for the production of enzymes pectinase and cellulase by *A. niger* NCIM 548 was performed on different substrates, such as wheat bran, corn bran, and kinnow peel [19]. In the study of Garrigues et al., *A. niger* was involved in the SBP degradation to assess the role of pectinolytic and hemicelulolytic enzyme regulators [49]. In the work of Berlowska and co-authors, the efficiency of lactic acid production using simultaneous enzymatic hydrolysis and the fermentation of SPB was analysed [10]. Gönen and co-authors optimised the SBP acid pretreatment under pressure and non-pressure conditions [50].

The literature data and the obtained results of one-factor experiments demonstrated that reducing sugar production during the fungal degradation of SBP can be influenced by various processing parameters.

3.3. The Optimisation of Degradation Processing and Mathematical Model Analysis

The effect of fungal hydrolysis time (*t*), temperature (*T*), and *s/w* ratio on the RS yield and cellulase activity was analysed via optimisation using CCD and RSM. The experiments were carried out according to the experimental design for three variables (Table 6), adopting the ranges of processing parameters based on the one-factor investigations (Section 3.2, Figures 1 and 2). The RS yield varied significantly from 27.12 to 38.08 g/100 g d.w., and the cellulase activity values varied from 3.49 to 6.57 U/g d.w., applying different *T*, *t*, and s/w ratio combinations (Table 6).

Table 6. Process variables and observed responses.

Exp. No.	Process Variables			Y1 (g/100 g d.w.)			Y2 (U/g d.w.)		
	X1 (s/w)	*X2* (*T*, °C)	*X3* (*t*, h)	Exp	Predicted	/RE/(%)	Exp	Predicted	/RE/(%)
1	1	1	1	33.18 ± 1.02	33.02	0.48	4.98	4.99	0.20
2	1	1	−1	31.21 ± 0.86	31.11	0.32	4.54	4.54	0.10
3	1	−1	1	27.12 ± 0.32	26.77	1.26	4.30	4.31	0.21
4	1	−1	−1	26.63 ± 0.64	26.62	0.04	3.49	3.52	0.80
5	−1	1	1	32.33 ± 0.68	32.31	0.07	5.14	5.13	0.21
6	−1	1	−1	30.24 ± 0.37	30.35	0.37	3.85	3.86	0.49
7	−1	−1	1	28.15 ± 0.35	28.20	0.17	4.30	4.31	0.25
8	−1	−1	−1	27.68 ± 1.12	28.00	1.15	3.64	3.63	0.11
9	1	0	0	32.58 ± 0.87	32.59	0.01	5.81	5.87	1.10
10	0	1	0	37.86 ± 0.57	38.31	1.18	6.12	6.20	1.27
11	0	0	1	36.96 ± 0.51	36.97	0.04	6.13	6.11	0.25
12	−1	0	0	33.08 ± 0.33	32.92	0.47	5.91	5.99	1.40
13	0	−1	0	34.26 ± 1.02	34.01	0.73	5.15	5.17	0.37
14	0	0	−1	35.91 ± 0.28	35.92	0.01	5.59	5.65	1.13
15	0	0	0	38.06 ± 0.86	37.91	0.40	6.54	6.55	0.08
16	0	0	0	38.04 ± 1.14	37.91	0.34	6.56	6.55	0.14
17	0	0	0	37.96 ± 0.87	37.91	0.13	6.57	6.55	0.31

T: temperature; *t*: time; *s/w*: solid-to-water ratio; RE: relative error.

The quartic order regression models (Table S1), describing the relationships between the RS yield (Y1), cellulase activity (Y2), and independent variables, were created in Equations (2) and (3):

$$Y1 = 37.909 - 0.167X1 + 2.150X2 + 0.527X3 - 5.152X1^2 - 1.747X2^2 - 1.462X3^2 - 0.01X1X3 + 0.535X1X2 + 0.44X2X3, \quad (2)$$

$$Y2 = 6.5285 - 0.0601X1 + 0.5148X2 + 0.23X3 - 0.6164X1^2 - 0.8661X2^2 - 0.6651X3^2 - 0.1997X1X2 + 0.1961X1X3 - 0.0823X2X3, \quad (3)$$

where Y1 is the RS yield, Y2 is cellulase activity, and X1, X2, and X3 are the values of the solid/water ratio, temperature, and time, respectively.

The relative errors (RE) between the experimental data and theoretical predictions varied between 0.01 and 1.26% for Y1 and between 0.08 and 1.40% for Y2 (Table 6), indicating that the models can be used to estimate the responses for the optimisation.

Based on ANOVA (Table S2), the mathematical models were significant for Y1 and Y2 and fitted well to the experimental data: Fisher values were 6.81 and 53.61 ($p < 0.0001$), and determination coefficients were $R^2 = 0.9859$ and $R^2 = 0.9889$ for Y1 and Y2, respectively.

The ANOVA of the models confirmed that the *s/w* ratio (X1) ($p = 0.045$ and $p = 0.013$), processing temperature (X2) ($p = 0.0001$), time (X3) ($p = 0.035$ and $p = 0.0001$), and their quadratic effects ($p = 0.0001$–0.005 and $p = 0.0001$) significantly affected the production of cellulase activity and RS (model $p < 0.0001$) (Table S2). The significant effects of the linear interaction between time and temperature (X2X3) ($p = 0.049$ and 0.004, respectively), and between the *s/w* ratio and temperature (X1X2) ($p = 0.051$ and $p = 0.0001$) were found. The *p*-value calculated for *'lack-of-fit'* was not significant ($p = 0.088$ and 0.985, respectively); hence, the models were satisfactory in explaining the obtained data at a 95% confidence level. The predicted R^2 values (0.9241 and 0.9699) for the experimental design are close to the adjusted R^2 values (0.9732 and 0.9789), indicating that up to 92, and 97% of the data can be described by these regression models, respectively.

3.4. Optimisation and Prediction of Process Parameters

Based on the obtained mathematical models, two-dimensional plots were constructed to predict the relationship between independent variables. Predictive plots were constructed by varying the factors from the low to high level of the optimum conditions based on the increase in the RS yield or cellulase activity. Figure 3 presents the relationships between the response values (Y1 and Y2) and the independent variables (time, temperature, and *s/w* ratio).

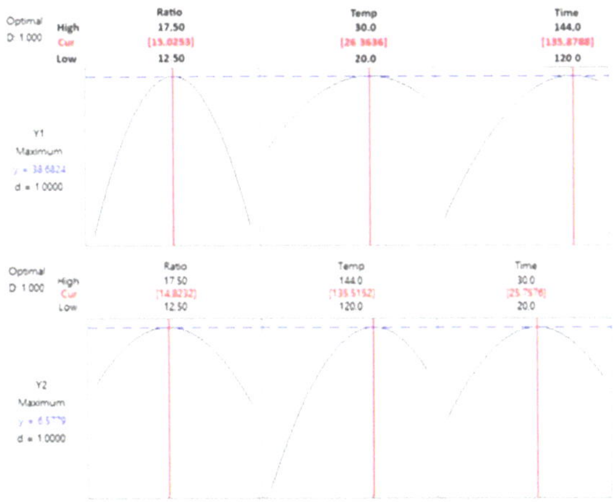

Figure 3. The RSM plots of the interactions between temperature and time, *s/w* and time, and *s/w* and temperature for the RS yield (Y1) and cellulase activity (Y2).

The elliptical contours and relatively sharp slopes in most cases of the surface responses indicate strong interactions between process parameters and a significant influence of them on the response values (Figure S2). Based on the RSM results, the optimal process conditions for the most effective production of RS (38.68 g/100 g d.w.) and cellulase activity (6.62 U/g d.w.) during the SBP degradation with *A. niger* CCF 3264 were set as follows: temperature 26 °C, time 136 h, and *s/w* ratio 1:15.

The confirmatory experiment was performed in triplicate under the conditions suggested by mathematical models to validate them. The RS concentration was produced after 136 h of fermentation at 26 °C, the optimum *s/w* ratio was 39.15 ± 0.16 U/100 g d.w., and cellulase activity was 6.65 ± 0.08 U/g d.w. These values with RE 1.22 and 0.45%, respectively, were found to be in agreement with the RSM model's prediction.

In summary, the fungal degradation of biomass can be characterised as a sustainable process with the potential to produce reducing sugars from agrobiomass for further specific applications in the industry.

4. Conclusions

This study presents a concept for the biorefinery approach to pretreat and ensure the sustainable utilisation of waste sugar beet pulp biomass. An efficient, low-cost substrate and fungal strains are the key points addressed for enhanced saccharification and enzyme production at an industrial scale. The biomass after fungal hydrolysis can be used as a source of carbohydrates for yeast fermentation or microbial cultivation. Low-energy biological pretreatments overall are considered environmentally friendly processes since they do not involve chemical materials compared to most of the other pretreatments used. However, the efficiency of biological lignocellulose hydrolysis is relatively low. Thus, emerging technology, such as ultrasound, might be considered for evaluation in further experiments for biomass pretreatment in combination with fungal degradation to enhance reducing sugar yield for further applications. The results of this study provide knowledge for further research on the use of sugar beet pulp biomass for yeast fermentation or feed-value protein production, contributing to the goals of sustainable sugar beet processing.

Supplementary Materials: The following supporting information can be downloaded at: https://www.mdpi.com/article/10.3390/polym16091178/s1, S1: Chemical and microbiological analysis; S2: Determination of soluble and insoluble dietary fibre; S3: Lignin determination; Table S1: Coded coefficients of the mathematical models for Y1 and Y2 responses; Table S2: ANOVA of mathematical models; Figure S1: Hydrolytic activity of tested fungal strains on CMC-based agar; Figure S2: Response surface plots of the interactions between process parameters for RS yield (Y1) and cellulase activity (Y2). References [51–53] are cited in the Supplementary Materials.

Author Contributions: Methodology, formal analysis, validation, writing—original draft, writing—review and editing, Z.G.; Methodology, data curation, R.Z.; Methodology, formal analysis, software, M.S.; Data curation, writing—review and editing, L.B.; Conceptualisation, resources, data curation, writing—original draft, writing—review and editing, supervision, D.Z. All authors have read and agreed to the published version of the manuscript.

Funding: This research received no external funding.

Institutional Review Board Statement: Not applicable.

Informed Consent Statement: Not applicable.

Data Availability Statement: Data are contained within the article and Supplementary Materials. The data presented in this article are part of first author's PhD work.

Acknowledgments: The authors would like to acknowledge the Department of Botany of Charles University (Prague, Czech Republic) for kindly providing fungal strains for PhD research purposes.

Conflicts of Interest: The authors declare no conflicts of interest.

References

1. Cárdenas-Fernández, M.; Bawn, M.; Hamley-Bennett, C.; Bharat, P.K.V.; Subrizi, F.; Suhaili, N.; Ward, D.P.; Bourdin, S.; Dalby, P.A.; Hailes, H.C.; et al. An integrated biorefinery concept for conversion of sugar beet pulp into value-added chemicals and pharmaceutical intermediates. *Faraday Discuss.* **2017**, *202*, 415–431. [CrossRef] [PubMed]
2. Muir, B.M.; Anderson, A.R. Development and diversification of sugar beet in Europe. *Sugar Tech* **2022**, *24*, 992–1009. [CrossRef]
3. Dygas, D.; Kręgiel, D.; Berłowska, J. Sugar beet pulp as a biorefinery substrate for designing feed. *Molecules* **2023**, *28*, 2064. [CrossRef] [PubMed]

4. Ubando, A.T.; Felix, C.B.; Chen, W.-H. Biorefineries in circular bioeconomy: A comprehensive review. *Bioresour. Technol.* **2020**, *299*, 122585. [CrossRef] [PubMed]
5. Usmani, Z.; Sharma, M.; Diwan, D.; Tripathi, M.; Whale, E.; Jayakody, L.N.; Moreau, B.; Thakur, V.K.; Tuohy, M.; Gupta, V.K. Valorization of Sugar Beet Pulp to Value-Added Products: A Review. *Bioresour. Technol.* **2022**, *346*, 126580. [CrossRef]
6. Joanna, B.; Michal, B.; Piotr, D.; Agnieszka, W.; Dorota, K.; Izabela, W. Sugar beet pulp as a source of valuable biotechnological products. In *Advances in Biotechnology for Food Industry*; Holban, A.M., Grumezescu, A.M., Eds.; Academic Press: London, UK, 2018; pp. 359–392.
7. Glaser, S.J.; Abdelaziz, O.Y.; Demoitié, C.; Galbe, M.; Pyo, S.-H.; Hati-Kaul, R. Fractionation of sugar beet pulp polysaccharides into component sugars and pre-feasibility analysis for further valorisation. *Biomass. Conv. Bioref.* **2022**, *14*, 3575–3588. [CrossRef]
8. Leijdekkers, A.G.; Bink, J.P.; Geutjes, S.; Schols, H.A.; Gruppen, H. Enzymatic saccharification of sugar beet pulp for the production of galacturonic acid and arabinose; a study on the impact of the formation of recalcitrant oligosaccharides. *Bioresour. Technol.* **2013**, *12*, 518–525. [CrossRef] [PubMed]
9. Bibra, M.; Samanta, D.; Sharma, N.K.; Singh, G.; Johnson, G.R.; Sani, R.K. Food waste to bioethanol: Opportunities and challenges. *Fermentation* **2023**, *9*, 8. [CrossRef]
10. Berlowska, J.; Cieciura-Wloch, W.; Kalinowska, H.; Kregiel, D.; Borowski, S.; Pawlikowska, E.; Binczarski, M.; Witonska, I. Enzymatic conversion of sugar beet pulp: A comparison of simultaneous saccharification and fermentation and separate hydrolysis and fermentation for lactic acid production. *Food Technol. Biotechnol.* **2018**, *56*, 188. [CrossRef] [PubMed]
11. Cavalaglio, G.; Gelosia, M.; D'Antonio, S.; Nicolini, A.; Pisello, A.L.; Barbanera, M.; Cotana, F. Lignocellulosic ethanol production from the recovery of stranded driftwood residues. *Energies* **2016**, *9*, 634. [CrossRef]
12. Li, C.; Lin, F.; Zhou, L.; Qin, L.; Li, B.; Zhou, Z.; Jin, M.; Chen, Z. Cellulase hyper-production by *Trichoderma reesei* mutant SEU-7 on lactose. *Biotechnol. Biofuels* **2017**, *10*, 228. [CrossRef] [PubMed]
13. Gabelle, J.C.; Jourdier, E.; Licht, R.B.; Ben Chaabane, F.; Henaut, I.; Morchain, J.; Augier, F. Impact of rheology on the mass transfer coefficient during the growth phase of *Trichoderma reesei* in stirred bioreactors. *Chem. Eng. Sci.* **2012**, *75*, 408–417. [CrossRef]
14. Yang, B.; Wyman, C.E. Effect of xylan and lignin removal by batch and flowthrough pretreatment on the enzymatic digestibility of corn stover cellulose. *Biotechnol. Bioeng.* **2004**, *86*, 88–95. [CrossRef] [PubMed]
15. Andlar, M.; Rezić, T.; Marđetko, N.; Kracher, D.; Ludwig, R.; Šantek, B. Lignocellulose degradation: An overview of fungi and fungal enzymes involved in lignocellulose degradation. *Eng. Life Sci.* **2018**, *18*, 768–778. [CrossRef] [PubMed]
16. Kumar, A.; Chandra, R. Ligninolytic enzymes and its mechanisms for degradation of lignocellulosic waste in environment. *Heliyon* **2020**, *6*, e03170. [CrossRef] [PubMed]
17. Arantes, V.; Goodell, B. Current understanding of brown-rot fungal biodegradation mechanisms: A review. In *Deterioration and Protection of Sustainable Biomaterials*; Schultz, T.P., Goodell, B., Nicholas, D.D., Eds.; American Chemical Society: Washington, DC, USA, 2014; pp. 4–21.
18. Shirkavand, E.; Baroutian, S.; Gapes, D.J.; Young, B.R. Combination of fungal and physicochemical processes for lignocellulosic biomass pretreatment—A review. *Renew. Sust. Energ. Rev.* **2016**, *54*, 217–234. [CrossRef]
19. Kumar, S.; Sharma, H.; Sarkar, B. Effect of substrate and fermentation conditions on pectinase and cellulase production by *Aspergillus niger* NCIM 548 in submerged (SmF) and solid state fermentation (SSF). *Food Sci. Biotechnol.* **2011**, *20*, 1289–1298. [CrossRef]
20. Marzo, C.; Díaz, A.B.; Caro, I.; Blandino, A. Conversion of exhausted sugar beet pulp into fermentable sugars from a biorefinery approach. *Foods* **2020**, *9*, 1351. [CrossRef] [PubMed]
21. Aberkane, A.; Cuenca-Estrella, M.; Gomez-Lopez, A.; Petrikkou, E.; Mellado, E.; Monzón, A.; Rodriguez-Tudela, J.L. Comparative evaluation of two different methods of inoculum preparation for antifungal susceptibility testing of filamentous fungi. *J. Antimi. Chemother.* **2002**, *50*, 719–722. [CrossRef]
22. Maki, M.L.; Idrees, A.; Leung, K.T.; Qin, W. Newly isolated and characterized bacteria with great application potential for decomposition of lignocellulosic biomass. *J. Mol. Microbiol. Biotechnol.* **2012**, *22*, 156–166. [CrossRef]
23. Namnuch, N.; Thammasittirong, A.; Thammasittirong, S.N. Lignocellulose hydrolytic enzymes production by *Aspergillus flavus* KUB2 using submerged fermentation of sugarcane bagasse waste. *Mycology* **2020**, *12*, 119–127. [CrossRef] [PubMed]
24. Gasparotto, J.M.; Werle, L.B.; Foletto, E.L.; Kuhn, R.C.; Jahn, S.L.; Mazutti, M.A. Production of cellulolytic enzymes and application of crude enzymatic extract for saccharification of lignocellulosic biomass. *Appl. Biochem. Biotechnol.* **2015**, *175*, 560–572. [CrossRef] [PubMed]
25. Ghose, T.K. Measurement of cellulase activities. *Pure Appl. Chem.* **1987**, *59*, 257–268. [CrossRef]
26. Miller, G.L. Use of Dinitrosalicylic acid reagent for determination of reducing sugar. *Anal. Chem.* **1959**, *31*, 426–428. [CrossRef]
27. Verchot, L.V.; Borelli, T. Application of para-nitrophenol (pNP) enzyme assays in degraded tropical soils. *Soil Biol. Biochem.* **2005**, *37*, 625–633. [CrossRef]
28. David, W.W.; Stout, T.R. Disc plate method of microbiological antibiotic assay. I. Factors influencing variability and error. *Appl. Microbiol.* **1971**, *22*, 659–665.
29. Akintunde, O.; Chukwudozie, C. Hydrolytic and inhibitory activity of two closely related *Bacillus* isolates. *J. Appl. Environ. Microbiol.* **2021**, *9*, 5–8. [CrossRef]
30. Vaithanomsat, P.; Songpim, M.; Malapant, T.; Kosugi, A.; Thanapase, W.; Mori, Y. Production of β-glucosidase from a newly isolated *Aspergillus* species using response surface methodology. *Int. J. Microbiol.* **2011**, *2011*, 949252. [CrossRef] [PubMed]

31. Mrudula, S.; Murugamnal, R. Production of cellulase by *Aspergillus niger* under submerged and solid state fermentation using Coir waste as a substrate. *Braz. J. Microbiol.* **2011**, *42*, 1119–1127. [CrossRef]
32. Moran-Aguilar, M.G.; Costa-Trigo, I.; Calderón-Santoyo, M.; Domínguez, J.M.; Aguilar-Uscanga, M.G. Production of cellulases and xylanases in solid-state fermentation by different strains of *Aspergillus niger* using sugarcane bagasse and brewery spent grain. *Biochem. Eng. J.* **2021**, *172*, 108060. [CrossRef]
33. Kumar, A.; Gautam, A.; Dutt, D. Co-Cultivation of *Penicillium* sp. AKB-24 and *Aspergillus nidulans* AKB-25 as a cost-effective method to produce cellulases for the hydrolysis of pearl millet stover. *Fermentation* **2016**, *2*, 12. [CrossRef]
34. de Oliveira Simões, L.C.; da Silva, R.R.; de Oliveira Nascimento, C.E.; Boscolo, M.; Gomes, E.; da Silva, R. Purification and physicochemical characterization of a novel thermostable xylanase secreted by the fungus *Myceliophthora heterothallica* F.2.1.4. *Appl. Biochem. Biotechnol.* **2019**, *188*, 991–1008. [CrossRef] [PubMed]
35. Leite, R.S.R.; Alves-Prado, H.F.; Cabral, H.; Pagnocca, F.C.; Gomes, E.; Da-Silva, R. Production and characteristics comparison of crude β-glucosidases produced by microorganisms *Thermoascus aurantiacus* e *Aureobasidium pullulans* in agricultural wastes. *Enzyme Microb. Technol.* **2008**, *43*, 391–395. [CrossRef]
36. Kumari, D.; Singh, R. Pretreatment of lignocellulosic wastes for biofuel production: A critical review. *Renew. Sustain. Energy Rev.* **2018**, *90*, 877–891. [CrossRef]
37. Rajasree, K.P.; Mathew, G.M.; Pandey, A.; Sukumaran, R.K. Highly glucose tolerant β-glucosidase from *Aspergillus unguis* NII08123 for enhanced hydrolysis of biomass. *J. Ind. Microbiol. Biotechnol.* **2013**, *40*, 967–975. [CrossRef] [PubMed]
38. Knesebeck, M.; Schäfer, D.; Schmitz, K.; Rüllke, M.; Benz, J.P.; Weuster-Botz, D. Enzymatic one-pot hydrolysis of extracted sugar beet press pulp after solid-state fermentation with an engineered *Aspergillus niger* strain. *Fermentation* **2023**, *9*, 582. [CrossRef]
39. Rezić, T.; Oros, D.; Marković, I.; Kracher, D.; Ludwig, R.; Šantek, B. Integrated hydrolyzation and fermentation of sugar beet pulp to bioethanol. *J. Microbiol. Biotechnol.* **2013**, *23*, 1244–1252. [CrossRef] [PubMed]
40. Sandri, I.G.; Lorenzoni, C.M.T.; Fontana, R.C.; da Silveira, M.M. Use of pectinases produced by a new strain of *Aspergillus niger* for the enzymatic treatment of apple and blueberry juice. *LWT* **2013**, *51*, 469–475. [CrossRef]
41. Park, H.-C.; Kim, Y.-J.; Lee, C.-W.; Rho, Y.-T.; Kang, J.W.; Lee, D.-H.; Seong, Y.-J.; Park, Y.-C.; Lee, D.; Kim, S.-G. Production of d-ribose by metabolically engineered *Escherichia coli*. *Process Biochem.* **2017**, *52*, 73–77. [CrossRef]
42. Metreveli, E.; Khardziani, T.; Elisashvili, V. The carbon source controls the secretion and yield of polysaccharide-hydrolyzing enzymes of basidiomycetes. *Biomolecules* **2021**, *11*, 1341. [CrossRef]
43. Passamani, F.R.F.; Hernandes, T.; Alves Lopes, N.; Carvalho Bastos, S.; Douglas Santiago, W.; Das Graças Cardoso, M.; Batist, L.R. Effect of temperature, water activity, and pH on growth and production of ochratoxin a by *Aspergillus niger* and *Aspergillus carbonarius* from Brazilian grapes. *J. Food Prot.* **2014**, *77*, 1947–1952. [CrossRef]
44. Hui, Y.; Berlin Nelson, J.R. Effect of temperature on *Fusarium solani* and *F. tricinctum* growth and disease development in soybean. *Can. J. Plant Pathol.* **2020**, *42*, 527–537.
45. Azhar, S.H.M.; Abdulla, R.; Jambo, S.A.; Marbawi, H.; Gansau, J.A.; Mohd Faik, A.A.; Rodrigues, K.F. Yeasts in sustainable bioethanol production: A review. *Biochem. Biophys. Rep.* **2017**, *10*, 52–61.
46. de Vries, R.P.; Patyshakuliyeva, A.; Garrigues, S.; Agarwal-Jans, S. The current biotechnological status and potential of plant and algal biomass degrading/modifying enzymes from ascomycete fungi. In *Grand Challenges in Fungal Biotechnology*; Nevalainen, H., Ed.; Springer: Cham, Switzerland, 2020; pp. 81–120.
47. van den Brink, J.; de Vries, R.P. Fungal enzyme sets for plant polysaccharide degradation. *Appl. Microbiol. Biotechnol.* **2011**, *91*, 1477–1492. [CrossRef]
48. Frisvad, J.C.; Møller, L.L.; Larsen, T.O.; Kumar, R.; Arnau, J. Safety of the fungal workhorses of industrial biotechnology: Update on the mycotoxin and secondary metabolite potential of *Aspergillus niger*, *Aspergillus oryzae*, and *Trichoderma reesei*. *Appl. Microbiol. Biotechnol.* **2018**, *102*, 9481–9515. [CrossRef]
49. Cubells, S.G.; Kun, R.S.; Peng, M.; Bauer, D.; Keymanesh, K.; Lipzen, A.; Ng, V.; Grigoriev, I.V.; de Vries, R.P. Unraveling the regulation of sugar beet pulp utilization in the industrially relevant fungus *Aspergillus niger*. *iScience* **2022**, *25*, 104065.
50. Gönen, Ç.; Akter Önal, N.; Deveci, E.Ü. Optimization of sugar beet pulp pre-treatment with weak and strong acid under pressure and non-pressure conditions via RSM. *Biomass Conv. Bioref.* **2023**, *13*, 9213–9226. [CrossRef]
51. AOAC. *Official Methods of Analysis*, 17th ed.; Methods 920.152, 942.05, 942.20, 2002.02, 985.29, 991.43, 996.11; The Association of Official Analytical Chemists: Gaithersburg, MD, USA, 2018.
52. Sluiter, A.; Hames, B.; Ruiz, R.; Scarlata, C.; Sluiter, J.; Templeton, D.; Crocker, D. Determination of Structural Carbohydrates and Lignin in Biomass. Laboratory Analytical Procedure (LAP). 2004. Available online: https://api.semanticscholar.org/CorpusID:100361490 (accessed on 5 April 2024).
53. Bouaziz, F.; Koubaa, M.; Ben Jeddou, K.; Kallel, F.; Helbert, C.B.; Khelfa, A.; Ghorbel, R.E.; Chaabouni, S.E. Water-soluble polysaccharides and hemicelluloses from almond gum: Functional and prebiotic properties. *Int. J. Biol. Macromol.* **2016**, *93 Pt A*, 359–368. [CrossRef]

polymers

Article

Development of a Biopolymer-Based Anti-Fog Coating with Sealing Properties for Applications in the Food Packaging Sector

Masoud Ghaani [1,2,*], Maral Soltanzadeh [2], Daniele Carullo [2] and Stefano Farris [2]

1 Department of Civil, Structural & Environmental Engineering, School of Engineering, Trinity College Dublin, College Green, 2 Dublin, Ireland
2 Food Packaging Laboratory, Department of Food, Environmental and Nutritional Sciences—DeFENS, University of Milan, Via Celoria 2, 20133 Milan, Italy; maral.soltanzadeh@unimi.it (M.S.); daniele.carullo@unimi.it (D.C.); stefano.farris@unimi.it (S.F.)
* Correspondence: masoud.ghaani@tcd.ie

Abstract: The quest for sustainable and functional food packaging materials has led researchers to explore biopolymers such as pullulan, which has emerged as a notable candidate for its excellent film-forming and anti-fogging properties. This study introduces an innovative anti-fog coating by combining pullulan with poly (acrylic acid sodium salt) to enhance the display of packaged food in high humidity environments without impairing the sealing performance of the packaging material—two critical factors in preserving food quality and consumers' acceptance. The research focused on varying the ratios of pullulan to poly (acrylic acid sodium salt) and investigating the performance of this formulation as an anti-fog coating on bioriented polypropylene (BOPP). Contact angle analysis showed a significant improvement in BOPP wettability after coating deposition, with water contact angle values ranging from ~60° to ~17° for formulations consisting only of poly (acrylic acid sodium salt) (P0) or pullulan (P100), respectively. Furthermore, seal strength evaluations demonstrated acceptable performance, with the optimal formulation (P50) achieving the highest sealing force (~2.7 N/2.5 cm) at higher temperatures (130 °C). These results highlight the exceptional potential of a pullulan-based coating as an alternative to conventional packaging materials, significantly enhancing anti-fogging performance.

Keywords: biopolymers; anti-fog coating; pullulan; poly (acrylic acid); seal strength; sustainable packaging

Citation: Ghaani, M.; Soltanzadeh, M.; Carullo, D.; Farris, S. Development of a Biopolymer-Based Anti-Fog Coating with Sealing Properties for Applications in the Food Packaging Sector. *Polymers* **2024**, *16*, 1745. https://doi.org/10.3390/polym16121745

Academic Editor: Ick-Soo Kim

Received: 26 April 2024
Revised: 16 June 2024
Accepted: 17 June 2024
Published: 20 June 2024

1. Introduction

As a novel source of innovative materials, biopolymers have garnered significant interest over the past few decades, becoming a critical area of study in the materials science field [1]. In particular, the food packaging sector is rapidly evolving with the commercialization of these materials with enhanced and unprecedented functional properties [2]. Among others (e.g., barrier performance against gases and vapours, thermal stability, active attributes, optical properties, etc.), the anti-fog feature of packaging films has received considerable attention. This feature refers to the material's ability to prevent the formation of tiny water droplets on the inner surface of packaging, a phenomenon that typically occurs due to environmental changes in temperature and humidity [3]. The newly formed droplets produce a foggy layer that alters the optical properties of the material, hindering the clear display of the food product by scattering light in various directions. This phenomenon occurs especially in fresh food and minimally processed vegetables during cold storage (i.e., refrigeration), whereby the transparency of the film significantly influences consumer choices [4,5].

Pullulan is a non-ionic exopolysaccharide (derived from *Aureobasidium pullulans*) with commendable characteristics, such as non-toxicity, biodegradability, and superior film-forming capabilities [6]. Pullulan's distinctive structural attributes, such as the $\alpha(1-6)$ linkage between maltotriose units, yield peculiar characteristics of this biopolymer, such as high flexibility and solubility in water. Its molecular structure, adorned with hydroxyl groups, facilitates extensive intermolecular hydrogen bonding and super-hydrophilicity. These features were first harnessed to develop coatings with excellent oxygen barrier properties [7,8], even in high relative humidity environments [9]. Later on, one of the main drawbacks associated with pullulan, moisture sensitivity [10], has proved advantageous for the new generation of anti-fog coatings [11].

However, the effectiveness of anti-fog pullulan coatings for practical uses also depends on their ability to adequately seal coated material, as the coating is intended for the inner surface of the packaging film, typical in products such as fresh salads. While pullulan inherently possesses some sealing attributes, it does not rival the sealing strength of more conventional materials such as thermoplastic polymers (polyethylene or polypropylene). Therefore, the exploration and development of innovative pullulan-based films or coatings that concurrently offer anti-fogging efficacy and adequate sealing capabilities are needed.

This study introduces a ground-breaking anti-fog coating that combines the biopolymer pullulan with an acrylic component—specifically, poly (acrylic acid) sodium salt. This component was chosen for its negative charges (COO^-) in an aqueous medium, which significantly enhances the sealing properties of the coating. Polyacrylic acid is characterised by sequences of negatively charged carboxylate groups (COO^-). Similar to well-known ionomers, the possibility of ionic interactions culminates in its strength potential, including hydrogen bonding and chain entanglement [12].

Unlike previous studies that primarily focused on anti-fogging or sealing properties, this research integrates both functionalities into a single coating. This dual-functional coating offers a more compelling, functional alternative to existing market offerings, addressing critical concerns like fogging while preserving the visual appeal and integrity of the packaged food throughout its shelf life.

2. Materials and Methods

2.1. Materials

Pullulan powder (PF-20 grade, molecular weight ~200.000 Da), was procured from Hayashibara Biochemical Laboratories Inc., Okayama, Japan. Poly acrylic acid (PAA) sodium salt (molecular weight 2100 Da) was obtained from Sigma-Aldrich (Burlington, MA, USA). A primer solution containing 0.5 wt % aziridine (Michem® Flex P2300) was acquired from Michelman International (Aubange, Belgium). The substrate for coating deposition, corona-treated bi-oriented polypropylene (BOPP) with a thickness of 20 ± 0.5 μm, was sourced from Taghleef Industries S.p.A. (S. Giorgio di Nogaro, Italy). The choice of this material accounts for its widespread use as a packaging material for fresh and ready-to-eat salads. Milli-Q water, exhibiting a resistivity of 18.3 MΩ cm, was employed throughout the experiment.

2.2. Film Preparation

Various concentrations of pullulan were added to PAA (22 wt% in water) solution under stirring conditions (600 rpm for 1 h) to achieve pullulan/PAA ratios of 1:3.78, 1:1.78, and 1:1.11. These formulations were designated as P25, P50, and P75, respectively, as reported in Table 1. Additionally, a solution of pure pullulan was labelled P100, and a solution solely consisting of PAA was identified as P0. Both P100 and P0 served as control specimens.

The surface of rectangular BOPP stripes (24×18 cm^2) underwent corona treatment (Arcotec, Ülm, Germany) for surface activation to bolster the adhesion between the substrate and the coatings. To optimise this process, a primer layer consisting of 0.5 wt% aziridine homopolymer in a water/ethanol solution was uniformly applied to the BOPP surface after corona treatment using an automatic film applicator (model 1137, Sheen Instruments, Kingston, UK). This applicator was equipped with a steel rod and an engraved pattern, which facilitated the deposition of the primer layer with a wet thickness of 4.0 μm. The subsequent evaporation of the solvent was achieved by directing a steady, perpendicular flow of mild air (25.0 ± 0.3 °C for 2 min) at a distance of 40 cm from the applicator. Following this, a secondary layer comprising different pullulan/poly (acrylic acid sodium salt) solutions was deposited using a rod, ensuring a consistent wet layer thickness of 4.0 μm. For the initial four formulations (P0 to P75), a singular coating layer was applied. By contrast, for the P100 formulation (with a dry substance content of 11%), the coating process was repeated to attain a final layer thickness comparable to the P0 formulation, which had a dry substance content of 22%.

Table 1. Composition and dry matter content of pullulan/PAA coating solutions.

Code	Pullulan (g)	PAA Solution (g)	Dry Matter of the Coating Solution (%)	Pullulan/PAA
P0	0	10	22	0:1
P25	0.55	9.45	26.29	1:3.78
P50	1.1	8.9	30.58	1:1.78
P75	1.65	8.35	34.87	1:1.113
P100	1.1	0	11	1:0

The application of both layers was conducted at a uniform speed of 2.5 mm s^{-1}, following ASTM D823-18(2022)-Practice C [13]. The coated films were subsequently stored under controlled conditions (25.0 ± 0.3 °C in a desiccator) for a duration of 15 days before any analytical evaluations. This step ensured a controlled and gradual removal of residual moisture while avoiding any potential moisture uptake from the surrounding environment before the following tests. The coated films were uniform and clear, with no visible defects or irregularities. This uniform appearance indicates the successful application and adhesion of the pullulan/PAA coating on the BOPP substrate.

2.3. Thickness Determination

To ascertain the thickness of the layers coating the plastic films, a systematic approach was employed. Initially, a sample measuring 10 × 10 cm^2 was sectioned, and its weight (M_1) was accurately determined. Subsequently, the coating was removed from the base film using hot water at a temperature of 80 °C, and the weight of the stripped base film was recorded as M_2. Adequate coating thickness was deduced using Equation (1):

$$l = \frac{M_1 - M_2}{\rho} \times 100 \tag{1}$$

where, M_1 represents the total mass per unit area of the plastic film inclusive of the coating (expressed in g/dm^2); M_2 signifies the mass per unit area of the plastic film alone (also in g/dm^2); ρ denotes the density of the aqueous solution (measured in g/cm^3). The thickness of the coating layer, l, in micrometers (μm), is determined by the known values of $M_1 - M_2$ [14].

2.4. Seal Strength Determination

Film strips measuring 2.5 cm × 10 cm^2 were heat-sealed using a Polikrimper TX/08 thermal heat sealer (Alipack, Pontecurone, Italy), equipped with smooth plates. The temperature of the sealing plates varied from 80 °C to 130 °C in 10 °C increments. A microprocessor consistently regulated the preset temperature of each bar during the experiment. The sealing pressure was maintained at 4.5 bar with a dwell time of 1 s. The seal strength of the heat-sealed samples was quantitatively assessed via the T-peel test, conforming to the standardised protocol delineated in ASTM F88-07a [15]. This evaluation was conducted employing a dynamometer (model Z005, Zwick Roell, Ulm, Germany), equipped with a 5 kN load cell and interfaced with two clamps spaced 10 cm apart. The test was executed at a crosshead speed of 300 mm/min. Critical parameters, such as the peak load (indicating the maximum force required to break the joints) and strain energy (represented by the area under the load (N)—deformation (mm) curve up to the point of breakage), were accurately computed and extracted using TestXpert V10.11 Master software (Zwick Roell, Ulm, Germany), which provided a quantitative measure of seal strength.

2.5. Contact Angle Measurements

Contact angle analysis was conducted using an advanced optical contact angle apparatus (OCA 15 Plus, Data Physics Instruments GmbH, Filderstadt, Germany) outfitted with a high-resolution CCD camera and a high-performance digitizing adapter. SCA20 software (version 1.0) facilitated the contact angle measurements. Rectangular specimens, each measuring 5×2 cm^2, were securely positioned and maintained in a flat orientation throughout the analysis using a specialised sample holder equipped with parallel clamping jaws.

The static contact angle of water in the air (θ, degrees) was meticulously determined by employing the sessile drop method. This process involved dispensing a droplet of Milli-Q water (18.3 MΩ cm) with a volume of 4.0 ± 0.5 µL onto the coated surface. The procedure followed was the so-called pick-up technique: A droplet, suspended from a needle, was gently positioned on the solid surface by elevating the sample stage until the solid/liquid interface was established. The measurements were conducted under controlled conditions at a temperature of 23 ± 1 °C and $50 \pm 2\%$ relative humidity (RH). To ensure uniform measurements, all droplets (ten per sample) were dispensed at a consistent height of 1 cm above the surface. The static contact angle was then recorded immediately following the droplet's deposition (time zero, t0) and after a duration of three seconds (t3). This angle was identified as the one formed between the drop baseline and the tangent at the drop's periphery.

2.6. Coefficient of Friction Measurements

The static and dynamic coefficient of friction (COF) values for the films were quantified in compliance with the ASTM D1894-14 standard using a dynamometer (model Z005, Zwick Roell, Ulm, Germany) equipped with a 100 N load cell [16]. The COF assessments, encompassing both coating-to-coating and coating-to-metal interactions, were conducted in a controlled laboratory environment, maintaining a stable temperature of 23 ± 2 °C and relative humidity (RH) of $50 \pm 2\%$. These values represent the mean of three individual measurements to ensure accuracy and repeatability.

2.7. Optical Properties

Haze measurements were performed across a wavelength spectrum of 780–380 nm, adhering to the guidelines of ASTM D1003-00 [17]. These measurements were facilitated by a UV–Vis high-performance spectrophotometer (Lambda 650, PerkinElmer, Waltham, MA, USA) coupled with a 150 mm integrated sphere, allowing diffuse transmitted light to be captured. Haze is quantified as the percentage of light that, upon transmission, deviates by an angle of more than 2.5° from the incident beam's direction. This parameter is of significant commercial relevance as it directly influences the contrast and visibility between objects viewed through the film. These values represent the mean of three individual measurements, ensuring statistical robustness and reliability of the data.

2.8. Statistical Analysis

Determining statistical significance among the mean values was carried out via one-way analysis of variance (ANOVA) using Statgraphics Plus 4.0 software. Where applicable, the mean values were compared using Student's *t*-test with a predefined significance threshold set at $p < 0.05$. This methodological approach ensured a rigorous statistical evaluation of the data and discerned significant differences in mean values across the groups.

3. Results and Discussion

3.1. Thickness Analysis

Accurate thickness analysis ensures uniformity and consistency in coatings, which are crucial for maintaining barrier performance, mechanical strength, and optical clarity. Moreover, precise thickness measurement is particularly important for determining haze, as variations can significantly impact the optical properties of the coatings. Table 2 outlines experimental thickness measurements derived using the methodology described in Section 2.3, exhibited deviations from the theoretical thickness (*TT*) values, and calculated based on the proportion:

$$4{:}100 = TT{:}DW \tag{2}$$

where, 4 μm symbolises the thickness of the coating applied with a steel horizontal rod to achieve a wet coating thickness of approximately 4 μm, and *DW* represents the dry weight content of the solution deposited on the substrate. The comparative analysis between nominal (theoretical) and actual (experimental) thickness values for each sample is presented below:

Table 2. Nominal and actual coating thicknesses for samples P0, P25, P50, P75, and P100.

Sample	Nominal Thickness (μm)	Actual Thickness (μm)
P0	0.88	$0.26^{a} \pm 0.03$
P25	1.05	$0.98^{bc} \pm 0.05$
P50	1.22	$1.15^{c} \pm 0.07$
P75	1.39	$1.37^{d} \pm 0.06$
P100	0.88	$0.88^{b} \pm 0.06$

Data for actual thickness are the mean of triplicate measurements ± SD values. Superscripts in this column indicate the results of a statistical comparison using a post hoc test. Samples with the same or overlapping superscript letters are not significantly different ($p > 0.05$).

According to the statistical analysis, significant differences in thickness were observed across various samples. However, no considerable variance was noted between the P25 and P50 formulations, and between the P25 and P100 formulations. Figure 1A graphically represents the divergence of actual thickness from theoretical values, indicating that experimentally obtained grammage values tend to align more closely with theoretical grammage values when the coatings have an increased pullulan concentration. This finding can be explained based on coating solutions' rheological properties. Similarly, Figure 1B reveals that the percentage difference between theoretical and experimental thickness diminishes and eventually nullifies with an increasing pullulan/acrylic lacquer ratio.

The trend between nominal and actual coating layer thickness has two plausible explanations. On the one hand, by increasing the pullulan concentration, the coating solution can leave the engraved pattern of the deposition rod easier than formulations based on higher amounts of acrylic, whose charged nature explains higher interactions with the metallic rod, leading to a lower amount of coating solution deposited on the plastic substrate. For practical purposes, this aspect must be considered, as it would require a different engraved pattern (i.e., able to deposit an extra amount of coating solution) or increasing the solids content of the coating solution to achieve the target grammage. On the other hand, the negative charges of the acrylic component engage in electrostatic interactions with the positively charged primer (aziridine), fostering an attraction that makes coating removal from the substrate difficult; consequently, the coating thickness is underestimated (see Equation (1)). By contrast, pullulan, a molecule devoid of electric charge, forms weaker interactions with the substrate, easing its removal. This characteristic is particularly evident in the near-complete detachment of the pure pullulan coating (P100), where the experimental grammage values closely match the theoretical ones.

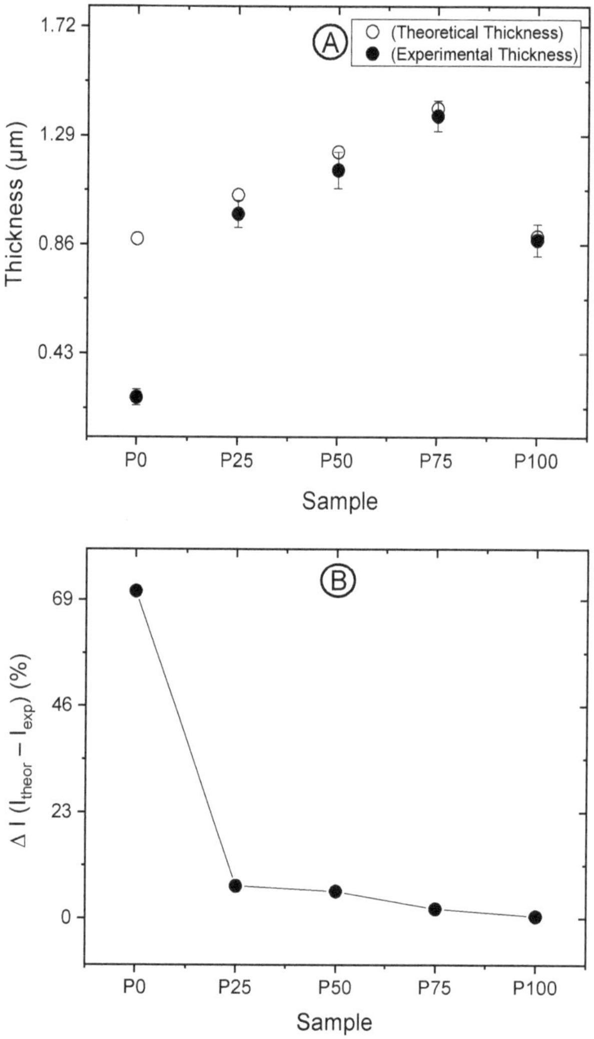

Figure 1. (**A**) Comparison between theoretical and experimental coating thicknesses for samples P0, P25, P50, P75, and P100. (**B**) Percentage variation between theoretical and experimental coating thicknesses for the different samples.

3.2. Sealing Analysis

As illustrated in Figure 2, the P100 formulation exhibits a markedly lower maximum sealing force than other formulations. This observation underscores the fact that pullulan, in its pure form and devoid of additives, lacks inherent sealing properties—a characteristic that remains unchanged even under elevated temperatures. Despite these thermal conditions, the sealing performance of the P100 formulation cannot surpass even the least effective performance of the P0 formulation, highlighting the pivotal role of the acrylic component in conferring sealability.

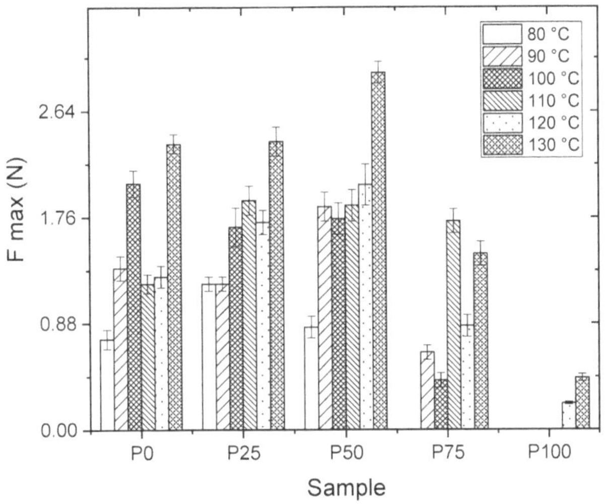

Figure 2. Seal strength comparison at various temperatures for samples P0, P25, P50, P75, and P100.

Sealing is a macroscopic phenomenon where molecular dynamics play a crucial role. At the glass transition temperature (T_g), there is a significant increase in molecules' mobility, marking the transition from a glassy, rigid state to a more flexible state. This mobility enhancement, as temperatures escalate towards the melting temperature (T_m), allows molecules to blend, diffuse and, upon cooling, solidify into a new position, facilitating the formation of a stable seal. The interactions between ionic charges help maintain this stability, especially in materials where ionic bonds contribute to the structure. While repulsive forces between like charges exist, their impact is often overshadowed by cohesive forces that drive the sealing process. This balance of forces ensures the material's integrity and functionality at the macroscopic level [12,18].

Furthermore, Figure 2 delineates a general pattern where higher seal strength is achieved with coatings subjected to a sealing temperature of 130 °C, implying that sealing properties are enhanced at elevated temperatures. The same graph also indicates that optimal performance (~2.7 N/2.5 cm) is attributed to the P50 formulation, which aligns with the highest sealing force values. Notably, up to a temperature of 110 °C, no statistically significant differences ($p < 0.05$) are discernible between this formulation and the others, suggesting a nuanced interaction between temperature, formulation composition, and the resulting sealing efficacy. The enhanced seal strength indicates adequate adhesion between the coating and the BOPP substrate mediated by the presence of aziridine homopolymer. Additionally, uncoated BOPP seal strength was measured to be ~5.5 N/2.5 cm, providing a reference for comparison. This finding indicates that pullulan/PAA coatings still demonstrate adequate sealing performance despite a lower seal strength than uncoated BOPP.

3.3. Contact Angle Analysis

The data presented in Table 3 elucidate specific trends and relationships between contact angle measurements at the initial time (t = 0 s) and after three seconds (t = 3 s). The deposition of the coating yielded a decrease in the water contact angle of the plastic substrate (BOPP), which is approximately 80° [19]. The significant reduction in water contact angles after coating deposition indicates enhanced wettability and suggests an even distribution of the coating on the BOPP substrate.

Table 3. Contact angle measurements at the initial time (t = 0 s) and after three seconds (t = 3 s) for various pullulan concentrations in the coating formulations.

Sample	Contact Angle (°)	
	t = 0 s	t = 3 s
P0	$61.19^{a} \pm 1.79$	$58.83^{d} \pm 1.43$
P25	$59.00^{a} \pm 1.89$	$55.17^{e} \pm 2.13$
P50	$56.35^{b} \pm 1.05$	$53.30^{f} \pm 0.95$
P75	$60.87^{a} \pm 1.04$	$37.62^{g} \pm 1.89$
P100	$44.25^{c} \pm 5.68$	$16.81^{h} \pm 1.24$

Data are the mean of triplicate measurements ± SD. Different letters as superscripts in each column represent significant differences between means ($p < 0.05$) using a post hoc test.

An inverse correlation between the pullulan concentration in the coating and the contact angle is evident, notably at t = 3 s, as proven by statistically significant differences ($p < 0.05$) among the various formulations. Additionally, the contact angle at t = 0 s invariably exceeded t = 3 s. The correlation mentioned above can be rationalised by investigating the chemical composition of the coating constituents, which can shed light on why pullulan and polyacrylic acid behave differently when in contact with water.

Pullulan, characterised by an abundance of hydroxyl (OH) groups along its chain, exhibits pronounced hydrophilicity. These OH groups engage in hydrogen bonding with water droplets on the coating, promoting uniform distribution and resulting in reduced contact angle values. Conversely, polyacrylic acid, consisting primarily of a non-polar hydrocarbon chain, exhibits a predominantly hydrophobic nature, except for its polar carboxylate groups. This configuration results in water droplets minimizing their surface area when in contact with a prevalently hydrophobic coating. They adopt a more spherical geometry and, consequently, higher contact angle values.

Figure 3 presents a detailed visualization of how a droplet's geometry evolves over time. This evolution, marked by a gradual reduction in the contact angle, is primarily influenced by two fundamental interactions at the interface of the solid surface and liquid: absorption and spreading [11]. The morphology of the droplet may theoretically be influenced by additional factors, including the evaporation of the liquid from the droplet and the physical expansion of the substrate due to moisture absorption. However, their effects were deemed negligible in this specific analysis due to a lack of experimental confirmation under the conditions being studied.

Figure 3. Water contact angle of (**a**) polypropylene film neutral (no surface treatment); (**b**) anti-fog coating (formulation P100) immediately after depositing the water droplet; (**c**) anti-fog coating (formulation P100) after ~10 s of depositing the water droplet.

The absorption process is intricately linked to the physical structure of the substrate, notably the presence of microscopic channels or capillaries that enable liquid droplets to permeate the substrate. This infiltration leads to a notable decrease in the droplet's overall volume and the area of its base, which is in direct contact with the substrate. Spreading, however, describes the dynamic process by which the droplet extends outward across the substrate surface. This spreading is significantly influenced by the physical characteristics of the substrate, such as its smoothness and chemical affinity for water, resulting in the droplet adopting a more flattened shape. This change increases the contact area of the droplet's base with the substrate but does not affect its volume. Notably, on pullulan-based substrates, a biopolymer known for its high hydrophilicity, spreading occurs at a markedly faster rate than absorption [11]. Given the negligible role of absorption in this context, spreading emerges as the sole mechanism driving change at the contact angle, underscored by pullulan's exceptional water-attracting capabilities and the substrate surface's specific morphological features designed to facilitate this interaction.

3.4. Coefficient of Friction (COF)

The COF analysis has practical relevance for assessing a material's suitability to run smoothly on packaging equipment during converting and packaging operations. Too high COF values can determine the blocking effects of the reels, eventually leading to unwanted ruptures of the webs and severely impacting the overall throughput of the process.

The COF results obtained in this work revealed insightful patterns in the interaction between coatings and metal surfaces. The COF values for coating/metal interfaces were higher in formulations containing the acrylic component than in the P100 formulation, which solely consists of pullulan. According to the statistical evaluation presented in Table 4, COF values on metal substrates did not exhibit statistically significant differences among acrylic-containing formulations, except for dynamic coating/metal COF measurements. Notably, the P100 formulation demonstrated significantly lower static and dynamic COF values than its counterparts.

Table 4. Coefficient of friction (COF) measurements for coating/metal and coating/coating interfaces at static and dynamic states across various formulations.

Sample	COF (Coating/Metal)		COF (Coating/Coating)	
	Static	Dynamic	Static	Dynamic
P0	$0.45^{a} \pm 0.03$	$0.31^{cd} \pm 0.04$	$0.71^{e} \pm 0.02$	$0.61^{g} \pm 0.03$
P25	$0.45^{a} \pm 0.01$	$0.33^{d} \pm 0.02$	$0.68^{e} \pm 0.02$	$0.61^{g} \pm 0.03$
P50	$0.46^{a} \pm 0.02$	$0.32^{cd} \pm 0.02$	$0.72^{e} \pm 0.06$	$0.62^{g} \pm 0.01$
P75	$0.45^{a} \pm 0.02$	$0.35^{d} \pm 0.01$	$0.77^{e} \pm 0.03$	$0.62^{g} \pm 0.02$
P100	$0.39^{b} \pm 0.01$	$0.28^{c} \pm 0.01$	$1.59^{f} \pm 0.68$	$0.72^{h} \pm 0.04$

Data are the mean of triplicate measurements ± SD. Different letters as superscripts in each column represent significant differences between means ($p < 0.05$) using a post hoc test.

The presence of the acrylic component markedly influenced COF values, which can be attributed to the interaction between negative charges of the polyacrylic acid, which somehow promote interactions with the metallic surface. These interactions restrict the movement of the sled covered with the coated film that slides on the metal surface. By contrast, pullulan, being an uncharged molecule, does not form such interactions with the metal surface, resulting in lower COF values.

A contrasting trend is observed when comparing coating/metal and coating/coating COF values. The P100 formulation exhibits significantly higher COF values in coating/coating interactions than acrylic-containing formulations. This difference is due to the unique characteristics of acrylic-based formulations, where the presence of negatively charged particles creates a repelling effect, leading to lower friction between the coating layers. This repelling action allows acrylic formulations to have lower COF values. On the other hand, the P100 formulation, comprised solely of pullulan, lacks such repelling forces and has a smoother surface. This lack of repulsion, combined with a smoother surface, increases resistance when coatings touch each other, raising COF values. This contrast highlights the significant impact of acrylic's negative charge and the coatings' physical qualities on the friction observed in coating/coating interactions.

3.5. Optical Properties

The data presented in Table 5 indicate no notable statistical differences in the haze values among the different formulations, except for the P75 and P100 formulations. Additionally, all measured values are below the 3% threshold, a value widely recognised for plastic materials in food packaging to ensure sufficient transparency for visual inspection of the contents [20]. The normalised haze values presented in Figure 4 correspond to an assumed coating thickness of 1 μm based on the premise that haze variation is directly proportional to thickness. The calculation of opacity values for a 1 μm thickness was executed using the following proportions:

$$\text{Haze}_s\text{:Thickness}_s = \text{Haze}_n\text{:1 μm} \tag{3}$$

Table 5. Haze percentage measurements for samples P0, P25, P50, P75, and P100.

Sample	Haze (%)
P0	1.15 ^{ab} ± 0.14
P25	1.92 ^{ab} ± 0.96
P50	1.88 ^{ab} ± 0.09
P75	2.40 ^{b} ± 1.56
P100	0.82 ^{a} ± 0.12

Data are mean of triplicate measurements ± SD. Different letters as superscripts in each column represent significant differences between means ($p < 0.05$) using a post hoc test.

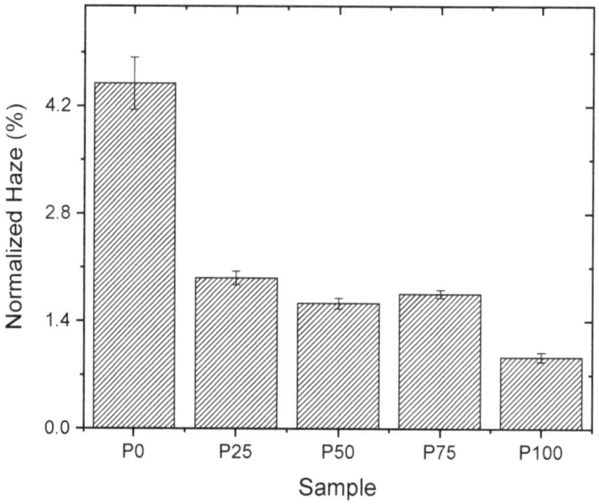

Figure 4. Normalised haze percentages at 1 μm coating thickness for samples P0, P25, P50, P75, and P100.

Here, $Haze_s$ signifies the experimentally determined opacity value (%), $Thickness_s$ denotes the coating's thickness as measured by the weight method (μm), and $Haze_n$ represents the normalised opacity value for a 1 μm thick coating (%). Consequently, the formula to deduce the normalised haze is articulated as follows:

$$Haze_n = \frac{(Haze_s \times 1)}{Thickness_s} \tag{4}$$

Figure 4 reveals a direct correlation between the increase in haze values and the augmented acrylic content in the coating. This trend is potentially attributable to the emergence of scattering centres within the polyacrylic acid formed by the clustering of acrylic molecules, which scatter the incident light and increase opacity. The P75 formulation exhibits peculiar behaviour due to its composition, featuring less acrylic content but higher pullulan content, rendering it more resistant to dissolution. Therefore, it is reasonable to infer that the formation of partially solubilised pullulan aggregates, observable in optical analyses, contributes to heightened opacity and serves as scattering centres. For reference, the haze of uncoated OPP is measured at 0.71 ± 0.05%. This baseline measurement demonstrates that while the addition of pullulan/PAA coatings increase haze, coatings still maintain transparency levels acceptable for food packaging applications.

4. Conclusions

This study introduces a pioneering approach to solving the problem of fog formation on food packaging films by combining pullulan, a naturally derived biopolymer, with poly (acrylic acid sodium salt) to forge a novel anti-fog coating. The research demonstrated that this unique combination not only mitigates the fogging phenomenon by leveraging pullulan's inherent hydrophilicity but also elevates the seal strength critical for maintaining food freshness, as evidenced by the enhanced sealing properties observed at varying temperatures. Notably, the investigation highlighted the formulation's superior anti-fogging performance, which was quantitatively affirmed through reduced contact angles, indicating more efficient moisture management on the packaging surface. Additionally, including poly (acrylic acid sodium salt) introduced negative charges that improved the coating's adhesion and sealing capabilities, a notable advancement over conventional materials. Optical properties analysis further validated the formulation's efficacy and maintained transparency within acceptable limits, thus ensuring product visibility without compromising sustainability. This research not only advances sustainable biopolymer-based packaging but also highlights the importance of innovation in solving the food industry's challenges, setting a foundation for future advancements in eco-friendly and functional packaging solutions. Future work will include comprehensive anti-fogging tests under simulated conditions and on real food products to thoroughly validate the coating's effectiveness. The effectiveness and clarity of this research will enhance its commercial applicability to food packaging.

Author Contributions: Methodology, formal analysis, validation, writing—original draft, writing—review and editing, M.G.; Data curation, writing—review and editing, M.S.; Data curation, writing—review and editing, D.C.; Conceptualization, resources, data curation, writing—original draft, writing—review and editing, supervision, S.F. All authors have read and agreed to the published version of the manuscript.

Funding: This research received no external funding.

Institutional Review Board Statement: Not applicable.

Data Availability Statement: Data are contained within the article.

Conflicts of Interest: The authors declare no conflicts of interest.

References

1. Dirpan, A.; Ainani, A.F.; Djalal, M. A Review on Biopolymer-Based Biodegradable Film for Food Packaging: Trends over the Last Decade and Future Research. *Polymers* **2023**, *15*, 2781. [CrossRef] [PubMed]
2. Ncube, L.K.; Ude, A.U.; Ogunmuyiwa, E.N.; Zulkifli, R.; Beas, I.N. Environmental Impact of Food Packaging Materials: A Review of Contemporary Development from Conventional Plastics to Polylactic Acid Based Materials. *Materials* **2020**, *13*, 4994. [CrossRef] [PubMed]
3. Matheus, J.R.V.; Dalsasso, R.R.; Rebelatto, E.A.; Andrade, K.S.; de Andrade, L.M.; de Andrade, C.J.; Monteiro, A.R.; Fai, A.E.C. Biopolymers as Green-Based Food Packaging Materials: A Focus on Modified and Unmodified Starch-Based Films. *Compr. Rev. Food Sci. Food Saf.* **2023**, *22*, 1148–1183. [CrossRef] [PubMed]
4. Grosu, G.; Andrzejewski, L.; Veilleux, G.; Ross, G.G. Relation between the Size of Fog Droplets and Their Contact Angles with CR39 Surfaces. *J. Phys. D Appl. Phys.* **2004**, *37*, 3350. [CrossRef]
5. Min, T.; Zhu, Z.; Sun, X.; Yuan, Z.; Zha, J.; Wen, Y. Highly Efficient Antifogging and Antibacterial Food Packaging Film Fabricated by Novel Quaternary Ammonium Chitosan Composite. *Food Chem.* **2020**, *308*, 125682. [CrossRef] [PubMed]
6. Agrawal, S.; Budhwani, D.; Gurjar, P.; Telange, D.; Lambole, V. Pullulan Based Derivatives: Synthesis, Enhanced Physicochemical Properties, and Applications. *Drug Deliv.* **2022**, *29*, 3328–3339. [CrossRef] [PubMed]

7. Rashid, A.; Qayum, A.; Liang, Q.; Kang, L.; Ekumah, J.N.; Han, X.; Ren, X.; Ma, H. Exploring the Potential of Pullulan-Based Films and Coatings for Effective Food Preservation: A Comprehensive Analysis of Properties, Activation Strategies and Applications. *Int. J. Biol. Macromol.* **2024**, *260*, 129479. [CrossRef] [PubMed]
8. Kumar, N.; Petkoska, A.T.; AL-Hilifi, S.A.; Fawole, O.A. Effect of Chitosan–Pullulan Composite Edible Coating Functionalized with Pomegranate Peel Extract on the Shelf Life of Mango (Mangifera Indica). *Coatings* **2021**, *11*, 764. [CrossRef]
9. Uysal Unalan, I.; Boyaci, D.; Ghaani, M.; Trabattoni, S.; Farris, S. Graphene Oxide Bionanocomposite Coatings with High Oxygen Barrier Properties. *Nanomaterials* **2016**, *6*, 244–254. [CrossRef] [PubMed]
10. Ding, Z.; Chang, X.; Fu, X.; Kong, H.; Yu, Y.; Xu, H.; Shan, Y.; Ding, S. Fabrication and Characterization of Pullulan-Based Composite Films Incorporated with Bacterial Cellulose and Ferulic Acid. *Int. J. Biol. Macromol.* **2022**, *219*, 121–137. [CrossRef] [PubMed]
11. Farris, S.; Introzzi, L.; Biagioni, P.; Holz, T.; Schiraldi, A.; Piergiovanni, L. Wetting of Biopolymer Coatings: Contact Angles Kinetics and Image Analysis Investigation. *Langmuir* **2011**, *27*, 7563–7574. [CrossRef] [PubMed]
12. Bamps, B.; Buntinx, M.; Peeters, R. Seal materials in flexible plastic food packaging: A review. *Packag. Technol. Sci.* **2023**, *36*, 507–532. [CrossRef]
13. *ASTM D823-18(2022)*; Standard Practices for Producing Films of Uniform Thickness of Paint, Coatings and Related Products on Test Panels. ASTM International: West Conshohocken, PA, USA, 2022.
14. Brown, N. *Plastics in Food Patkaging: Properties: Design and Fabrication*; Dekker: New York, NY, USA, 1992; pp. 200–202.
15. *ASTM F88-07a*; Standard Test Method for Seal Strength of Flexible Barrier Materials. ASTM International: West Conshohocken, PA, USA, 2010.
16. *ASTM D1894-14*; Standard Test Method for Static and Kinetic Coefficients of Friction of Plastic Film and Sheeting. ASTM International: West Conshohocken, PA, USA, 2023.
17. *ASTM D1003-00*; Standard Test Method for Haze and Luminous Transmittance of Transparent Plastics. ASTM International: West Conshohocken, PA, USA, 2010.
18. Ilhan, I.; Turan, D.; Gibson, I.; Klooster, R. Understanding the factors affecting the seal integrity in heat sealed flexible food packages: A review. *Packag. Technol. Sci.* **2021**, *34*, 321–337. [CrossRef]
19. Cozzolino, C.A.; Campanella, G.; Türe, H.; Olsson, R.T.; Farris, S. Microfibrillated Cellulose and Borax as Mechanical, O_2-Barrier, and Surface-Modulating Agents of Pullulan Biocomposite Coatings on BOPP. *Carbohydr. Polym.* **2016**, *143*, 179–187. [CrossRef] [PubMed]
20. Uysal Unalan, I.; Boyaci, D.; Trabattoni, S.; Tavazzi, S.; Farris, S. Transparent Pullulan/Mica Nanocomposite Coatings with Outstanding Oxygen Barrier Properties. *Nanomaterials* **2017**, *7*, 281. [CrossRef] [PubMed]

polymers

Article

Generation of Microplastics from Biodegradable Packaging Films Based on PLA, PBS and Their Blend in Freshwater and Seawater

Annalisa Apicella, Konstantin V. Malafeev, Paola Scarfato * and Loredana Incarnato

Department of Industrial Engineering, University of Salerno, Via Giovanni Paolo II n. 132, 84084 Fisciano, SA, Italy; anapicella@unisa.it (A.A.); kmalafeev@unisa.it (K.V.M.); lincarnato@unisa.it (L.I.)
* Correspondence: pscarfato@unisa.it

Abstract: Biodegradable polymers and their blends have been advised as an eco-sustainable solution; however, the generation of microplastics (MPs) from their degradation in aquatic environments is still not fully grasped. In this study, we investigated the formation of bio-microplastics (BMPs) and the changes in the physicochemical properties of blown packaging films based on polylactic acid (PLA), polybutylene succinate (PBS) and a PBS/PLA 70/30 wt% blend after degradation in different aquatic media. The tests were carried out in two temperature/light conditions to simulate degradation in either warm water, under sunlight exposure (named Warm and Light—W&L), and cold deep water (named Cold and Dark—C&D). The pH changes in the aqueous environments were evaluated, while the formed BMPs were analyzed for their size and shape alongside with variations in polymer crystallinity, surface and mechanical properties. In W&L conditions, for all the films, the hydrolytic degradation led to the reorganization of the polymer crystalline phases, strong embrittlement and an increase in hydrophilicity. The PBS/PLA 70/30 blend exhibited increased resistance to degradation with respect to the neat PLA and PBS films. In C&D conditions, no microparticles were observed up to 12 weeks of degradation.

Keywords: biodegradable microplastics; PBS/PLA blend; degradation; fresh water; seawater

Citation: Apicella, A.; Malafeev, K.V.; Scarfato, P.; Incarnato, L. Generation of Microplastics from Biodegradable Packaging Films Based on PLA, PBS and Their Blend in Freshwater and Seawater. *Polymers* **2024**, *16*, 2268. https://doi.org/10.3390/polym16162268

Academic Editors: Masoud Ghaani and Stefano Farris

Received: 18 July 2024
Revised: 2 August 2024
Accepted: 6 August 2024
Published: 10 August 2024

1. Introduction

The environmental pollution generated by accumulation of synthetic plastic wastes and debris in terrestrial and aquatic ecosystems has raised numerous concerns in the last years and has recently been the subject of numerous studies [1–5].

The main source of plastic waste derives from packaging materials, including flexible films and rigid containers. In fact, the plastic packaging market is the largest one for plastic consumption worldwide [6–9]. In particular, plastics represent the largest, most harmful and most persistent fraction of aquatic litter: it is estimated that every day, the equivalent of more than 2000 plastic-filled garbage trucks are dumped into our oceans, rivers and lakes [7,10].

Currently, the replacement of fossil fuel-based plastics by biodegradable polymers, such as polylactic acid (PLA), polybutylene succinate (PBS), polyhydroxyalkanoates (PHA), poly (butylene adipate-co-terephthalate) (PBAT), cellulose acetate, polycaprolactone (PCL) and poly(vinyl alcohol) (PVOH) has been encouraged as a possible solution to reduce environmental issues [11–13]. Today, they are increasingly used as disposable packaging, mainly in the form of films [14–16]. However, unlike conventional fossil-derived polymers, biodegradable polymers often lack functional performance suitable for large-scale applications, such as in food packaging [17–21]. To improve their functional properties and expand their commercial uptake in the packaging field, biodegradable polymers are often mixed with the appropriate additives (i.e., active and smart agents, inorganic and organic nanoparticles, processing aids, metals, biomolecules from food wastes valorization, etc.) [22–29] or blended with other biodegradable polymers [30–34].

Proper management of post-consumer biodegradable polymers and their blends involves disposal through industrial composting sites, which break them down into low-molecular-weight compounds (water, methane, carbon dioxide) in a relatively short time under specific biotic and abiotic conditions (temperature, humidity, pH, solar radiation, bio-surfactants, presence of microorganisms and enzymes) [35–37]. The degradation rate of biodegradable polymers also depends on intrinsic material factors, such as number and type of functional groups, molecular weight, crystallinity degree, shape and size [38,39]. When biodegradable polymers enter the natural environment, their behavior becomes like that of non-biodegradable materials: this entails long-term decomposition through photodegradation, biodegradation and hydrolysis mechanisms, and generation of large amounts of bio-microplastics (BioMPs, $\leq$5 mm) which can be persistent and remain for decades, posing a real ecological pollution risk [35,40–42]. BioMPs can influence plant development, change the soil microbiota and affect the antioxidant system of fish and shellfish [43–47]. For biodegradable polymer blends, factors such as phase compatibility, interfacial interactions, as well as changes in morphology, wettability, mechanical properties, and surface charge induced by blending may result in an increased or decreased resistance to degradation and formation of BMPs [30,48,49]. As a matter of fact, the degradation behavior of a polymer blend may differ substantially from the degradation pathways of the pure components, as interactions between the different species in the blends and between the degradation products can occur during degradation. For this reason, the additive rule cannot often be applied in the case of degradation of polymer blends and, therefore, it is hard to foresee the degradative behavior of a polymer blend based on the properties of the pure components [50].

To our best knowledge, almost no work in the literature investigated the degradation of biodegradable polymer blends and the formation of BMPs from biodegradable packaging films in natural ecosystems, especially in aquatic ones. Among the studies on biodegradable films, previous literature focused on the degradation of neat films made by poly(ε-caprolactone) in a buffer–enzymatic solution [35,51,52], or on the thermal degradation of PLA-based blends [53,54]. Among biodegradable blends, only Zhao et al. addressed the photodegradation of a commercial blend based on polylactic acid/poly(butylene adipate-co-terephthalate)/thermoplastic starch (PLA/PBAT/TPS) in seawater; however, they prepared pristine microparticles of the blend by mechanical grinding, eventually subjected to UV irradiation in seawater [48].

In this scenario, the article aims at exploring the potential formation of microplastics from biodegradable packaging films based on PLA, PBS and a PBS/PLA 70/30wt% blend in different aquatic environments, and to understand the underlying mechanisms. PLA and PBS were chosen as they are among the most widely used biopolymers for flexible food packaging, thanks to their good processability and chemical–physical and functional properties, which are further enhanced by blending. The tests were carried out in four aquatic media, namely natural seawater, fresh water, sterilized seawater and sterilized fresh water, in the attempt to discriminate the contribution of water composition and microbial content on the degradation and BMPs formation. Two different conditions of temperature and light exposure were established during the tests: in the first condition, the films were kept at a temperature equal to 28 °C and exposed to a LED plant culture lamp for 14 h/day in order to simulate the ageing of plastic items immersed in water and exposed to sunlight; in the second condition, the films were kept at a temperature of 4 °C in darkness to simulate a deep-water environment. The tests were carried out up to 12 weeks. The pH changes in the aqueous environments were evaluated, while the morphology, size, wettability, crystallinity and mechanical properties of the formed microplastics were studied, analyzing the changes induced by polymer blending in the degradation behavior with respect to the neat polymers.

2. Materials and Methods

2.1. Materials, Water Collection and Analysis

PLA 4032D (semicrystalline, D-isomer content = 1.5 wt%, Mw ~155,000 g/mol, density = 1.24 g/cm^3, Tm = 155–170 °C) was supplied by NatureWorks (Minnesota, MN, USA).

BioPBS FZ91PM (MFR (190° C, 2.16 kg) = 5 g/10 min, density = 1.26 g/cm^3, Tm = 115 °C) was supplied by Mitsubishi Chemical Co. (Tokyo, Japan). Both the biodegradable polymers are suitable for food contact in conformity with regulations established by the U.S. Food and Drug Administration (FDA) and the European Union (EU). They also comply with EN13432 and ASTM D 6400 standards regarding compostability under controlled composting conditions.

Tap water from the system of the University of Salerno was used as freshwater (FW). Seawater (SW) was collected from the Tyrrhenian Sea in Vietri sul Mare, province of Salerno, Italy. Their main chemical–physical parameters were measured and are reported in Table S1. pH and conductivity measurements were conducted using the GPL 21 benchtop pH meter (Crison Instruments, Barcelona, Spain) equipped with a Sension+ 5010T pH electrode (Hach Lange, Düsseldorf, Germany) and an Orion Star™ A212 benchtop conductivity meter (Thermo Scientific, Waltham, Massachusetts, USA), respectively. Alkalinity, hardness and chlorides were determined by the Lovibond® MD610 benchtop photometer (Tintometer GmbH, Dortmund, Germany) equipped with reagent drop test kits. Total dissolved solids were calculated according to Rusydi [55]. Sterilized fresh water (SFW) and sterilized seawater (SSW) were obtained by autoclave sterilization at 121 °C for 20 min. All other chemicals used were of analytical grade and supplied by Sigma-Aldrich (Milan, Italy).

2.2. Blend and Films Preparation

Before processing, PLA and PBS pellets were vacuum dried at 70 °C for 14 h. The PBS and PLA4032D blend, with mass ratio 70/30 by weight, was prepared in a co-rotating twin screw extruder (Collin ZK25, with screw diameter equal to 25 mm and L/D = 42), with a screw speed equal to 150 rpm and a temperature profile from 150 °C to 190 °C from the hopper to the die. Monolayer films consisting of neat PLA, neat PBS and PBS/PLA blend were made by a lab-scale film blowing plant equipped by a single screw extruder (GIMAC, D = 12 mm, L/D = 24), setting the same temperature profile as described above and the blow-up ratio at 2.5. The collection speed was fixed at 3 m/min, yielding samples with an average thickness of 50 ± 2 μm.

2.3. Degradation Experiments

Degradation tests were carried out on 12.7 mm × 150 mm film cut strips that were placed in Greiner T-75 cell culture flasks with a canted neck, a ventilated cap, a cross-section surface area of 75 cm^2 and a capacity of 250 mL, as shown in Figure 1a. Each flask was filled with 12 film strips and 100 mL of aquatic medium, either fresh water (FW), fresh sterilized water (SFW), seawater (SW), or sterilized seawater (SSW).

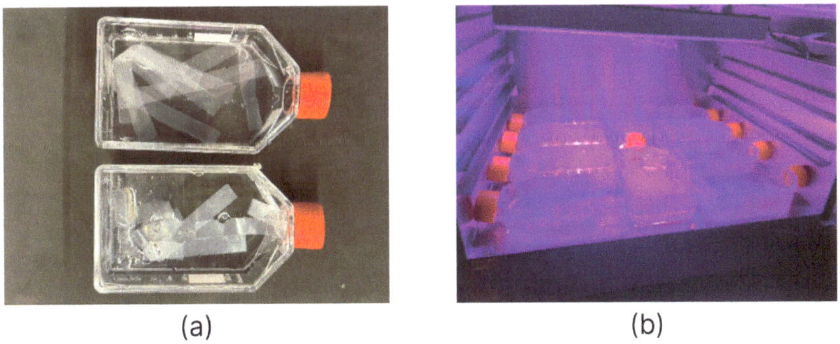

Figure 1. Experimental conditions: (**a**) samples in the flask; (**b**) flasks in the climate chamber.

Two different conditions of temperature and light exposure were established during the tests. In the first one (named Warm&Light—W&L), the flasks were kept in a conditioned chamber at 50% RH and 28 ± 1 °C under LED plant culture lamps (MOSOTON MT-102 25 W LED panel, Mosoton, Guangzhou, China) with 14 h of light and 10 h of darkness each day (Figure 1b) in order to simulate the ageing of the plastic items immersed in water and exposed to sunlight. The flasks were shaken twice a week to improve oxygen supply in the medium and exchanged positions to ensure homogeneity of light exposure for all samples. In the second condition (named Cold&Dark—C&D), the flasks were kept in a refrigerator for 12 weeks at 4 ± 1 °C, and culture flasks were wrapped in aluminum foil to avoid light detection, in order to simulate a deep-water environment.

The tests were conducted for up to 12 weeks, and the samples were analyzed at time 0 and after 5, 7 and 12 weeks of storage for the W&L condition, and after 12 weeks for the C&D condition. At each time interval, three replicate flasks for each type of biodegradable film and degradation condition were withdrawn. Then, the exposed film strips were removed, washed in an ultrasonic bath for 5 min with Tergazyme detergent at 3% in warm water to remove the biofilm, rinsed with distilled water, patted dry with a lint-free cloth and equilibrated in a climate chamber at controlled conditions (23 °C and 50% RH) for 24 h before the analyses. The remaining culture flasks with water were stored in a refrigerator at a temperature of 4 ± 1 °C before further tests.

2.4. Characterizations

pH changes occurring in the different aqueous media during degradation were measured using a GPL 21 benchtop pH meter (Crison Instruments, Barcelona, Spain) equipped with a Sension+ 5010T pH electrode (Hach Lange, Düsseldorf, Germany), with pH accuracy equal to 0.02.

The shape and average size of the formed microplastics were analyzed using an optical microscope (Zeiss Axioskop 40, Carl Zeiss, Oberkochen, Germany) equipped with a Axiocam 208 color camera. The analyses were performed on droplets (V = 0.5 mL) of the different aqueous media, taken from the flasks, pipetted on microscope slides and dried at room temperature for 1 day in a vacuum oven.

Thermal analyses of the samples were carried out by a differential scanning calorimeter DSC PT 1000 (Linseis, Selb, Germany). The heating/cooling thermal cycle was set at a rate of 10 °C/min in the range from 25 to 200 °C under a nitrogen gas flow (20 mL/min). The crystallinity degree of each component *i* (i.e., PBS and PLA) of the films was calculated according to the following equation:

$$X_{c_i} = (\Delta H_{m_i} - \Delta H_{cc_i})/(\phi_i \times \Delta H^0_{m_i}) \quad (1)$$

where ΔH_{m_i} and ΔH_{cc_i} are the melting and the cold crystallization enthalpies of component *i* in the film, respectively; $\Delta H^0_{m_i}$ is the melting enthalpy of component *i* 100% crystalline (equal to 110.3 J/g for PBS and 93.7 J/g for PLA); and ϕ_i is the weight fraction of component *i* in the film.

The film strips were submitted to tensile testing according to the ASTM D 882-91 procedure, using a Sans CMT6000 dynamometer (Shenzhen, China) equipped with a 100 N load cell. Tensile tests were carried out at 23 °C and 50% RH, setting the crossbar speed at 3 mm/min to evaluate the elastic modulus, and at 300 mm/min to evaluate strength and elongation at break. The results were expressed as the average of at least seven measurements taken for each type of film and water medium.

Static contact angle measurements using the sessile drop method were recorded and analyzed at room temperature using an FTA 1000 analyzer (First Ten Angstroms, Inc., Portsmouth, VA, USA) according to ASTM D 7490. A 2 ± 0.5 µL drop of distilled water was applied to the sample surface using a syringe. The drop's image was captured by a video camera immediately after deposition, and the contact angle was calculated from the drop's shape using proprietary image analysis software. The reported contact angles are the average of at least ten replicate measurements.

3. Results and Discussion

3.1. pH Changes of the Aquatic Media

The pH changes of the aquatic media with time, both in W&L and C&D conditions, were monitored as an indirect indication of the film degradation progress in the different aqueous environments. It is known, in fact, that the hydrolysis of the ester bonds of aliphatic polyesters decreases the molecular weight because of molecular chain fracture and leads to the formation of short-chain monomers and oligomers with a weakly acidic character [56]. The results of the measurements are reported in Figure 2 and Table 1.

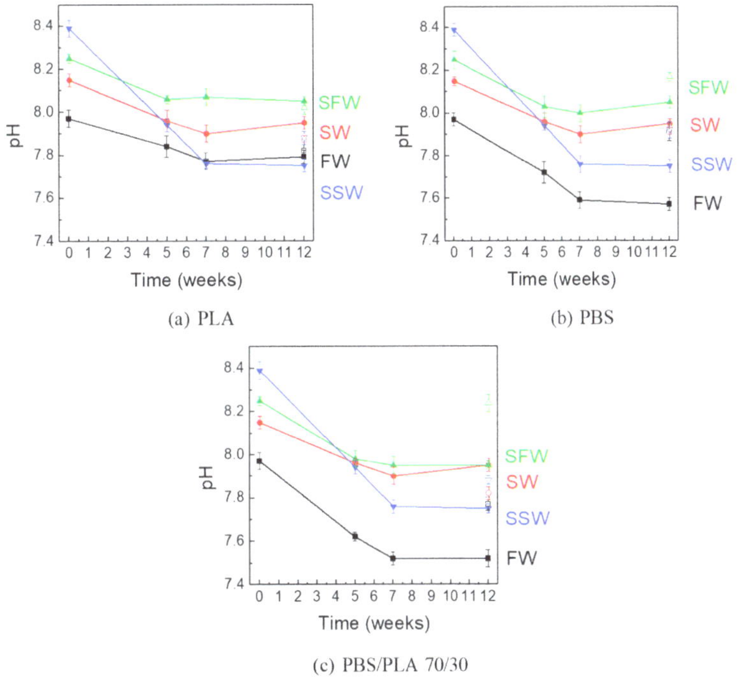

Figure 2. Measured pH values in the aqueous media (FW, SW, SFW and SSW) in the presence of films made of (**a**) PLA, (**b**) PBS and (**c**) PBS/PLA 70/30 (full symbols: Warm and Light conditions; empty symbols: Cold and Dark conditions).

Table 1. Differences between initial pH and pH after 12 weeks of the four aqueous media: fresh water (FW), seawater (SW), sterilized fresh water (SFW) and sterilized seawater (SSW), without film (control) and with PLA, PBS and PBS/PLA 70/30 films, stored under Warm and Light (W&L) and Cold and Dark (C&D) conditions.

Samples	Conditions	ΔpH ($pH_{week\ 0}$–$pH_{week\ 12}$)			
		FW	SFW	SW	SSW
Control	W&L	0.03	0.01	0.04	0.02
	C&D	0.02	0.01	0.03	0.02
PLA	W&L	0.18	0.20	0.20	0.65
	C&D	0.15	0.23	0.28	0.52
PBS	W&L	0.40	0.20	0.30	0.70
	C&D	0.05	0.08	0.22	0.48
PBS/PLA 70/30	W&L	0.45	0.30	0.28	0.70
	C&D	0.20	0.15	0.33	0.51

At the beginning, all aqueous media had weak alkaline pH values. The chemical potential of water plays a significant role in the hydrolysis of polyesters: in fact, hydrolysis in alkaline conditions is faster than in acidic ones [57]. FW initially had a pH of 7.97, while SW showed a pH equal to 8.15; these values are consistent with those reported in the literature [58]. Sterilized freshwater (SFW) and seawater (SW) had an initial pH equal to 8.25 and 8.39, respectively. In all cases, the pH values remained essentially constant with time in control samples.

In the W&L condition, a decrease in pH with respect to the initial value was observed during exposure to all the aqueous media with all the film samples. After 12 weeks, for FW, SW and SFW, the average pH decrease ranged within 0.18–0.20 units for the PLA film, 0.20–0.40 units for the PBS film and 0.28–0.45 units for the PBS/PLA 70/30 film. For all the tested films, the most consistent pH decrease was observed in SSW: at the end of the test, pH decreased by 0.65 units for the PLA film and by 0.70 units for PBS and PBS/PLA 70/30 samples. The observed acidification of the aqueous environments can be reasonably attributed to the release of monomer and oligomer byproducts, such as lactic acid as well as 4-hydroxybutyl succinate and succinic acid formed during the hydrolytic degradation of the PLA and PBS polymers, respectively [59]. The largest decrease in pH obtained for SSW could be explained by the fact that the sterilization of seawater destroys microorganisms capable of metabolizing PLA and PBS, as reported elsewhere [60–64]. Similar outcomes were gained for the tests in C&D conditions, although to a slightly smaller extent than in W&L conditions, as expected. In fact, the degradation kinetic of biopolymers is made faster by increasing temperature and the presence of light radiation, as largely reported in literature [59,65].

Overall, the obtained results are consistent with those reported by Romera-Castillo et al. [66], who observed an average pH decrease of 0.49 units due to biodegradable and non-biodegradable plastic leaching in seawater and emphasized the risk of exacerbating the harmful effects on marine organisms of acidification resulting from anthropogenic CO_2 emissions [67].

3.2. Optical Microscopy Analysis of the Formed Microplastics

Optical microscopy analysis was used to obtain information on the size and shape of the microparticles formed from PLA, PBS and PBS/PLA 70/30 films, after immersion in the four different aquatic environments in W&L conditions for 5, 7 and 12 weeks. Since the microscopy observations revealed no fragment formation in any case even after 12 weeks of exposure in C&D conditions and no appreciable differences between the same aquatic media in W&L conditions, sterilized and non-sterilized, only the images taken on the non-sterile fresh and seawater in W&L conditions are reported below.

Figures 3–5 show the photos of microplastics from PLA, PBS and PBS/PLA 70/30 films, respectively, after different immersion times in FW and SW.

For the PLA film (Figure 3), a large number of fragments was observed after 5 weeks in all the aqueous environments investigated, with similar shape and average size. In particular, the presence of either fibers, which constituted the most abundant fraction, with a length up to 700 μm, or platelets, with an average length up to 1.5 mm, was evident. After 7 weeks, BMPs of similar shape and size were observed compared to those obtained after 5 weeks. After 12 weeks, some longer fiber-like structures, up to 1.5 mm, were also found, while the average size of the platelets decreased to 300 μm. It was not possible to discriminate significant differences in the production of BMPs in the different aqueous media due to the presence of microorganisms in the unsterilized media or to differences in ionic strength. This result suggests that the PLA films predominantly underwent an abiotic degradation in all investigated aqueous media, with essentially comparable kinetics; as also reported by [59,68], in fact, hydrolytic degradation is predominant for aliphatic polyesters, such as PLA and PBS, following a bulk erosion mechanism for thicknesses between 0.5 and 2 mm.

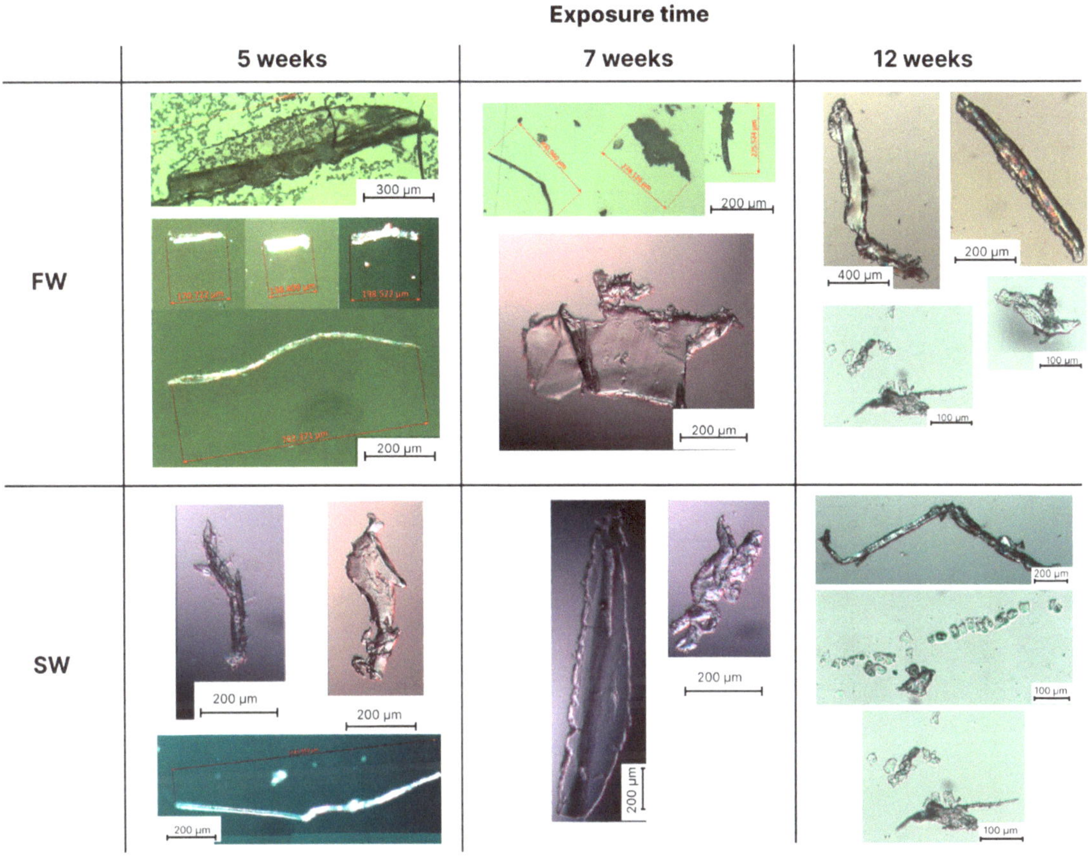

Figure 3. Optical microscopy images of the formed microplastics formed after 5, 7 and 12 weeks of degradation of PLA films in fresh water (FW) and seawater (SW). Black bars represent the reference size. Red bars represent the average measured particle length.

As regards the PBS film (Figure 4), after 5 weeks, the production of few acicular microplastics in FW and SFW was observed, having an average size of 20 µm. By increasing the exposure time to 7 and 12 weeks, the number of these particles steadily increased. Interestingly, a large number of BMPs entangled together and formed large fragments, with a length of up to 500 µm. This peculiar morphology was only found for FW and SFW. For SW and SSW, mostly fibers with 1 mm maximum length, and platelets with 500 µm maximum size were noted after 5 weeks; after 7 and 12 weeks, the maximum size for both fibers and platelets decreased to 300 µm. Additionally, in this case, no difference was appreciable due to the presence of microorganisms.

It is worth to note that, from the conducted observations, most of the plastic microparticles obtained for PBS and PLA films fell into the fiber category. BMPs with fiber-like structure and size up to 2 mm were also observed by Wei et al. for PBAT films after 10 weeks of exposure in artificial seawater and Milli-Q water [35]. In accordance with what reported by Hebner et al., the preferential formation of microplastic fibers rather than platelets can be attributed to the processing of the polymer films, which imparts orientation to the macromolecules [69]. The penetration of water in thin films further exacerbates the creation of oriented cracks that lead to the specimen's breakage along the direction of alignment of the polymer macromolecules, resulting in fiber-like particles rather than platelets.

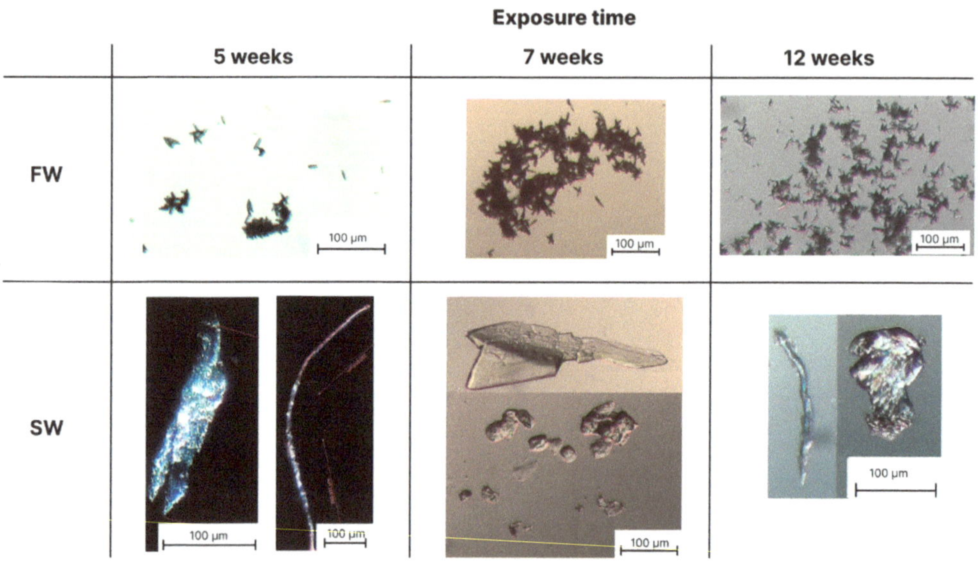

Figure 4. Optical microscopy images of microplastics formed after 5, 7 and 12 weeks of degradation of PBS films in fresh water (FW) and seawater (SW). Black bars represent the reference size. Red bars represent the average measured particle length.

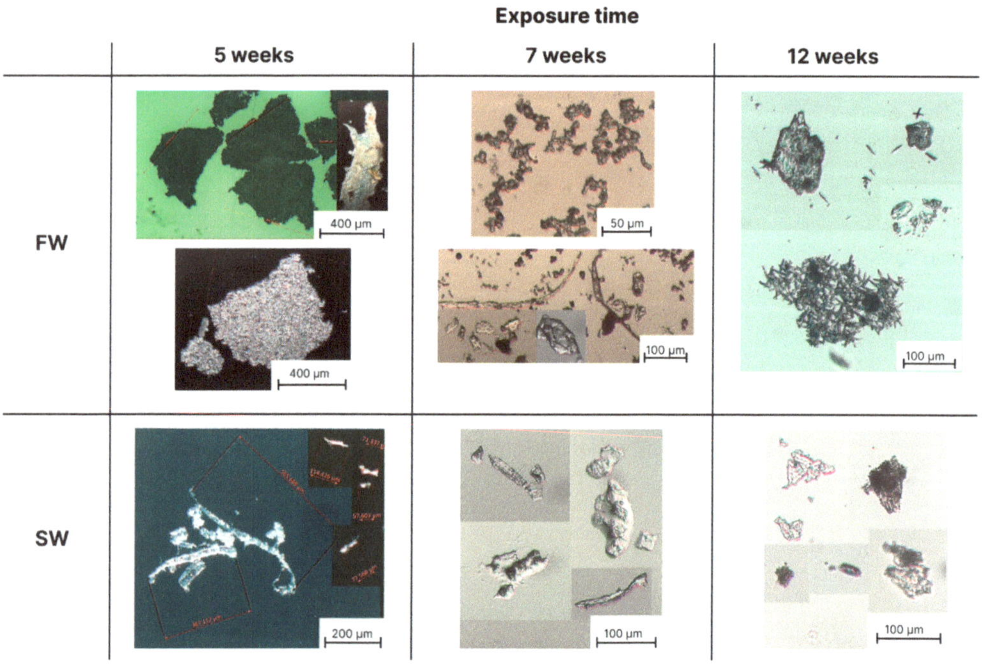

Figure 5. Optical microscopy images of microplastics formed after 5, 7 and 12 weeks of degradation of PBS/PLA 70/30 films in fresh water (FW) and seawater (SW). Black bars represent the reference size. Red bars represent the average measured particle length.

As regards the PBS/PLA 70/30 film (Figure 5), the formation of microplastics was much more heterogeneous than in the pure films. In FW and SFW, large heterogeneous particles with size up to 800 μm were observed after 5 weeks of exposure, as well as platelets up to 1 mm, consisting of aggregates of smaller particles like in the case of pure PBS (Figure 5). After 7 weeks of exposure, similarly to neat PBS film, the micrographs also revealed the presence of 20 μm-long acicular particles, alone or aggregated in bigger clusters. After 12 weeks, fibers up to 500 μm and acicular clusters up to 100 μm were observed. In SW and SSW, after 5 weeks of exposure, fiber-shaped particles up to 600 μm in length were found, as well as small particles of inhomogeneous shape up to 100 μm (Figure 5). After 7 and 12 weeks, BMPs mainly consisted of platelets up to 500 μm, fibers up to 200 μm and inhomogeneous particles up to 100 microns (Figure 5).

The resulting outcomes, compared with those achieved for pure PLA and PBS films, suggest a slower degradation rate for the PBS/PLA 70/30 blend, at least initially, with the formation of larger and more heterogeneous particles. As exposure time increased, the formation of microplastics having the characteristic shape and size of both the PBS and PLA phases was observed. The degradation rate of the blend was largely influenced by the composition of the blend, phase compatibility, degree of crystallinity, wettability and mechanical properties achieved. These characteristics will be discussed further in the article. For PBS/PLA mixtures, studies in the literature reveal very different degradation behaviors depending on these parameters. For example, Luzi et al. in their study observed a decreased degradation rate for the PLA/PBS 80/20 blend compared to that of neat polymers because of the higher degree of crystallinity induced by PBS [70]. On the other hand, Wang et al. found that immiscibility between PLA and PBS induces gaps in the mixture, providing channels for water penetration and enhancing hydrolytic degradation [71].

3.3. Melting Behavior, Crystallinity and Glass Transition Temperature

DSC measurements were carried out to investigate the initial film morphology and crystallinity, which deeply affect the degradation rate, and to study the changes in the thermal behavior of the polymer phases during the degradation processes.

Figure 6 displays the heating and cooling thermograms of the PLA, PBS and PBS/PLA 70/30 blend films, undegraded and after 12 weeks of degradation in the W&L condition in the four aqueous media; the corresponding thermal parameters are shown in Table 2.

Regarding the PLA film, before degradation, it exhibited a glass transition temperature at 67 °C, a broad cold crystallization peak in the range 100–120 °C, a second small exothermic event just before melting that, according to the literature, can be related to the reorganization of the PLA α′-mesophase into more stable α-crystals [72], and a melting peak at 169 °C. The crystallinity degree was equal to 12.8%. After 12 weeks of exposure in aqueous media, the Tg slightly raised from 68 up to a maximum of 71 °C. This increase of Tg values has been previously reported by other authors [73,74] and can be attributable to the stabilized packing of the chains in the amorphous region by annealing in the presence of water molecules. This raise in Tg leads to further embrittlement of the PLA sample, which is one of the most detrimental effects of degradation and accelerates film fragmentation. The specific interactions between water and PLA polymer chains do not only promote order and stability in the amorphous region but also in the crystalline region, as highlighted by the slight increase in the melting temperature and the crystallinity of the PLA films, up to maximum Tm = 173 °C and Xc = 19.5% after 12 weeks in SFW. These changes can be related to the formation of thicker, more perfect crystals with a higher melting temperature [75]. No peaks were found in the cooling scans in all cases, as expected, due to the low crystallization rate of PLA [76].

Concerning the PBS film, before degradation, the DSC trace showed a small cold crystallization peak at 96 °C and a melting transition at 118 °C, giving a crystallinity degree equal to 56.4%. After degradation experiments, the cold crystallization temperatures tended to increase. In fact, the lowering of the molecular weight caused by molecular chain fracture affected the cold crystallization kinetic, which became slower, as found also by others [77,78]. However, the corresponding ΔH_{cc} values remained almost the same. No relevant changes in the crystallinity degree of the films after degradation were detected, too. In the cooling scans, the crystallization peak became broader and shifted towards lower temperatures for all the films biodegraded in the aqueous media. A similar trend has been reported in the literature for dynamic crystallization from melt of a different polyester with decreasing molecular weights and is coherent with the hydrolytic degradation of PBS [79]. The changes are more relevant in both the seawater media, suggesting that the PBS hydrolysis rate can be enhanced in high-ionic-strength media, such as marine solutions [35,80].

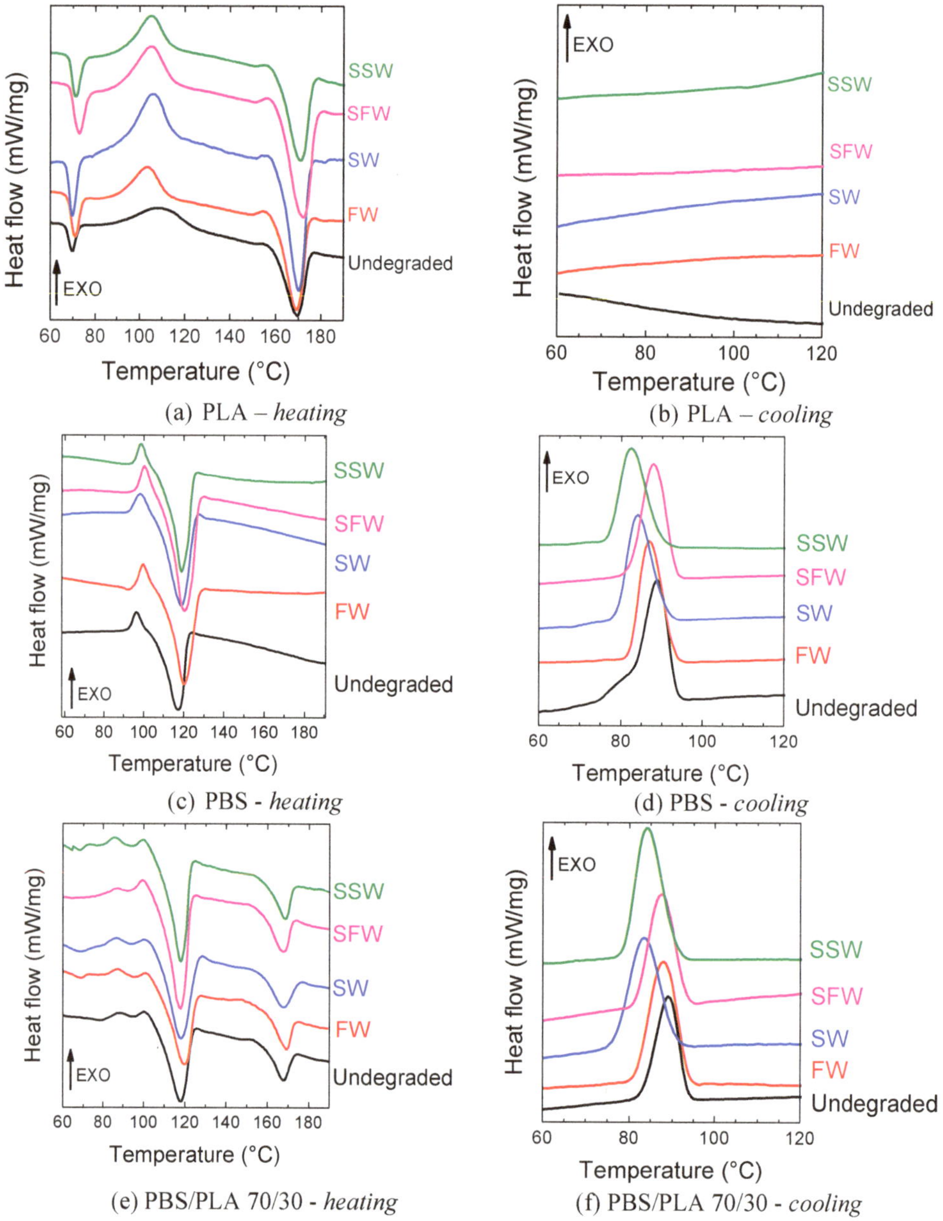

Figure 6. DSC thermograms of the PLA, PBS and PBS/PLA 70/30 films as produced and after 12 weeks of degradation in aqueous media in the Warm and Light condition.

Table 2. Thermal parameters of the PLA, PBS and PBS/PLA 70/30 films undegraded and after 12 weeks of degradation, calculated from the first heating and the cooling scans.

Film Sample	Degradation Medium	Heating											Cooling	
		T_g, [°C]	$T_{cc\,PBS}$, [°C]	$\Delta H_{cc\,PBS}$, [mWs/mg]	$T_{cc\,PLA}$, [°C]	$\Delta H_{cc\,PLA}$, [mWs/mg]	$T_{m\,PBS}$, [°C]	$\Delta H_{m\,PBS}$, [mWs/mg]	X_{cPBS}, [%]	$T_{m\,PLA}$, [°C]	$\Delta H_{m\,PLA}$, [mWs/mg]	X_{cPLA}, [%]	T_{cr}, [°C]	ΔH_{cr}, [mWs/mg]
PLA	Undegraded	68			109	25.2				169	37.2	12.8		
	FW	69			104	20.4				169	37.9	18.7		
	SW	69			106	19.6				170	36.4	17.9		
	SFW	71			105	18.6				173	36.9	19.5		
	SSW	70			106	21.7				172	37.8	17.2		
PBS	Undegraded		97	7.6			118	69.8	56.4				89	63.8
	FW		100	7.4			120	66.4	56.6				88	55.1
	SW		99	7			119	63.5	55.8				84	57.9
	SFW		100	6.9			119	70.6	60.4				88	57.5
	SSW		99	5.7			119	65.4	54.1				82	51.4
PBS/PLA 70/30	Undegraded		88	4.1	100		118	49.7	59.1	167	14.7	52.3	89	38.2
	FW		88	5.1	101		120	51.6	60.2	169	15.3	54.5	88	40.1
	SW		87	5.6	100		118	48.4	55.4	168	12.8	45.6	84	37.1
	SFW		87	3.4	100		118	51.5	62.3	168	14.4	51.2	88	39.5
	SSW		87	4.5	100		118	51.4	60.7	168	14.3	51.0	84	37.8

As for the PBS/PLA 70/30 blend, before degradation, the thermogram showed a first small cold crystallization peak at 88 °C, related to PBS, another small exotherm transition at ca. 100 °C attributable to PLA just before the PBS melting peak at 118 °C, and a second melting peak at 167 °C due to PLA. The two well-distinct melting peaks confirmed the complete phase separation of the two polymers. Due to the overlapping of the cold crystallization peak of PLA with the melting peak of PBS, it was difficult to determine the cold crystallization enthalpy and thus the crystallinity of PLA in the blend. However, following the same approach described in [32], it was possible to calculate this value with a slight bias: the crystallinity degrees of PBS and PLA phases in the blends, equal to 64.4% and 52.3%, respectively, revealed a substantial increase in the crystallinity degree of the PLA phase with respect to the neat PLA film, ascribable to the higher mobility of the PLA chains in the presence of PBS, with an increased ability to crystallize.

This suggests an increase in the resistance to hydrolytic degradation of the PBS/PLA 70/30 blend compared to that of pure PLA and PBS films, as already observed by Luzi et al., and is corroborated by optical microscope observations, which revealed the presence of larger and more heterogeneously shaped fragments compared to the microplastics produced by pure polymer films [70]. With regard to the cooling thermograms, similarly to what was observed for the pure PBS film, a shift of the crystallization temperatures ascribable to the PBS phase towards lower values was observed after degradation exposure in SW and SSW (Tc equal to 89, 84 and 82 °C for the initial film and after exposure in SW and SSW, respectively), ascribable also in this case to the shortening of the PBS chains due to hydrolytic degradation.

3.4. Mechanical Properties

Tensile tests were carried out to reveal changes in the mechanical properties of the PLA, PBS and PBS/PLA 70/30 films during exposure in aquatic environments. This represents an effective methodology for investigating the degradation rate of the biodegradable films [60]. The results, expressed as elastic modulus E, maximum strength σ_m and elongation at break ε_b, are displayed in Figure 7 for the W&L condition in the four aqueous media. It is worth pointing out that, since the PLA films experienced fast degradation kinetics, as also underlined by optical microscopy and DSC analyses, their tensile properties could be measured only up to 7 weeks of exposure; after 12 weeks, their advanced state of degradation did not allow to subject them to tensile testing. For this reason, in order to make a comparison between the mechanical properties of PLA, PBS and PBS/PLA 70/30 films, the results presented were collected after 7 weeks of exposure for all the samples. The results of tensile tests on PBS and PBS/PLA 70/30 film strips after 12 weeks of degradation in W&L condition are available in the Supplementary Material.

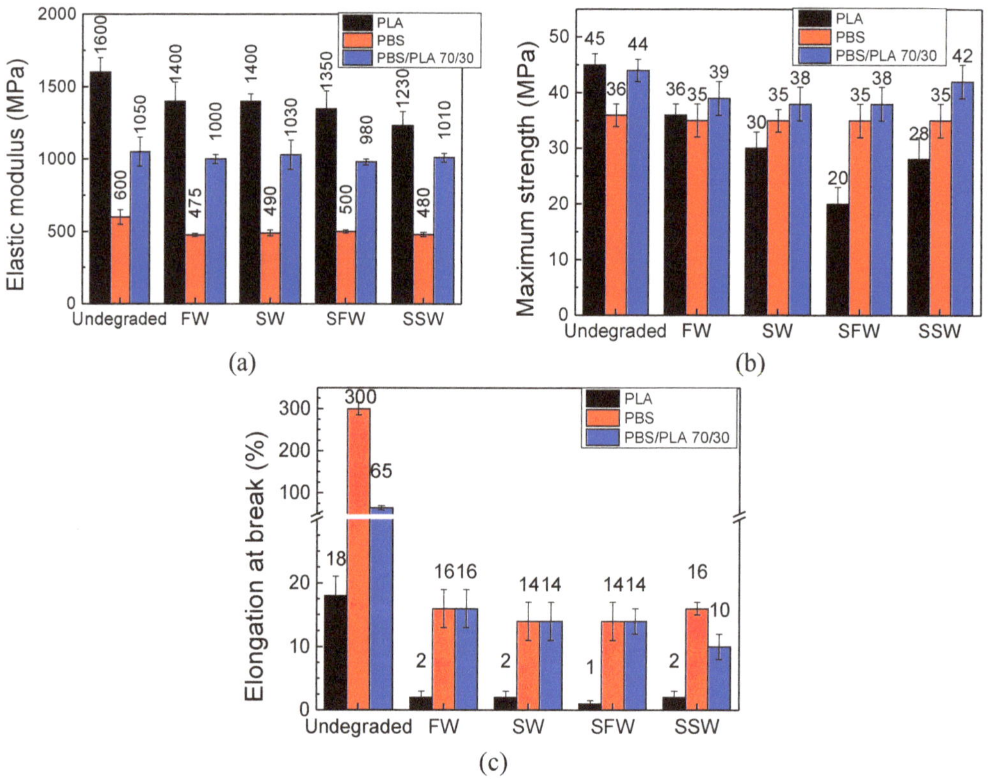

Figure 7. Mechanical properties of films after 7 weeks of degradation in Warm and Light (W&L) conditions: (**a**) elastic modulus E; (**b**) maximum strength σ_m; (**c**) elongation at break ε_b.

The mechanical properties obtained at the initial time confirmed the inherent stiffness and brittleness of the PLA film (E = 1600 ± 100 MPa, σ_m = 45 ± 3 MPa, ε_b = 18 ± 3%), while PBS showed very ductile behavior (E = 600 ± 50 MPa, σ_m = 36 ± 2 MPa, ε_b = 300 ± 20%), as also reported by other authors [81,82]. The PBS/PLA 70/30 film exhibited a good balance between stiffness and ductility (E = 1050 ± 70 MPa, σ_m = 44 ± 3 MPa, ε_b = 65 ± 5%).

A decrease in the tensile properties was observed for all the films after exposure in the aqueous media, especially regarding elongation at break, which is the most sensitive index for monitoring the cleavage of polymer chains [83]. However, some relevant differences could be noticed based on the composition of the films, rather than on the presence of microorganisms or ionic strength.

After 7 weeks, the PLA film underwent the most severe embrittlement and decrease in stiffness: with respect to initial time, in the four aqueous media, the decrease in elastic modulus, elongation at break and maximum tensile strength ranged from 13% to 23%, from 88% to 95% and from 20% to 55%, respectively. The results obtained for the PBS film also demonstrated a transition from ductile to brittle behavior upon hydrolytic degradation; however, to a lesser extent than in the PLA sample: the decreases in E, ε_b and σ_m ranged from 16% to 21%, from 71% to 94% and from 2% to 3%, respectively. As underlined by

DSC analyses, the increased resistance to hydrolysis can be attributable to the higher crystallinity degree of PBS with respect to PLA. The PBS/PLA 70/30 film exhibited the lowest embrittlement and stiffness reduction: the decrease in E, ε_b and σ_m ranged from 3% to 6%, from 77% to 84% and from 2% to 13%, respectively. These results further confirm the outcomes gained by DSC and optical microscopy analyses, which underlined the increased resistance to hydrolytic degradation of the PBS/PLA blend with respect to that of their parent polymers. Different outcomes were obtained by Zhou et al., who observed that PBS/PLA blends lost their tensile properties earlier than their parent polymers with the proceeding of hydrolysis [84]. Other authors analyzed the effect of hydrolytic degradation on the mechanical properties of fossil-based polyesters such as polyethylene terephthalate and highlighted that during hydrolysis, an embrittlement of the polymer is observed that leads to a large decrease in both strain and stress at break, whereas Young's modulus remains almost constant [85].

The mechanical properties of the specimens after 12 weeks under C&D conditions are shown in Figure 8. It can be noted that under these conditions, the specimens retained their strength and modulus of elasticity regardless of the water environment; however, an increase in the brittleness of the specimens was observed, but to a lower extent than that measured in W&L conditions. These results are in accordance with the outcomes gained from the optical microscopy analysis and could be attributable to slower water diffusion into the samples under these conditions: this points out that, in cold deep water, biodegradable plastics can be highly persistent and could behave like non-degradable polymers.

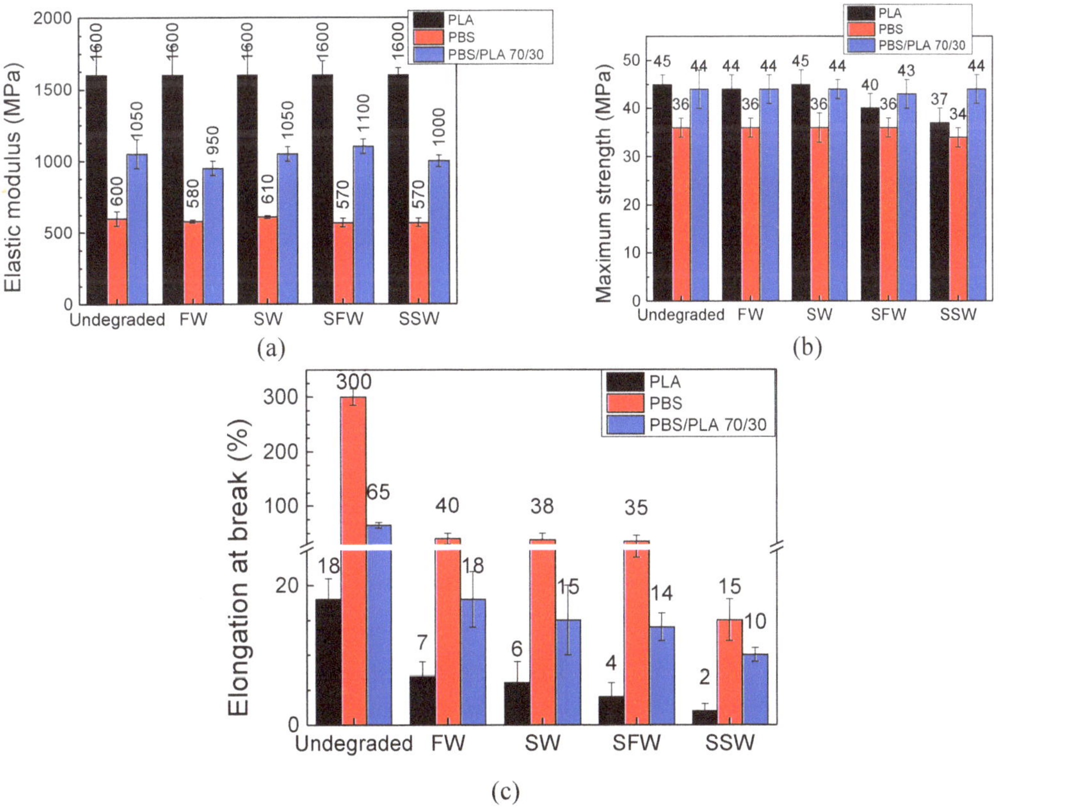

Figure 8. Mechanical properties of films after the 12 weeks of degradation in Cold and Dark (C&D) conditions: (**a**) elastic modulus E; (**b**) maximum strength σ_m; (**c**) elongation at break ε_b.

3.5. Surface Wettability

Water contact angle measurements were carried out to evaluate changes in surface wettability of the biodegradable films during degradation in the aqueous media in Warm and Light (W&L) conditions. Table 3 shows the values of the water contact angle for PLA, PBS and PBS/PLA 70/30 films, both undegraded and after 12 weeks of degradation in FW, SW, SFW and SSW.

Table 3. Water contact angle values of PLA, PBS and PBS/PLA 70/30 films, both undegraded and after 12 weeks of degradation in different aqueous media in Warm and Light (W&L) conditions.

Sample	Water Contact Angle [°]				
	Undegraded	**FW**	**SW**	**SFW**	**SSW**
PLA	63.7 ± 2.5	41.0 ± 6.1	45.8 ± 6.0	53.5 ± 6.4	56.7 ± 4.0
PBS	72.5 ± 2.1	60.6 ± 2.7	53.9 ± 2.9	60.1 ± 4.8	50.2 ± 2.7
PBS/PLA 70/30	65.5 ± 3.1	50.4 ± 5.9	54.0 ± 3.9	57.5 ± 4.2	57.4 ± 5.1

For all the materials and the aqueous media, the wetting angle decreased after 12 weeks of exposure, implying an increase in hydrophilicity. The PLA film exhibited the maximum decrease in contact angle equal to 36% in FW, while for PBS and PBS/PLA 70/30, maximum decreases equal to 31% in SSW and 23% in FW, respectively, were registered. The increase in hydrophilicity could be attributed to both the increase of surface roughness due to swelling and erosion phenomena and to the cleavage of the ester bonds occurring during hydrolytic degradation, with the accumulation of polar low-molecular-weight substances on the surface of the films. In particular, PLA degradation entails the formation of as carboxylic acids (e.g., lactic acid) and hydroxyl groups [86], while the main products of PBS hydrolytic degradation are 4-hydroxybutyl succinate, succinic acid and butane-1,4-diol [87]. Compared to that of the PLA film, the smaller decrease in water contact angle of the PBS and PBS/PLA 70/30 films could represent an indirect measurement of the increased resistance to hydrolytic degradation of these samples, which involved the generation of lower amounts of water-soluble oligomers and monomers.

4. Conclusions

In this work, the risk associated with the production of BMPs from biodegradable packaging was investigated by monitoring the deterioration behaviors of PLA, PBS and PBS/PLA 70/30 blend films for 12 weeks in different aquatic media and two temperature/light conditions.

Among the different biodegradable polymers, PLA underwent the most rapid degradation, particularly in W&L tests. In this condition, the PLA film, due to water swelling that increased the chain mobility, underwent to both solvent-induced crystallization, shown by DSC results, and hydrolytic degradation, evidenced by the dramatic film embrittlement (with a ε_b decrease in the range ca. 88–94%), which is the most sensitive index for monitoring the cleavage of polymer chains. Therefore, a high number of BMPs was generated already after 5 weeks, mostly consisting of fibers up to 700 μm long. After 12 weeks of exposure, the PLA film completely lost its integrity. On the other hand, the PBS film, thanks to its intrinsic high crystallinity degree, had a much slower degradation rate, exhibiting only slight and insignificant modifications of its thermal, tensile and surface properties, and produced only few acicular BMPs after 5 weeks, and more fibrous BMPs having 300 to 500 μm average size for longer exposure times. Lastly, the PBS/PLA 70/30 blend exhibited the highest resistance to deterioration, attributable to the higher crystallinity degree of the PLA phase in the blend (52.3%) with respect to that of the neat PLA film (12.8%). Compared to its parent polymers, the blend showed the lowest embrittlement (maximum εb decrease equal to 84%) and the lowest increase in hydrophilicity, and its hydrolytic degradation led to the formation of large and quite heterogeneous BMPs, consisting of acicular clusters, fibers and inhomogeneous platelets, whose sizes ranged from 100 to 800 μm. These findings pointed out that polymer blends may pose a greater risk to aquatic environments than neat polymers, and caution should be adopted when encouraging biodegradable polymer mixtures.

Only minor changes were detectable in degradation behavior of the films due to the different ionic strength of aqueous media and to the presence of microorganisms. However, all the aqueous environments incurred acidification during the degradation experiments, and a larger decrease in pH was observed in sterilized media (i.e., SFW and SSW) than in unsterilized ones (i.e., FW and SW), suggesting that microorganisms could absorb acidic degradation byproducts generated during polyester hydrolysis.

Finally, the results pointed out that temperature and light exposure severely affected the degradation rate of the bioplastics, which can be highly persistent in cold deep water, behaving like non-degradable polymers.

Supplementary Materials: The following supporting information can be downloaded at: https://www.mdpi.com/article/10.3390/polym16162268/s1, Table S1: Chemical-physical parameters of the freshwater and seawater; Figure S1: Mechanical properties of PBS and PBS/PLA 70/30 films after the 12 weeks of degradation in Cold and Dark (C&D) conditions: a—strength at break, b—maximum strength, c—elongation at break.

Author Contributions: Conceptualization, K.V.M. and A.A.; methodology, validation and formal analysis, K.V.M. and A.A.; investigation and data curation, K.V.M. and P.S.; resources, L.I. and P.S.; writing—original draft preparation, K.V.M. and A.A.; writing—review and editing, P.S. and L.I.; visualization, P.S. and L.I.; supervision, A.A., P.S. and L.I.; project administration and funding acquisition, L.I. All authors have read and agreed to the published version of the manuscript.

Funding: This research received no external funding.

Institutional Review Board Statement: Not applicable.

Data Availability Statement: The data presented in this study are available on request from the corresponding author.

Conflicts of Interest: The authors declare no conflicts of interest.

References

1. Rhodes, C.J. Plastic Pollution and Potential Solutions. *Sci. Prog.* **2018**, *101*, 207–260. [CrossRef] [PubMed]
2. Li, P.; Wang, X.; Su, M.; Zou, X.; Duan, L.; Zhang, H. Characteristics of Plastic Pollution in the Environment: A Review. *Bull. Environ. Contam. Toxicol.* **2021**, *107*, 577–584. [CrossRef] [PubMed]
3. Barnes, S.J. Understanding plastics pollution: The role of economic development and technological research. *Environ. Pollut.* **2019**, *249*, 812–821. [CrossRef] [PubMed]
4. Mong, G.R.; Tan, H.; Sheng, D.D.C.V.; Kek, H.Y.; Nyakuma, B.B.; Woon, K.S.; Othman, M.H.D.; Kang, H.S.; Goh, P.S.; Wong, K.Y. A review on plastic waste valorisation to advanced materials: Solutions and technologies to curb plastic waste pollution. *J. Clean. Prod.* **2024**, *434*, 140180. [CrossRef]
5. Suzuki, G.; Uchida, N.; Tuyen, L.H.; Tanaka, K.; Matsukami, H.; Kunisue, T.; Takahashi, S.; Viet, P.H.; Kuralmochi, H.; Osako, M. Mechanical recycling of plastic waste as a point source of microplastic pollution. *Environ. Pollut.* **2022**, *303*, 119114. [CrossRef] [PubMed]
6. Siracusa, V.; Blanco, I. Bio-Polyethylene (Bio-PE), Bio-Polypropylene (Bio-PP) and Bio-Poly(ethylene terephthalate) (Bio-PET): Recent Developments in Bio-Based Polymers Analogous to Petroleum-Derived Ones for Packaging and Engineering Applications. *Polymers* **2020**, *12*, 1641. [CrossRef]
7. United Nations. Fast Facts—What is Plastic Pollution?—United Nations Sustainable Development. 2023. Available online: https://www.un.org/sustainabledevelopment/blog/2023/08/explainer-what-is-plastic-pollution/ (accessed on 18 July 2024).
8. Resource Futures and Nextek. Eliminating Avoidable Plastic Waste by 2042: A Use-Based Approach to Decision and Policy Making. Resourcing the Future Conference. 2018. Available online: https://www.circularonline.co.uk/wp-content/uploads/2019/06/Eliminating-avoidable-plastic-waste-by-2042-a-use-based-approach-to-decision-and-policy-making.pdf (accessed on 18 July 2024).
9. Bandaru, S.; Ravipati, M.; Busi, K.B.; Phukan, P.; Bag, S.; Chandu, B.; Dalapati, G.K.; Biring, S.; Chakrabortty, S. A Review on the Fate of Microplastics: Their Degradation and Advanced Analytical Characterization. *J. Polym. Environ.* **2023**, *32*, 2532–2550. [CrossRef]

10. Issac, M.N.; Kandasubramanian, B. Effect of microplastics in water and aquatic systems. *Environ. Sci. Pollut. Res.* **2021**, *28*, 19544–19562. [CrossRef]
11. Haider, T.P.; Völker, C.; Kramm, J.; Landfester, K.; Wurm, F.R. Plastics of the Future? The Impact of Biodegradable Polymers on the Environment and on Society. *Angew. Chem. Int. Ed.* **2019**, *58*, 50–62. [CrossRef]
12. Dörnyei, K.R.; Uysal-Unalan, I.; Krauter, V.; Weinrich, R.; Incarnato, L.; Karlovits, I.; Colelli, G.; Chrysochou, P.; Fenech, M.C.; Pettersen, M.K.; et al. Sustainable food packaging: An updated definition following a holistic approach. *Front. Sustain. Food Syst.* **2023**, *7*, 1119052. [CrossRef]
13. Li, B.; Ma, Y.; Li, H. A new journey of plastics: Towards a circular and low carbon future. *Giant* **2022**, *11*, 100115. [CrossRef]
14. European Bioplastics, e.V. Bioplastics Market Development Update 2023. 2023. Available online: https://www.european-bioplastics.org/bioplastics-market-development-update-2023-2/ (accessed on 18 July 2024).
15. Scarfato, P.; Di Maio, L.; Incarnato, L. Recent advances and migration issues in biodegradable polymers from renewable sources for food packaging. *J. Appl. Polym. Sci.* **2015**, *132*, 42597. [CrossRef]
16. Siracusa, V.; Rocculi, P.; Romani, S.; Rosa, M.D. Biodegradable polymers for food packaging: A review. *Trends Food Sci. Technol.* **2008**, *19*, 634–643. [CrossRef]
17. Carullo, D.; Casson, A.; Rovera, C.; Ghaani, M.; Bellesia, T.; Guidetti, R.; Farris, S. Testing a coated PE-based mono-material for food packaging applications: An in-depth performance comparison with conventional multi-layer configurations. *Food Packag. Shelf Life* **2023**, *39*, 101143. [CrossRef]
18. Leneveu-Jenvrin, C.; Apicella, A.; Bradley, K.; Meile, J.; Chillet, M.; Scarfato, P.; Incarnato, L.; Remize, F. Effects of maturity level, steam treatment, or active packaging to maintain the quality of minimally processed mango (*Mangifera indica* cv. José). *J. Food Process. Preserv.* **2021**, *45*, e15600. [CrossRef]
19. Apicella, A.; Scarfato, P.; Di Maio, L.; Incarnato, L. Oxygen absorption data of multilayer oxygen scavenger-polyester films with different layouts. *Data Brief* **2018**, *19*, 1530–1536. [CrossRef] [PubMed]
20. Viscusi, G.; Bugatti, V.; Vittoria, V.; Gorrasi, G. Antimicrobial sorbate anchored to layered double hydroxide (LDH) nano-carrier employed as active coating on Polypropylene (PP) packaging: Application to bread stored at ambient temperature. *Futur. Foods* **2021**, *4*, 100063. [CrossRef]
21. Apicella, A.; Scarfato, P.; Di Maio, L.; Garofalo, E.; Incarnato, L. Evaluation of performance of PET packaging films based on different copolyester O_2-scavengers. *AIP Conf. Proc.* **2018**, *1981*, 020130. [CrossRef]
22. Benbettaïeb, N.; Karbowiak, T.; Debeaufort, F. Bioactive edible films for food applications: Influence of the bioactive compounds on film structure and properties. *Crit. Rev. Food Sci. Nutr.* **2019**, *59*, 1137–1153. [CrossRef]
23. Bhargava, N.; Sharanagat, V.S.; Mor, R.S.; Kumar, K. Active and intelligent biodegradable packaging films using food and food waste-derived bioactive compounds: A review. *Trends Food Sci. Technol.* **2020**, *105*, 385–401. [CrossRef]
24. Apicella, A.; Scarfato, P.; D'Arienzo, L.; Garofalo, E.; Di Maio, L.; Incarnato, L. Antimicrobial biodegradable coatings based on LAE for food packaging applications. *AIP Conf. Proc.* **2018**, *1981*, 020010. [CrossRef]
25. Bugatti, V.; Brachi, P.; Viscusi, G.; Gorrasi, G. Valorization of tomato processing residues through the production of active bio-composites for packaging applications. *Front. Mater.* **2019**, *6*, 34. [CrossRef]
26. Boccalon, E.; Viscusi, G.; Lamberti, E.; Fancello, F.; Zara, S.; Sassi, P.; Marinozzi, M.; Nocchetti, M.; Gorrasi, G. Composite films containing red onion skin extract as intelligent pH indicators for food packaging. *Appl. Surf. Sci.* **2022**, *593*, 153319. [CrossRef]
27. Bugatti, V.; Viscusi, G.; Gorrasi, G. Formulation of a Bio-Packaging Based on Pure Cellulose Coupled with Cellulose Acetate Treated with Active Coating: Evaluation of Shelf Life of Pasta Ready to Eat. *Foods* **2020**, *9*, 1414. [CrossRef]
28. Gorrasi, G.; Viscusi, G.; Gerardi, C.; Lamberti, E.; Giovinazzo, G. Physicochemical and Antioxidant Properties of White (Fiano cv) and Red (Negroamaro cv) Grape Pomace Skin Based Films. *J. Polym. Environ.* **2022**, *30*, 3609–3621. [CrossRef]
29. Aliotta, L.; Vannozzi, A.; Panariello, L.; Gigante, V.; Coltelli, M.-B.; Lazzeri, A. Sustainable Micro and Nano Additives for Controlling the Migration of a Biobased Plasticizer from PLA-Based Flexible Films. *Polymers* **2020**, *12*, 1366. [CrossRef] [PubMed]
30. Wei, X.F.; Nilsson, F.; Yin, H.; Hedenqvist, M.S. Microplastics Originating from Polymer Blends: An Emerging Threat? *Environ. Sci. Technol.* **2021**, *55*, 4190–4193. [CrossRef]
31. Anukiruthika, T.; Sethupathy, P.; Wilson, A.; Kashampur, K.; Moses, J.A.; Anandharamakrishnan, C. Multilayer packaging: Advances in preparation techniques and emerging food applications. *Compr. Rev. Food Sci. Food Saf.* **2020**, *19*, 1156–1186. [CrossRef]
32. Apicella, A.; Scarfato, P.; Incarnato, L. Tailor-made coextruded blown films based on biodegradable blends for hot filling and frozen food packaging. *Food Packag. Shelf Life* **2023**, *37*, 101096. [CrossRef]
33. Aliotta, L.; Gigante, V.; Pont, B.D.; Miketa, F.; Coltelli, M.-B.; Lazzeri, A. Tearing fracture of poly(lactic acid) (PLA)/poly(butylene succinate-co-adipate) (PBSA) cast extruded films: Effect of the PBSA content. *Eng. Fract. Mech.* **2023**, *289*, 109450. [CrossRef]
34. Aliotta, L.; Vannozzi, A.; Canesi, I.; Cinelli, P.; Coltelli, M.-B.; Lazzeri, A. Poly(lactic acid) (PLA)/Poly(butylene succinate-co-adipate) (PBSA) Compatibilized Binary Biobased Blends: Melt Fluidity, Morphological, Thermo-Mechanical and Micromechanical Analysis. *Polymers* **2021**, *13*, 218. [CrossRef] [PubMed]
35. Wei, X.F.; Bohlén, M.; Lindblad, C.; Hedenqvist, M.; Hakonen, A. Microplastics generated from a biodegradable plastic in freshwater and seawater. *Water Res.* **2021**, *198*, 117123. [CrossRef] [PubMed]
36. Pischedda, A.; Tosin, M.; Degli-Innocenti, F. Biodegradation of plastics in soil: The effect of temperature. *Polym. Degrad. Stab.* **2019**, *170*, 109017. [CrossRef]

37. Agarwal, S. Biodegradable Polymers: Present Opportunities and Challenges in Providing a Microplastic-Free Environment. *Macromol. Chem. Phys.* **2020**, *221*, 2000017. [CrossRef]
38. Gorrasi, G.; Pantani, R. Hydrolysis and Biodegradation of Poly(lactic acid). In *Advances in Polymer Science*; Springer: New York, NY, USA, 2018; pp. 119–151.
39. Tokiwa, Y.; Calabia, B.P.; Ugwu, C.U.; Aiba, S. Biodegradability of plastics. *Int. J. Mol. Sci.* **2009**, *10*, 3722–3742. [CrossRef] [PubMed]
40. Qin, M.; Chen, C.; Song, B.; Song, B.; Shen, M.; Cao, W.; Yang, H.; Zeng, G.; Gong, J. A review of biodegradable plastics to biodegradable microplastics: Another ecological threat to soil environments? *J. Clean. Prod.* **2021**, *312*, 127816. [CrossRef]
41. Sifuentes-Nieves, I.; Flores-Silva, P.C.; Ledezma-Perez, A.S.; Hernandez-Gamez, J.F.; Gonzalez-Morones, P.; Saucedo-Salazar, E.; Hernandez-Hernandez, E. Sustainable and Ecological Poly (Vinyl Alcohol)/Agave Fiber-Based Films: Structural Features post Composting Process. *J. Polym. Environ.* **2024**, *32*, 2448–2456. [CrossRef]
42. Kwon, S.; Zambrano, M.C.; Pawlak, J.J.; Ford, E.; Venditti, R.A. Aquatic Biodegradation of Poly(β-Hydroxybutyrate) and Polypropylene Blends with Compatibilizer and the Generation of Micro- and Nano-Plastics on Biodegradation. *J. Polym. Environ.* **2023**, *31*, 3619–3631. [CrossRef]
43. Green, D.S. Effects of microplastics on European flat oysters, *Ostrea edulis* and their associated benthic communities. *Environ. Pollut.* **2016**, *216*, 95–103. [CrossRef]
44. Malafeev, K.V.; Apicella, A.; Incarnato, L.; Scarfato, P. Understanding the Impact of Biodegradable Microplastics on Living Organisms Entering the Food Chain: A Review. *Polymers* **2023**, *15*, 3680. [CrossRef]
45. Serrano-Ruiz, H.; Martin-Closas, L.; Pelacho, A.M. Impact of buried debris from agricultural biodegradable plastic mulches on two horticultural crop plants: Tomato and lettuce. *Sci. Total Environ.* **2023**, *856*, 159167. [CrossRef] [PubMed]
46. Zhang, X.; Xia, M.; Su, X.; Yuan, P.; Li, X.; Zhou, C.; Wan, Z.; Zou, W. Photolytic degradation elevated the toxicity of polylactic acid microplastics to developing zebrafish by triggering mitochondrial dysfunction and apoptosis. *J. Hazard. Mater.* **2021**, *413*, 125321. [CrossRef] [PubMed]
47. Li, M.; Ma, Q.; Su, T.; Wang, Z.; Tong, H. Effect of Polycaprolactone Microplastics on Soil Microbial Communities and Plant Growth. *J. Polym. Environ.* **2023**, *32*, 1039–1045. [CrossRef]
48. Zhao, S.; Liu, L.; Li, C.; Zheng, H.; Luo, Y.; Pang, L.; Lin, Q.; Zhang, H.; Sun, C.; Chen, L.; et al. Photodegradation of biobased polymer blends in seawater: A major source of microplastics in the marine environment. *Front. Mar. Sci.* **2022**, *9*, 1046179. [CrossRef]
49. Narancic, T.; Verstichel, S.; Chaganti, S.R.; Morales-Gamez, L.; Kenny, S.T.; De Wilde, B.; Padamati, R.B.; O'connor, K.E. Biodegradable Plastic Blends Create New Possibilities for End-of-Life Management of Plastics but They Are Not a Panacea for Plastic Pollution. *Environ. Sci. Technol.* **2018**, *52*, 10441–10452. [CrossRef]
50. La Mantia, F.P.; Morreale, M.; Botta, L.; Mistretta, M.C.; Ceraulo, M.; Scaffaro, R. Degradation of polymer blends: A brief review. *Polym. Degrad. Stab.* **2017**, *145*, 79–92. [CrossRef]
51. Wei, X.F.; Capezza, A.J.; Cui, Y.; Li, L.; Hakonen, A.; Liu, B.; Hedenqvist, M.S. Millions of microplastics released from a biodegradable polymer during biodegradation/enzymatic hydrolysis. *Water Res.* **2022**, *211*, 118068. [CrossRef] [PubMed]
52. Weinstein, J.E.; Dekle, J.L.; Leads, R.R.; Hunter, R.A. Degradation of bio-based and biodegradable plastics in a salt marsh habitat: Another potential source of microplastics in coastal waters. *Mar. Pollut. Bull.* **2020**, *160*, 111518. [CrossRef]
53. Belioka, M.P.; Siddiqui, M.N.; Redhwi, H.H.; Achilias, D.S. Thermal degradation kinetics of recycled biodegradable and non-biodegradable polymer blends either neat or in the presence of nanoparticles using the random chain-scission model. *Thermochim. Acta* **2023**, *726*, 179542. [CrossRef]
54. Siddiqui, M.N.; Redhwi, H.H.; Belioka, M.P.; Achilias, D.S. Effect of nanoclay on the thermal degradation kinetics of recycled biodegradable/non-biodegradable polymer blends using the random chain-scission model. *J. Anal. Appl. Pyrolysis* **2024**, *177*, 106291. [CrossRef]
55. Rusydi, A.F. Correlation between conductivity and total dissolved solid in various type of water: A review. *IOP Conf. Ser. Earth Environ. Sci.* **2018**, *118*, 012019. [CrossRef]
56. Polyák, P.; Nagy, K.; Vértessy, B.; Pukánszky, B. Self-regulating degradation technology for the biodegradation of poly(lactic acid). *Environ. Technol. Innov.* **2023**, *29*, 103000. [CrossRef]
57. Min, K.; Cuiffi, J.D.; Mathers, R.T. Ranking environmental degradation trends of plastic marine debris based on physical properties and molecular structure. *Nat. Commun.* **2020**, *11*, 727. [CrossRef]
58. Radke, L. pH of Coastal Waterways—OzCoasts. 2002. Available online: https://ozcoasts.org.au/indicators/biophysical-indicators/ph_coastal_waterways/ (accessed on 18 July 2024).
59. Bher, A.; Mayekar, P.C.; Auras, R.A.; Schvezov, C.E. Biodegradation of Biodegradable Polymers in Mesophilic Aerobic Environments. *Int. J. Mol. Sci.* **2022**, *23*, 12165. [CrossRef] [PubMed]
60. Elsawy, M.A.; Kim, K.H.; Park, J.W.; Deep, A. Hydrolytic degradation of polylactic acid (PLA) and its composites. *Renew. Sustain. Energy Rev.* **2017**, *79*, 1346–1352. [CrossRef]
61. Lindström, A.; Albertsson, A.C.; Hakkarainen, M. Quantitative determination of degradation products an effective means to study early stages of degradation in linear and branched poly(butylene adipate) and poly(butylene succinate). *Polym. Degrad. Stab.* **2004**, *83*, 487–493. [CrossRef]
62. Zaaba, N.F.; Jaafar, M. A review on degradation mechanisms of polylactic acid: Hydrolytic, photodegradative, microbial, and enzymatic degradation. *Polym. Eng. Sci.* **2020**, *60*, 2061–2075. [CrossRef]

63. Liu, T.-Y.; Huang, D.; Xu, P.-Y.; Lu, B.; Wang, G.-X.; Zhen, Z.-C.; Ji, J. Biobased Seawater-Degradable Poly(butylene succinate-l-lactide) Copolyesters: Exploration of Degradation Performance and Degradation Mechanism in Natural Seawater. *ACS Sustain. Chem. Eng.* **2022**, *10*, 3191–3202. [CrossRef]
64. Hugenholtz, J. Citrate metabolism in lactic acid bacteria. *FEMS Microbiol. Rev.* **1993**, *12*, 165–178. [CrossRef]
65. Boskhomdzhiev, A.P.; Bonartsev, A.P.; Ivanov, E.A.; Makhina, T.; Myshkina, V.; Bagrov, D.; Filatova, E.; Bonartseva, G.; Iordanskii, A. Hydrolytic Degradation of Biopolymer Systems Based on Poly-3-hydroxybutyrate. Kinetic and Structural Aspects. *Int. Polym. Sci. Technol.* **2010**, *37*, 25–30. [CrossRef]
66. Romera-Castillo, C.; Lucas, A.; Mallenco-Fornies, R.; Briones-Rizo, M.; Calvo, E.; Pelejero, C. Abiotic plastic leaching contributes to ocean acidification. *Sci. Total Environ.* **2023**, *854*, 158683. [CrossRef]
67. Intergovernmental Panel on Climate Change (IPCC). Changing Ocean, Marine Ecosystems, and Dependent Communities. In *The Ocean and Cryosphere in a Changing Climate: Special Report of the Intergovernmental Panel on Climate Change*; Cambridge University Press: Cambridge, UK, 2022; pp. 447–588. [CrossRef]
68. Von Burkersroda, F.; Schedl, L.; Göpferich, A. Why degradable polymers undergo surface erosion or bulk erosion. *Biomaterials* **2002**, *23*, 4221–4231. [CrossRef]
69. Hebner, T.S.; Maurer-Jones, M.A. Characterizing microplastic size and morphology of photodegraded polymers placed in simulated moving water conditions. *Environ. Sci. Process Impacts* **2020**, *22*, 398–407. [CrossRef] [PubMed]
70. Luzi, F.; Fortunati, E.; Jiménez, A.; Puglia, D.; Pezzolla, D.; Gigliotti, G.; Kenny, J.; Chiralt, A.; Torre, L. Production and characterization of PLA_PBS biodegradable blends reinforced with cellulose nanocrystals extracted from hemp fibres. *Ind. Crops Prod.* **2016**, *93*, 276–289. [CrossRef]
71. Wang, Y.P.; Xiao, Y.J.; Duan, J.; Yang, J.H.; Wang, Y.; Zhang, C.L. Accelerated hydrolytic degradation of poly(lactic acid) achieved by adding poly(butylene succinate). *Polym. Bull.* **2016**, *73*, 1067–1083. [CrossRef]
72. Di Lorenzo, M.L.; Androsch, R. Influence of α'-/α-crystal polymorphism on properties of poly(l-lactic acid). *Polym. Int.* **2019**, *68*, 320–334. [CrossRef]
73. Tsuji, H.; Shimizu, K.; Sato, Y. Hydrolytic degradation of poly(L-lactic acid): Combined effects of UV treatment and crystallization. *J. Appl. Polym. Sci.* **2012**, *125*, 2394–2406. [CrossRef]
74. Vu, T.; Nikaeen, P.; Chirdon, W.; Khattab, A.; Depan, D. Improved Weathering Performance of Poly(Lactic Acid) through Carbon Nanotubes Addition: Thermal, Microstructural, and Nanomechanical Analyses. *Biomimetics* **2020**, *5*, 61. [CrossRef]
75. Pantani, R.; Sorrentino, A. Influence of crystallinity on the biodegradation rate of injection-moulded poly(lactic acid) samples in controlled composting conditions. *Polym. Degrad. Stab.* **2013**, *98*, 1089–1096. [CrossRef]
76. Saeidlou, S.; Huneault, M.A.; Li, H.; Park, C.B. Poly(lactic acid) crystallization. *Prog. Polym. Sci.* **2012**, *37*, 1657–1677. [CrossRef]
77. Muthuraj, R.; Misra, M.; Mohanty, A.K. Hydrolytic degradation of biodegradable polyesters under simulated environmental conditions. *J. Appl. Polym. Sci.* **2015**, *132*, 42189. [CrossRef]
78. Papageorgiou, G.Z.; Bikiaris, D.N.; Achilias, D.S. Effect of molecular weight on the cold-crystallization of biodegradable poly(ethylene succinate). *Thermochim. Acta* **2007**, *457*, 41–54. [CrossRef]
79. Wang, X.S.; Yan, D.; Tian, G.H.; Li, X.G. Effect of molecular weight on crystallization and melting of poly(trimethylene terephthalate). 1: Isothermal and dynamic crystallization. *Polym. Eng. Sci.* **2001**, *41*, 1655–1664. [CrossRef]
80. Oyama, H.T.; Kimura, M.; Nakamura, Y.; Ogawa, R. Environmentally safe bioadditive allows degradation of refractory poly(lactic acid) in seawater: Effect of poly(aspartic acid-co-l-lactide) on the hydrolytic degradation of PLLA at different salinity and pH conditions. *Polym. Degrad. Stab.* **2020**, *178*, 109216. [CrossRef]
81. Aliotta, L.; Seggiani, M.; Lazzeri, A.; Gigante, V.; Cinelli, P. A Brief Review of Poly(Butylene Succinate) (PBS) and Its Main Copolymers: Synthesis, Blends, Composites, Biodegradability, and Applications. *Polymers* **2022**, *14*, 844. [CrossRef] [PubMed]
82. Farah, S.; Anderson, D.G.; Langer, R. Physical and mechanical properties of PLA, and their functions in widespread applications—A comprehensive review. *Adv. Drug Deliv. Rev.* **2016**, *107*, 367–392. [CrossRef]
83. Tsuji, H. In vitro hydrolysis of blends from enantiomeric poly(lactide)s. Part 4: Well-homo-crystallized blend and nonblended films. *Biomaterials* **2003**, *24*, 537–547. [CrossRef]
84. Zhou, J.; Wang, X.; Hua, K.; Duan, C.; Zhang, W.; Ji, J.; Yang, X. Enhanced mechanical properties and degradability of poly(butylene succinate) and poly(lactic acid) blends. *Iran. Polym. J.* **2013**, *22*, 267–275. [CrossRef]
85. Arhant, M.; Le Gall, M.; Le Gac, P.Y.; Davies, P. Impact of hydrolytic degradation on mechanical properties of PET-towards an understanding of microplastics formation. *Polym. Degrad. Stab.* **2019**, *161*, 175–182. [CrossRef]
86. Tham, C.Y.; Hamid, Z.A.A.; Ahmad, Z.; Ismail, H. Surface modification of poly(lactic acid) (PLA) via alkaline hydrolysis degradation. *Adv. Mater. Res.* **2014**, *970*, 324–327. [CrossRef]
87. Taniguchi, I.; Nakano, S.; Nakamura, T.; El-Salmawy, A.; Miyamoto, M.; Kimura, Y. Mechanism of Enzymatic Hydrolysis of Poly(butylene succinate) and Poly(butylene succinate-co-L-lactate) with a Lipase from *Pseudomonas cepacia*. *Macromol. Biosci.* **2002**, *2*, 447–455. [CrossRef]

polymers MDPI

Article

Influence of Cross-Linkers on the Wash Resistance of Chitosan-Functionalized Polyester Fabrics

Tanja Pušić [1], Tea Bušac [1] and Julija Volmajer Valh [2,*]

[1] Faculty of Textile Technology, University of Zagreb, Prilaz Baruna Filipovića 28a, 10000 Zagreb, Croatia; tanja.pusic@ttf.unizg.hr (T.P.); tea.busac@ttf.unizg.hr (T.B.)
[2] Faculty of Mechanical Engineering, University of Maribor, Smetanova 17, 2000 Maribor, Slovenia
* Correspondence: julija.volmajer@um.si

Abstract: This study investigates the wash resistance of polyester fabrics functionalized with chitosan, a biopolymer known for its biocompatibility, non-toxicity, biodegradability and environmentally friendly properties. The interaction of chitosan with synthetic polymers, such as polyester, often requires surface treatment due to the weak natural affinity between the two materials. To improve the interaction and stability of chitosan on polyester, alkaline hydrolysis of the polyester fabric was used as a surface treatment method. The effectiveness of using cross-linking agents 1,2,3,4-butane tetracarboxylic acid (BTCA) and hydroxyethyl methacrylate (HEMA) in combination with ammonium persulphate (APS) to improve the stability of chitosan on polyester during washing was investigated. The wash resistance of polyester fabrics functionalized with chitosan was tested after 1, 5 and 10 washes with a standard ECE detergent. Staining tests were carried out to evaluate the retention of chitosan on the fabric. The results showed that polyester fabrics functionalized with chitosan without cross-linkers exhibited better wash resistance than the fabrics treated with crosslinkers.

Keywords: polyester; functionalization; chitosan; cross-linkers; stability; washing

Citation: Pušić, T.; Bušac, T.; Volmajer Valh, J. Influence of Cross-Linkers on the Wash Resistance of Chitosan-Functionalized Polyester Fabrics. *Polymers* **2024**, *16*, 2365. https://doi.org/10.3390/polym16162365

Academic Editors: Stefano Farris and Masoud Ghaani

Received: 22 July 2024
Revised: 11 August 2024
Accepted: 18 August 2024
Published: 21 August 2024

1. Introduction

After cellulose, chitosan is the most abundant biopolymer [1] and is used as a natural polysaccharide with exceptional biological and physico-chemical properties, as well as its environmental friendliness, in various areas [2] such as medicine, biomedicine [3], pharmacy, cosmetics, the textile, chemical and paper industries and agriculture. Chitosan has two important structural parameters, namely the degree of deacetylation (DD) and the molecular weight (MW). The degree of deacetylation determines the solubility and biodegradability of chitosan, which is closely related to the degree of crystallization [1,4,5].

Chitosan has primary amino groups with a pKa value of 6.3, the presence of which determines the properties of chitosan and its pH sensitivity. At a pH value below its pKa, chitosan is a polycation, while at a pH value ≤ 4, it is completely protonated. On the other hand, chitosan loses its electrical charge due to the deprotonation of the amine groups and becomes insoluble when the pH rises above 6. Therefore, the chitosan is prepared in an acidic medium consisting of inorganic and organic acids, preferably with organic acids [5–7].

Chitosan is of increasing interest when it comes to adding functionalities to textile surfaces [8]. However, the weak interaction of chitosan to textile fibers is the main problem in its application. Citric acid and other weak oxidizing agents have been shown to promote effective cross-linking between chitosan and textile substrates such as cotton and its blends [6,8–11].

The compatibility of two polymers in a mixture can be extended by modification depending on the composition [12]. Considering the partial compatibility of biopolymer chitosan with polyester as a synthetic polymer, modification of the materials should be carried out by classical or advanced methods to introduce functional groups [8,13–15].

The alkaline hydrolysis of polyester textiles with sodium hydroxide hydrolyzes the ester groups in the polymer chain so that modified fragments remain in the polyester structure. Alkaline-hydrolyzed polyesters have an improved reactivity and hydrophilicity of the textiles due to the introduction of carboxyl groups (-COOH) [8,14]. The reaction of alkaline hydrolysis can be catalyzed by cationic promoters, which can reduce the concentration of sodium hydroxide or short reaction time [11,13,16]. Alkaline hydrolysis reduces the tensile properties of polyester textiles, but functionalization with chitosan restores them and improves resistance to deformation, the wetting ability and hydrophilicity, antimicrobial properties, reduces charging by static electricity and reduces the proportion of released fragments from polyester textiles in the washing process [11,17].

Efficacy and durability of chitosan-functionalized polyester fabrics are certain limitations, so the aim of this study was to analyze the stability of three chitosan-functionalized polyester fabrics (chitosan itself, chitosan with BTCA and chitosan with HEMA/APS) during the washing process. The solution of chitosan (low molecular weight, LMW) was prepared in hydrochloric acid as a pH-controlling agent, which differs from the majority of available protocols referring to acetic acid [6,8,14,18]. Furthermore, the functionalization of polyester fabrics was carried out with a chitosan concentration of 1%, which is less than most protocols in the selected published papers [7,8,19,20]. The interaction of chitosan as well as chitosan cross-linkers with polyester was improved by alkaline hydrolysis of the polyester reference fabric samples. The effect of the functionalization of a polyester reference fabric with a biopolymer chitosan before and after one, three and five washing cycles was analyzed by a staining test with acid dyestuff. The staining test is a simple, effective and affordable method to confirm the presence of chitosan on fabric and does not require expensive research equipment.

2. Materials and Methods

2.1. Materials and Reagents

The research was carried out on the polyester reference fabric, supplied by Centre for Testmaterials, CFT, Vlaardingen, The Netherlands, specified by mass per unit area of 156 g/m^2, density in warp direction 27.7 cm^{-1} and 20 cm^{-1} threads in weft direction, fineness of the warp threads 30.4 tex and fineness of weft threads 31.9 tex. The ultrasonic cutter model TTS400, Sonowave S.r.l., Legnano MI, Italy was used to prepare 30 × 50 cm polyester reference fabric samples.

The following chemicals were used: sodium hydroxide (NaOH), supplied by Ivero d.d.; chitosane (low molecular weight (LMW) with an 85% degree of deacetylation); 1,2,3,4-butanetetracarboxylic acid (BTCA) (99%) and calcium carbonate ($CaCO_3$), supplied by Sigma Aldrich, St. Louis, USA; hydrochloric acid (HCl, 37%), supplied by GRAM-MOL, Zagreb, Croatia, 2-hydroxyethyl methacrylate (HEMA) (97%) and 4-methoxyphenol, supplied by Thermo Fisher (Kandel) GmbH, Karlsruhe, Germany; ammonium persulphate (APS), supplied by BDH Chemicals, Leuven, Belgium; standard ECE A detergent from SDL Atlas; and dye Telon® Blue M-GLW (C.I. Acid Blue 221) supplied by DyStar Colours Distribution GmbH, Raunheim, Germany.

2.2. Procedures

2.2.1. Alkali Hydrolysis

In order to improve the interaction between the surface of the synthetic polymer and the biopolymer chitosan, the polyester reference fabric was modified by treatments in solutions of NaOH. Alkaline hydrolysis of the fabric was carried out in solutions of 10 g/L (1AH), 20 g/L (2AH) and 30 g/L NaOH (3AH) with a bath ratio of 1:5. The polyester fabrics were inserted in a chamber of a total volume of 10 L, sodium hydroxide of a certain concentration prepared in soft water (50 ppm $CaCO_3$) was added and the hydrolysis started at 20 °C in a Polymat laboratory apparatus, Mathis, Switzerland. The chamber was heated by gradient of 4 °C/min until a temperature of 98 °C was achieved. The duration of alkali hydrolysis was 30 min. After the end of alkaline hydrolysis, the polyester fabrics were removed from the bath and rinsed four times due to removal of residual alkali, two times at 80 °C and two times with cold water. The so-prepared samples were air-dried.

2.2.2. Functionalization of Polyester Fabrics with Chitosan

The low-molecular-weight chitosan with a deacetylation degree of 85% was used to functionalize the reference polyester fabrics. The multiphase preparation of the stable chitosan solution (1%) is pH-sensitive and time-consuming. Hydrochloric acid (1 mol/L) was used to prepare the acidic chitosan solution with a pH value of 3.62 and to adjust the pH value after stirring overnight.

For the functionalization of the polyester reference fabric, the cross-linker BTCA was added to the prepared chitosan solution (1%) in amount of 5% per mass of chitosan. The HEMA stabilized with approximately 500 ppm 4-methoxyphenol in an amount of 5% and a catalyst APS in an amount of 1%, were added to the chitosan solution (1%) before functionalization of the polyester reference fabric.

The application of chitosan on polyester fabrics was carried out in three variants: with chitosan itself (Ch), chitosan with cross-linker 1,2,3,4-butanetetracarboxylic acid (Ch*BTCA) and chitosan with cross-linker 2-hydroxyethyl methacrylate with ammonium persulphate (Ch*HEMA/APS).

The functionalization of the reference polyester fabric without (U) and after the alkaline hydrolysis processes (1AH, 2AH, 3AH) with chitosan (Ch) solution, chitosan with BTCA (Ch*BTCA) solution and chitosan with HEMA/APS (Ch*HEMA/APS) solution was carried out on the padder and stenter, Ernst Benz, Rümlang-Zurich, Switzerland, at a pressure of 12.5 kg/cm. Thermal processes, drying at 90 °C for 40 s and curing at 130 °C for 20 s, followed the functionalization of the polyester fabrics.

2.2.3. Washing Process

The stability of chitosan-polyester fabrics in the washing process was tested according to HRN EN ISO 6330 (2A) procedure [21] using a standard ECE A detergent (1.25 g/L) at 60 °C with a bath ratio of 1:7 over 1, 3 and 5 cycles in the Rotawash laboratory machine, Atlas SDL. Washing in hard water was followed by rinsing over 4 cycles with a bath ratio of 1:8 and air-drying in a horizontal position. The labelling of the treatments and samples is shown in Table 1.

Table 1. Designation of polyester fabrics.

Polyester Sample	Label
Untreated	U
Alkali treated (NaOH) 10 g/L, 20 g/L, 30 g/L	1AH, 2AH, 3AH
Chitosan treated	Ch-U Ch-1AH, Ch-2AH, Ch-3AH
Chitosan treated with BTCA	Ch*BTCA-U Ch*BTCA-1AH, Ch*BTCA-2AH, Ch*BTCA-3AH
Chitosan treated with HEMA/APS	Ch*HEMA/APS-U Ch*HEMA/APS-1AH, Ch*HEMA/APS-2AH, Ch*HEMA/APS-3AH
Washed, cycles	−1×, −3×,−5×

2.2.4. Staining Test

The effect of the functionalization of a polyester reference fabric with a biopolymer chitosan in three variations (with chitosan itself, and with two cross-linkers (BTCA, HEMA/APS), and its stability in the washing process was analyzed by a staining test carried out using the dye Telon® Blue M-GLW (C.I. Acid Blue 221). The 5 × 5 cm samples were soaked in Petri dishes with the dye solution (1 %) for 15 min. After soaking, the samples were rinsed with water to remove excess non-fixed dye and air-dried.

All processes are presented in Scheme 1.

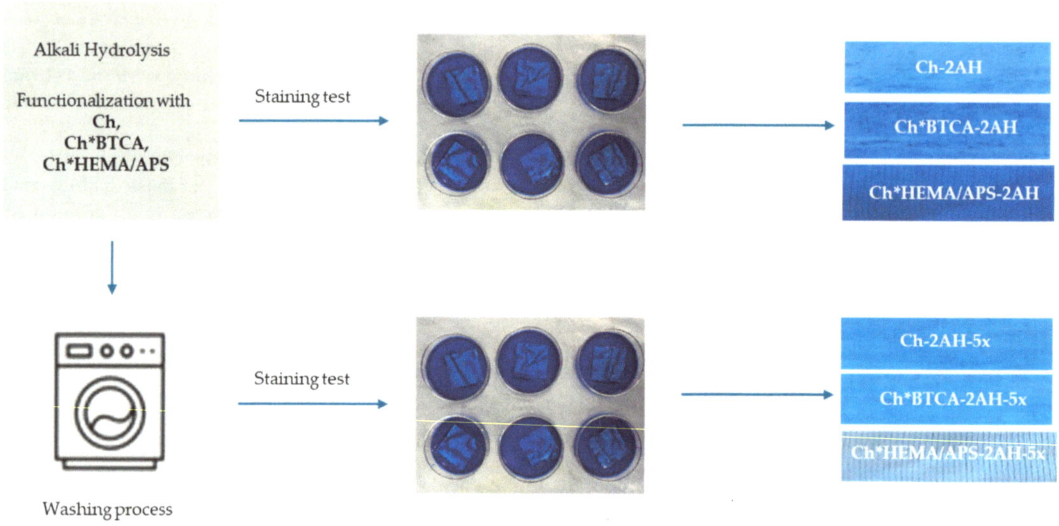

Scheme 1. Schematic representation of processes.

2.3. *Methods*

2.3.1. Gravimetric Analysis

Samples of the reference polyester fabrics before and after alkaline hydrolysis were weighed to determine the weight loss (Δm) as an indicator of the reaction with a surface and topochemical effect of alkaline hydrolysis.

2.3.2. Streaming Potential Method

The zeta potential of samples was determined by streaming potential method in the SurPASS electrokinetic analyzer equipped with a titration unit and an adjustable gap cell (AGC) controlled with Attract software 2.0 (all from Anton Paar GmbH, Graz, Austria). After stabilization of the measurement parameters in the electrokinetic analyzer, the streaming potential of the polyester fabric samples before and after alkaline hydrolysis in 20 g/L NaOH (sample 2AH) in the electrolyte solution (1 mM/L KCl) was measured as a function of pH, starting from alkaline (adjusted with 1 mol/L NaOH) to acidic (adjusted with 1 mol/L HCl). During the titration procedure, the streaming potential in mV and other parameters were recorded, from which the zeta potential was calculated as a function of pH according to the Helmholtz–Smoluchovsky equation [22,23].

2.3.3. Microscopic Observation

Photographs of stained polyester fabrics (micrographs) were taken using a DinoLite digital microscope, Premier IDCP B.V., Almere, The Netherlands, at 50× magnification.

2.3.4. Remission Spectrophotometry

The spectral values of polyester reference fabrics were determined using the DataColor SF300 spectrophotometer, aperture 2.2 cm, illumination D65 and geometry d/8°.

The whiteness of untreated and alkaline hydrolyzed polyester reference fabrics was evaluated according to AATCC test method 110 [24].

In accordance with [25], the total color difference, the ΔE, of all dyed samples was calculated as follows (Equation (1)):

$$DE^* = \sqrt{(DL^*)^2 + (DC^*)^2 + (DH^*)^2} \quad (1)$$

where

DL^*—difference in lightness ($DL^* = L^*$washed sample − L^*standard),
DC^*—difference in chroma ($DC^* = C^*$washed sample − C^*standard),
DH^*—difference in hue ($DH^* = H^*$washed sample − H^*standard).

Additionally, the evaluation of the change in color of washed samples to the appropriate standard based on greyscale values was evaluated according to ISO and AATCC [26,27].

Color strength (K/S value) was determined for selected chitosan-functionalized polyester samples before and after 5 washes.

3. Results

3.1. Alkaline Hydrolysis

The alkaline hydrolysis of polyester cleaved its ether bonds and formed -COOH groups, which increased its reactivity and hydrophilicity. The increase in polar -COOH groups after alkaline hydrolysis improved the wettability of polyester. However, this treatment also created craters that changed the porosity of the material and caused a reduction in mass.

3.1.1. Weight Loss

The weight losses of polyester reference fabrics resulting from the performed alkaline hydrolysis processes are presented in Table 2.

Table 2. Weight loss of polyester reference fabrics after alkaline hydrolysis.

Reference Polyester Fabric	Δm (%)
1AH	1.5 ± 0.04
2AH	4.2 ± 0.06
3AH	7.1 ± 0.01

The alkaline hydrolysis of polyester textiles is a topochemical process in sodium hydroxide solution at high temperatures, which should be maintained so that the weight loss does not exceed 5% [8]. The weight loss of polyester fibers brings with it many advantageous properties, such as increased absorbency, hydrophilicity, moisture recovery and dye absorption, while at the same time reducing the tendency to pilling and the generation of static charge.

The results in Table 2 show the effect of NaOH concentration on the weight loss of the polyester fabrics. Treatment with 30 g/L (3AH) resulted in a weight loss of 7.1%, which, according to [8], exceeds recommended value of 5%. Despite this fact, all polyester reference fabric samples (1AH, 2AH, 3AH) were further functionalized with chitosan solutions to analyze the influence of the surface modification on the uptake of the biopolymer chitosan as well as driven by cross-linkers applied in functionalization of polyester fabric.

3.1.2. Whiteness Quality

As the whiteness quality is an important criterion in all phases of pre-treatment, the whiteness degree (W_{CIE}), the basic whiteness (Y) and the change in tint (TV, TD) of the untreated (U) and the alkaline-hydrolyzed polyester samples (AH) were measured accordingly, as shown in Table 3.

Table 3. Whiteness degree of untreated and the alkaline-hydrolyzed polyester samples.

Samples	W	TV	TD	Y
U	68.1	0.1		81.8
1AH	56.7	−0.5	R1	80.3
2AH	72.2	0.0		82.2
3AH	71.6	0.0		81.7

The results in Table 3 show changes in whiteness quality that were not completely harmonized. Namely, the 1AH process reduced the whiteness by 10 units, with some reduction in the basic whiteness of (Y) and a deviation in tint (R1). The 2AH and 3AH processes, on the other hand, increased the degree of whiteness by 3–4 units. Enhanced whiteness quality proves a higher degree of alkaline cleaning and removal of preparations on the surface of the polyester reference fabric [28].

3.1.3. Streaming Potential

The level of surface modification of polyester fabric after alkaline hydrolysis was assessed by the streaming potential method. The comparison of the zeta potential curves of untreated (U) and alkali-treated polyester fabric (2AH) (Figure 1) proved a modification of the surface. More negative values of the zeta potential showed that the alkali-treated polyester fabric had more accessible groups.

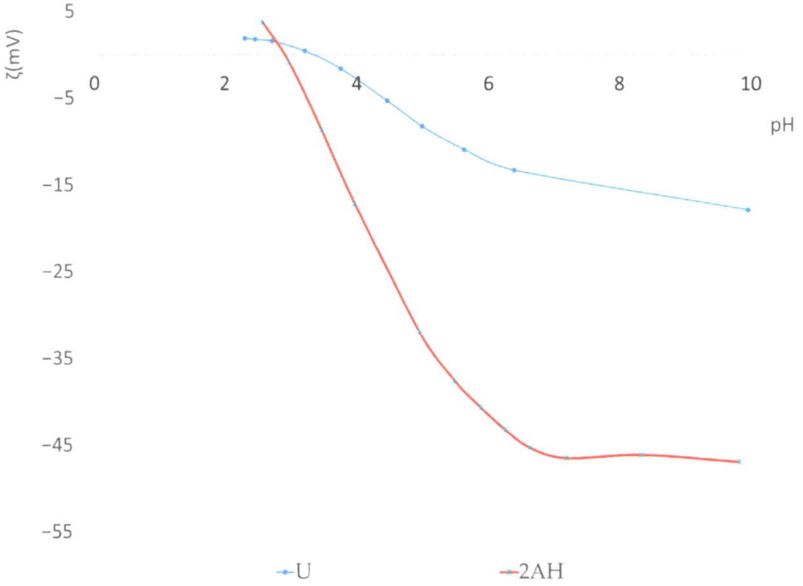

Figure 1. Zeta potential of polyester fabric before (U) and after (2AH) alkaline hydrolysis in variation of pH.

3.2. Stainability of Chitosan-Functionalized Polyester Fabrics

The staining test is a practical method for detecting the presence of chitosan. Depending on the substrate, different types of dyes (reactive, disperse, acid) can be used that react specifically with chitosan. Color intensity confirms how much chitosan is present on the surface or deposited from the surface. Both samples were colored blue, with the blank sample having a light color and the chitosan-functionalized polyester sample having a dark color [8,20,28,29]. According to [14], chitosan improves the stainability of the fabrics with acid dye, Telon Turquoise (C.I. Acid Blue 185), owing to the ionic interaction between protonated amino groups and sulfonic groups of the dye ions.

The color intensity and homogeneity of chitosan-functionalized polyester fabrics was tested by staining with Telon® Blue M-GLW (C.I. Acid Blue 221).

Untreated and alkaline hydrolyzed samples were also stained with the Telon Blue dyestuff, and the appearance of untreated and alkaline treated samples is shown in Table 4.

Table 4. Stained untreated and alkaline-hydrolyzed samples.

Stained Sample	U	1AH	2AH	3AH
Micrograph				

The differences between the samples in Table 4 indicate a different dye substantivity depending on the topography and the surface. The pale blue color of the untreated samples was according to the literature [29]. The colors of 1AH, 2AH and 3AH were consistent with the whiteness and roughness influenced by the topochemical reaction that was the result of alkaline hydrolysis. The alkaline hydrolysis impaired the binding of Telon® Blue M-GLW to the polyester samples.

In Table 5, the microscopic images of cthe olored polyester reference samples (without and after the alkaline hydrolysis processes) functionalized with chitosan, chitosan with BTCA and chitosan with HEMA/APS before and after 1, 3 and 5 wash cycles are shown.

The presence of chitosan on the surface of the polyester reference samples was detected by the blue coloration of the samples before and after washing (Table 5). Alkaline hydrolysis of polyester fabric improved its functionalization with chitosan by increasing the availability of reactive sites on the fiber surface. The treatment involved breaking ester linkages within the polyester polymer, leading to the formation of carboxyl (-COOH) and hydroxyl (-OH) groups on the fabric surface. These functional groups provided active sites for the bonding of chitosan, which was rich in amino groups (-NH_2). The degree of functionalization with chitosan was most pronounced when the polyester fabric was treated with a 30 g/L NaOH solution. According to this coloring of the samples in Table 5, the chitosan was preserved on surface of all washed samples.

An indicator for the stability of chitosan in the washing process are the spectral values and the fastness degree according to AATCC and ISO A05, shown in Tables 6–10.

The results in Tables 6–9 show that the wash cycles influenced the spectral values of the polyester reference fabrics. Just the first wash cycle reduced the spectral values, which continue to change over three cycles. The spectral values of all samples treated with chitosan were slightly reduced at five cycles compared to three cycles. The most favorable spectral values, confirming the good interaction of chitosan with polyester, were obtained from the chitosan-functionalized samples (Ch) and the chitosan alkaline-hydrolyzed samples (Ch-3AH). According to these indicators, it is desirable to carry out the alkaline hydrolysis of (3AH) as a preparatory phase of the topographical changes that favor compatibility with chitosan, as shown in Table 9.

Micrographs of untreated (U) and alkaline-hydrolyzed (1AH, 2AH, 3AH) polyester reference fabrics functionalized with chitosan and cross-linker BTCA before and after 1, 3 and 5 washes are shown in Table 10.

Table 5. Micrographs of stained polyester samples (U, 1AH, 2AH, 3AH) functionalized with chitosan (Ch) before and after 1, 3 and 5 washing cycles.

Samples	**Ch-U**	**Ch-1AH**	**Ch-2AH**	**Ch-3AH**
Micrograph				
Samples	Ch-U-1x	Ch-1AH-1x	Ch-2AH-1x	Ch-3AH-1x
Micrograph				
Samples	Ch-U-3x	Ch-1AH-3x	Ch-2AH-3x	Ch-3AH-3x
Micrograph				
Samples	Ch-U-5x	Ch-1AH-5x	Ch-2AH-5x	Ch-3AH-5x
Micrograph				

Table 6. Spectral values of the washed chitosan polyester reference fabric, Ch-U.

	Cycles	**ISO A05**	**AATCC**	**DE**	**DL***	**DC***	**DH***
Standard Ch-U	1	4	3-4	2.564	1.318	1.553	−1.258
	3	3-4	3	3.813	−0.041	3.461	−0.251
	5	3-4	3	2.986	2.344	1.068	−0.864

Table 7. Spectral values of washed chitosan alkaline-hydrolyzed polyester reference fabric, Ch-1AH.

	Cycles	**ISO A05**	**AATCC**	**DE**	**DL***	**DC***	**DH***
Standard Ch-1AH	1	3-4	3	3.780	0.919	3.234	−1.292
	3	3-4	3	3.778	0.738	3.503	−0.829
	5	3-4	3	3.590	2.352	2.200	−1.156

Table 8. Spectral values of washed chitosan alkaline-hydrolyzed polyester reference fabric, Ch-2AH.

	Cycles	ISO A05	AATCC	DE	DL*	DC*	DH*
Standard Ch-2AH	1	2-3	2	6.466	5.728	−1.235	−2.715
	3	3	3	3.979	2.988	1.982	−1.130
	5	2	2	7.207	6.639	−0.999	−2.548

Table 9. Spectral values of washed chitosan alkaline-hydrolyzed polyester reference fabric, Ch-3AH.

	Cycles	ISO A05	AATCC	DE	DL*	DC*	DH*
Standard Ch-3AH	1	4	3	3.302	−0.683	3.029	0.695
	3	4	4	2.118	1.359	1.311	0.280
	5	3-4	3-4	2.837	1.825	1.851	0.080

Microscopic images of colored Ch*BTCA polyester reference samples (U, 1AH, 2AH, 3AH) showed uneven coloration on the surface of all chitosan treated samples. BTCA, as a cross-linker during processing, influenced the localization of chitosan on the surface of the polyester reference fabric. The degree of this non-uniformity in coloration was particularly pronounced in the alkaline-hydrolyzed samples treated with chitosan and BTCA (Ch*BTCA-1AH, Ch*BTCA-2AH, Ch*BTCA-3AH) when compared to the untreated chitosan BTCA sample (Ch*BTCA-U). The non-uniform fragments were deposited during washing processes. The change in the intensity of the blue color showed that five washing cycles caused the deposition of chitosan from the surface of the polyester reference fabrics. The quantification of the differences between the samples before and after the washes is carried out using the spectral values and the color fastness evaluation in Tables 11–14.

The changes in the spectral values in Tables 11–14 show that the stability of chitosan with the cross-linker Ch*BTCA in the washing process was weaker than that of Ch itself. The differences in the spectral values in Tables 11–14 show the influence of the sodium hydroxide concentration and the washing process on the deposition of chitosan. The change in the spectral values of the samples evaluated using the ISO and AATCC greyscale values showed a lower deposition of chitosan from sample Ch*BTCA-2AH in the washing process. According to all spectral parameters, alkali hydrolysis at 20 g/L was the optimum preparation concentration for good and stable chitosan uptake on polyester fabrics.

Micrographs of untreated (U) and alkaline-hydrolyzed (1AH, 2AH, 3AH) polyester reference fabric functionalized with chitosan and cross-linker HEMA/APS before and after 1, 3 and 5 washes are shown in Table 15.

The intensity of the blue color of polyester standard fabrics functionalized with crosslinker HEMA/APS was slightly lighter than that of chitosan with cross-linker BTCA. However, the blue shade of all samples was retained over five wash cycles. The structural elements (warp and weft threads) of all samples washed with Ch*HEMA/APS were more visible compared to the samples washed with Ch and Ch*BTCA. This proves that the chitosan had almost completely separated from the surface of the Ch*HEMA/APS samples during the washing process.

The quantification of the changes in the samples functionalized with chitosan and the crosslinker HEMA/APS through the wash cycles is shown in Tables 16–19.

The changes in the spectral values in Tables 16–19 show that the stability of chitosan with the cross-linker HEMA/APS in the washing process was weaker than that of Ch itself and Ch*BTCA. Already after the first wash cycle, the persistence of color fastness was rated as 1, which confirms the further hydrolysis of fabrics treated with Ch*HEMA/APS in alkaline detergent solution. The spectral value, the total color difference (DE) was such that all samples were graded by the same category of changes, grade 1, although the lightness changes (DL*), chroma changes (DC*) and hue changes (DH*) of the samples differed. Within the analyzed series of chitosan-treated samples,

Ch*HEMA/APS and Ch*HEMA/APS-1AH retained more chitosan than Ch*HEMA/APS-2AH and Ch*HEMA/APS-3AH.

Taking into account the less use of chemicals and optimal effects, primary parameters weight loss and degree of whiteness, the alkaline hydrolysis (2AH) proved to be optimal. So, the results of the color strength (K/S) for all samples (2AH, Ch-2AH, Ch*BTCA-2AH, Ch*HEMA/APS-2AH) before and after five wash cycles (Ch-2AH-5x, Ch*BTCA-2AH-5x, Ch*HEMA/APS-2AH-5x) were selected for the final stability checking point, as shown in Table 20.

Table 10. Micrographs of stained polyester samples (U, 1AH, 2Ah, 3AH) with chitosan and cross-linker BTCA (Ch*BTCA) before and after 1, 3 and 5 washes.

Samples	**Ch*BTCA-U**	**Ch*BTCA-1AH**	**Ch*BTCA-2AH**	**Ch*BTCA-3AH**
Micrograph	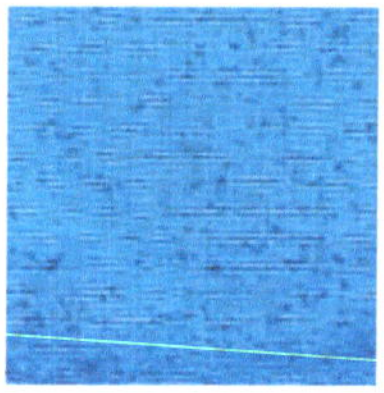			
Samples	Ch*BTCA-U-1x	Ch*BTCA-1AH-1x	Ch*BTCA-2AH-1x	Ch*BTCA-3AH-1x
Micrograph				
Samples	Ch*BTCA-U-3x	Ch*BTCA-1AH-3x	Ch*BTCA-2AH-3x	Ch*BTCA-3AH-3x
Micrograph				
Samples	Ch*BTCA-U-5x	Ch*BTCA-1AH-5x	Ch*BTCA-2AH-5x	Ch*BTCA-3AH-5x
Micrograph				

Table 11. Spectral values of the washed chitosan polyester reference fabric, Ch*BTCA.

	Cycles	ISO A05	AATCC	DE	DL*	DC*	DH*
Standard Ch*BTCA	1	2-3	2-3	4.665	3.973	0.077	−2.274
	3	2	2	6.321	5.744	−1.445	−2.189
	5	2-3	2-3	4.938	4.588	−0.648	−1.686

Table 12. Spectral values of the washed chitosan alkaline-hydrolyzed polyester reference fabric, Ch*BTCA-1AH.

	Cycles	ISO A05	AATCC	DE	DL*	DC*	DH*
Standard Ch*BTCA-1AH	1	2	2	6.338	5.512	0.502	−3.072
	3	2	2	7.217	6.581	−0.633	−2.764
	5	2	2	8.241	7.479	−1.491	−3.090

Table 13. Spectral values of washed chitosan alkaline-hydrolyzed polyester reference fabric, Ch*BTCA-2AH.

	Cycles	ISO A05	AATCC	DE	DL*	DC*	DH*
Standard Ch*BTCA-2AH	1	3	3	3.923	1.958	2.665	−1.722
	3	2-3	2-3	5.261	4.627	0.159	−2.418
	5	2-3	2-3	4.673	3.715	1.541	−2.262

Table 14. Spectral values of washed chitosan alkaline-hydrolyzed polyester reference fabric, Ch*BTCA-3AH.

	Cycles	ISO A05	AATCC	DE	DL*	DC*	DH*
Standard Ch*BTCA-3AH	1	2	2	6.749	5.811	0.089	−3.410
	3	2	1-2	8.526	7.465	−1.447	−3.840
	5	1-2	1-2	9.487	8.407	−2.075	−3.870

Table 15. Micrographs of stained polyester samples (U, 1AH, 2Ah, 3AH) with chitosan and cross-linker HEMA/APS (Ch*HEMA/APS) before and after 1, 3 and 5 washes.

Samples	**Ch*HEMA/APS-U**	**Ch*HEMA/APS-1AH**	**Ch*HEMA/APS-2AH**	**Ch*HEMA/APS-3AH**
Micrograph				
Samples	Ch*HEMA/APS-U-1x	Ch*HEMA/APS-1AH-1x	Ch*HEMA/APS-2AH-1x	Ch*HEMA/APS-3AH-1x
Micrograph				

Table 15. *Cont.*

Samples	Ch*HEMA/APS-U-3x	Ch*HEMA/APS-1AH-3x	Ch*HEMA/APS-2AH-3x	Ch*HEMA/APS-3AH-3x
Micrograph				
Samples	Ch*HEMA/APS-U-5x	Ch*HEMA/APS-1AH-5x	Ch*HEMA/APS-2AH-5x	Ch*HEMA/APS-3AH-5x
Micrograph			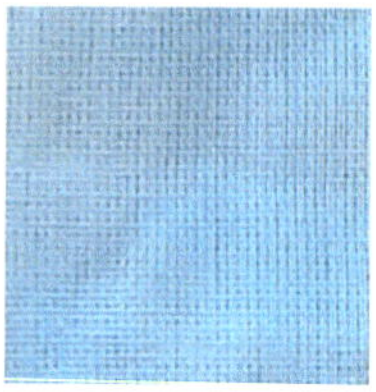	

Table 16. Spectral values of the washed chitosan polyester reference fabric, Ch*HEMA/APS-U.

	Cycles	ISO A05	AATCC	DE	DL*	DC*	DH*
Standard Ch*HEMA/APS-U	1	1	1	22.514	15.027	−14.805	−7.866
	3	1	1	22.023	14.794	−14.536	−7.379
	5	1	1	22.588	15.803	−14.250	−7.573

Table 17. Spectral values of the washed chitosan alkaline-hydrolyzed polyester reference fabric, Ch*HEMA/APS-1AH.

	Cycles	ISO A05	AATCC	DE	DL*	DC*	DH*
Standard Ch*BTCA-1AH	1	1	1	17.446	11.939	−11.246	−5.672
	3	1	1	23.543	17.406	−15.008	−6.460
	5	1	1	25.545	18.144	−16.470	−7.187

Table 18. Spectral values of the washed chitosan alkaline-hydrolyzed polyester reference fabric, Ch*HEMA/APS-2AH.

	Cycles	ISO A05	AATCC	DE	DL*	DC*	DH*
Standard Ch*BTCA-2AH	1	1	1	31.000	22.942	−18.379	−9.937
	3	1	1	33.907	25.103	−20.304	−10.360
	5	1	1	35.050	25.540	−21.510	−10.651

Table 19. Spectral values of the washed chitosan alkaline-hydrolyzed polyester reference fabric, Ch*HEMA/APS-3AH.

	Cycles	ISO A05	AATCC	DE	DL*	DC*	DH*
Standard Ch*BTCA-3AH	1	1	1	26.368	19.758	−14.951	−9.007
	3	1	1	30.535	22.440	−18.457	9.387
	5	1	1	31.422	22.929	−19.159	−9.720

Table 20. Color strength (K/S) of 2AH and chitosan-functionalized polyester samples before and after 5 wash cycles.

Samples	K/S	K/S, Checksum	Micrographs
2AH	0.02	0.69	
Ch-2AH	0.24	63.88	
Ch-2AH-5x	0.13	42.34	
Ch*BTCA-2AH	0.21	56.71	
Ch*BTCA-2AH-5x	0.13	40.95	
Ch*HEMA/APS-2AH	0.21	89.67	
Ch*HEMA/APS-2AH-5x	0.12	32.49	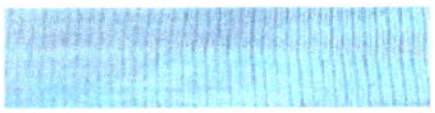

The results in Table 20 confirm the presence of chitosan on chitosan-functionalized polyester fabrics according to color strength and blue coloration; HEMA/APS provided the best results. However, stability analysis over five wash cycles gave priority to the treatment of polyester fabric with chitosan itself, without cross-linkers.

4. Conclusions

The chitosan solution, which was prepared by adding hydrochloric acid as a pH-sensitizing solution, confirmed the applicability of this acid in all variations, with and without cross-linking agents, due to its viscosity and time stability. Alkaline hydrolysis of the polyester fabric improved the degree of its functionalization with the biopolymer chitosan. Within the applied sodium hydroxide concentration, the optimal chitosan-functionalized polyester was obtained with 2% NaOH. The cross-linking agents 1,2,3,4-butanetetracarboxylic acid (BTCA) and hydroxyethyl methacrylate (HEMA) with ammonium persulphate (APS) used did not improve the degree of functionality of polyester and alkaline-hydrolyzed polyester with chitosan. The results confirmed the influence of the washing process on the reduction of the color strength of the chitosan-functionalized polyester, the chitosan-functionalized, alkaline-hydrolyzed polyester and, consequently, on the deposition of chitosan. The staining test for all polyester fabric samples before and after the washing process confirmed that the cross-linkers used did not improve the stability of the chitosan-functionalized polyester.

The staining test confirmed that five washing cycles had an impact on the reduction of spectral values, as quantified by the values of total color difference, fastness levels and color strength compared to unwashed polyester fabrics.

Author Contributions: Conceptualization, T.P. and J.V.V.; methodology, T.B. and J.V.V.; formal analysis, T.B., J.V.V. and T.P.; investigation, T.B. and T.P.; writing—original draft preparation, T.P. and J.V.V. All authors have read and agreed to the published version of the manuscript.

Funding: This research was funded by Croatian Science Foundation, grant numbers HRZZ-IP-2020-02-7575.

Institutional Review Board Statement: Not applicable.

Data Availability Statement: The original contributions presented in the study are included in the article; further inquiries can be directed to the corresponding author.

Acknowledgments: The part of the research was performed on equipment purchased by K.K.01.1.1.02.0024 project "Modernization of Textile Science Research Centre Infrastructure" (MI-TSRC).

Conflicts of Interest: The authors declare no conflicts of interest.

References

1. Takara, E.A.; Marchese, J.; Ochoa, N.A. NaOH treatment of Chitosan films: Impact on macromolecular structure and film properties. *Carbohydr. Polym.* **2015**, *132*, 25–30. [PubMed]
2. Morin-Crini, N.; Lichtfouse, E.; Torri, G.; Crini, G. Applications of chitosan in food, pharmaceuticals, medicine, cosmetics, agriculture, textiles, pulp and paper, biotechnology, and environmental chemistry. *Environ. Chem. Lett.* **2019**, *17*, 1667–1692.
3. Del Valle, L.; Diaz, A.; Puiggalí, J. Hydrogels for Biomedical Applications: Cellulose, Chitosan, and Protein/Peptide Derivatives. *Gels* **2017**, *3*, 217. [CrossRef]
4. Enescu, D. Use of chitosan in surface modification of textile materials. *Rom. Biotechnol. Lett.* **2008**, *13*, 4037–4048.
5. Saïed, N.; Aïder, M. Zeta Potential and Turbidimetry Analyzes for the Evaluation of Chitosan/Phytic Acid Complex Formation. *J. Food Res.* **2014**, *3*, 71–81.
6. Flinčec Grgac, S.; Tarbuk, A.; Dekanić, T.; Sujka, W.; Draczynski, Z. The Chitosan Implementation into Cotton and Polyester/Cotton Blend Fabrics. *Materials* **2020**, *13*, 1616. [CrossRef] [PubMed]
7. Joshi, M.; Ali, S.W.; Purwar, R. Ecofriendly antimicrobial finishing of textiles using bioactive agents based on natural products. *Indian J. Fibre Text. Res.* **2009**, *34*, 295–304.
8. Bhavsar, S.P.; Dalla Fontana, G.; Zoccola, M. Sustainable Superheated Water Hydrolysis of Black Soldier Fly Exuviae for Chitin Extraction and Use of the Obtained Chitosan in the Textile Field. *ACS Omega* **2021**, *13*, 8884–8893.
9. Luo, X.; Yao, M.Y.; Li, L. Application of chitosan in the form of textile: Production and sourcing. *Text. Res. J.* **2022**, *92*, 3522–3533.
10. Croisier, F.; Jerome, C. Chitosan-based biomaterials for tissue engineering. *Eur. Polym. J.* **2013**, *49*, 780–792.
11. Palacios-Mateo, C.; van der Meer, Y.; Seide, G. Analysis of the polyester clothing value chain to identify key intervention points for sustainability. *Environ. Sci. Eur.* **2021**, *33*, 2. [PubMed]
12. Vernaez, O.; Neubert, K.J.; Kopitzky, R.; Kabasci, S. Compatibility of Chitosan in Polymer Blends by Chemical Modification of Bio-based Polyesters. *Polymers* **2019**, *25*, 1939. [CrossRef]
13. Mitić, J.; Amin, G.; Kodrić, M.; Šmelcerović, M.; Đorđević, D. Polyester fibres structure modification using some organic solutions. *Tekstil* **2016**, *65*, 196–200.
14. Ferrero, F.; Periolatto, M. Chitosan Coating on Textile Fibers for Functional Properties. In *Handbook of Composites from Renewable Materials*; Thakur, V.K., Thakur, M.K., Kessler, M.R., Eds.; Wiley Online Library: Hoboken, NJ, USA, 2017; Volume 4, pp. 165–197.
15. Grancarić, A.M.; Pušić, T.; Kallay, N. Modifikacija poliesterskog vlakna alkalnom hidrolizom. *Polimeri* **1991**, *12*, 141–146.
16. Čorak, I.; Tarbuk, A.; Đorđević, D.; Višić, K.; Botteri, L. Sustainable alkaline hydrolysis of polyester fabric at low temperature. *Materials* **2022**, *15*, 1530. [CrossRef] [PubMed]
17. Periyasamy, A.P.; Tehrani-Bagha, A. A review on microplastic emission from textile materials and its reduction techniques. *Polym.Degrad. Stab.* **2022**, *199*, 109901.
18. Raza, Z.A.; Anwar, F.; Abid, S. Multi-response optimization in impregnation of chitosan nanoparticles on polyester fabric. *Polym. Bull.* **2019**, *76*, 3039–3058.
19. Hoque, M.-T.; Klinkhammer, K.; Mahltig, B. HT process for treatment of PET fabrics with chitosan containing recipes. Commun. *Dev. Assem. Text. Prod.* **2023**, *4*, 222–230.
20. Klinkhammer, K.; Hohenbild, H.; Hoque, M.T.; Elze, L.; Teshay, H.; Mahltig, B. Functionalization of Technical Textiles with Chitosan. *Textiles* **2024**, *4*, 70–90. [CrossRef]
21. *EN ISO 6330:2021*; Textiles—Domestic Washing and Drying Procedures for Textile Testing. European Committee for Standardization: Brussels, Belgium, 2021.
22. Bellmann, C.; Klinger, C.; Opfermann, A.; Böhme, F.; Adler, H.J. Evaluation of surface modification by electrokinetic measurements. *Prog. Org. Coat.* **2002**, *44*, 93–98.
23. Bišćan, J. Electrokinetic Data: Approaches, Interpretations and Applications. *Croat. Chem. Acta* **2007**, *80*, 357–365.
24. AATCC Test Method 110: Whiteness of Textiles. Available online: https://members.aatcc.org/store/tm110/521/ (accessed on 5 June 2024).
25. Parac-Osterman, Đ. *Osnove o Boji i Sustav Vrednovanja II. Izdanje*; Sveučilište u Zagrebu Tekstilno-tehnološki fakultet: Zagreb, Croatia, 2013.
26. *ISO 105-A03*; Textiles—Tests for Colour Fastness Part A03: Grey Scale for Assessing Staining. International Organization for Standardization (ISO): Geneva, Switzerland, 2019.
27. AATCC Evaluation Procedure 7 Instrumental Assessment of the Change in Color of a Test Specimen. Available online: https://members.aatcc.org/store/ep7/464/ (accessed on 5 June 2024).

28. Pušić, T.; Kaurin, T.; Liplin, M.; Budimir, A.; Čurlin, M.; Grgić, K.; Sutlović, A.; Valh, J.V. The Stability of the Chitosan Coating on Polyester Fabric in the Washing Process. *Tekstilec* **2023**, *66*, 85–104.
29. De Smet, D.; Vanneste, M. Application of Biobased and Biodegradable Materials in Textile Coating. In Proceedings of the International Federation of Associations of Textile Chemists and Colorists, IFATCC 2018, Greenville, SC, USA, 6–8 March 2018; pp. 1–8. Available online: https://www.ifatcc.org/wp-content/uploads/2018/01/O30.pdf (accessed on 5 June 2024).

polymers

MDPI

Article

Strategic Use of Vegetable Oil for Mass Production of 5-Hydroxyvalerate-Containing Polyhydroxyalkanoate from δ-Valerolactone by Engineered *Cupriavidus necator*

Suk-Jin Oh [1], Yuni Shin [1], Jinok Oh [1], Suwon Kim [1], Yeda Lee [1], Suhye Choi [1], Gaeun Lim [1], Jeong-Chan Joo [2], Jong-Min Jeon [3], Jeong-Jun Yoon [3], Shashi Kant Bhatia [1,4], Jungoh Ahn [5], Hee-Taek Kim [6,*] and Yung-Hun Yang [1,*]

1 Department of Biological Engineering, College of Engineering, Konkuk University, Seoul 05029, Republic of Korea; equal73@naver.com (S.-J.O.); sdbsdl0526@naver.com (Y.S.); xmfvm@naver.com (J.O.); rlatn990@naver.com (S.K.); karecurry@konkuk.ac.kr (Y.L.); suhye0823@konkuk.ac.kr (S.C.); lge0919@naver.com (G.L.); shashikonkukuni@konkuk.ac.kr (S.K.B.)
2 Department of Chemical Engineering, Kyung Hee University, Yongin-si 17104, Gyeonggi-do, Republic of Korea; jcjoo@khu.ac.kr
3 Department of Green & Sustainable Materials R&D, Research Institute of Clean Manufacturing System, Korea Institute of Industrial Technology (KITECH), Cheonan-si 31056, Chungcheongnam-do, Republic of Korea; j2pco@kitech.re.kr (J.-M.J.); jjyoon@kitech.re.kr (J.-J.Y.)
4 Institute for Ubiquitous Information Technology and Application, Konkuk University, Seoul 05029, Republic of Korea
5 Biotechnology Process Engineering Center, Korea Research Institute Bioscience Biotechnology (KRIBB), Cheongju-si 28116, Chungcheongbuk-do, Republic of Korea; ahnjo@kribb.re.kr
6 Department of Food Science and Technology, Chungnam National University, Daejeon 34134, Republic of Korea
* Correspondence: heetaek@cnu.ac.kr (H.-T.K.); seokor@konkuk.ac.kr (Y.-H.Y.); Tel.: +82-42-821-6722 (H.-T.K.); +82-2-450-2-3936 (Y.-H.Y.)

Abstract: Although efforts have been undertaken to produce polyhydroxyalkanoates (PHA) with various monomers, the low yield of PHAs because of complex metabolic pathways and inhibitory substrates remains a major hurdle in their analyses and applications. Therefore, we investigated the feasibility of mass production of PHAs containing 5-hydroxyvalerate (5HV) using δ-valerolactone (DVL) without any pretreatment along with the addition of plant oil to achieve enough biomass. We identified that $PhaC_{BP\text{-}M\text{-}CPF4}$, a PHA synthase, was capable of incorporating 5HV monomers and that *C. necator* PHB^{-4} harboring *phaC*$_{BP\text{-}M\text{-}CPF4}$ synthesized poly(3HB-*co*-3HHx-*co*-5HV) in the presence of bean oil and DVL. In fed-batch fermentation, the supply of bean oil resulted in the synthesis of 49 g/L of poly(3HB-*co*-3.7 mol% 3HHx-*co*-5.3 mol%5HV) from 66 g/L of biomass. Thermophysical studies showed that 3HHx was effective in increasing the elongation, whereas 5HV was effective in decreasing the melting point. The contact angles of poly(3HB-*co*-3HHx-*co*-5HV) and poly(3HB-*co*-3HHx) were 109 and 98°, respectively. In addition, the analysis of microbial degradation confirmed that poly(3HB-*co*-3HHx-*co*-5HV) degraded more slowly (82% over 7 days) compared to poly(3HB-*co*-3HHx) (100% over 5 days). Overall, the oil-based fermentation strategy helped produce more PHA, and the mass production of novel PHAs could provide more opportunities to study polymer properties.

Keywords: polyhydroxyalkanoates; 5-hydroxyvalerate; δ-valerolactone; plant oil

Citation: Oh, S.-J.; Shin, Y.; Oh, J.; Kim, S.; Lee, Y.; Choi, S.; Lim, G.; Joo, J.-C.; Jeon, J.-M.; Yoon, J.-J.; et al. Strategic Use of Vegetable Oil for Mass Production of 5-Hydroxyvalerate-Containing Polyhydroxyalkanoate from δ-Valerolactone by Engineered *Cupriavidus necator*. *Polymers* **2024**, *16*, 2773. https://doi.org/10.3390/polym16192773

Academic Editors: Masoud Ghaani and Stefano Farris

Received: 21 August 2024
Revised: 25 September 2024
Accepted: 26 September 2024
Published: 30 September 2024

1. Introduction

Polyhydroxyalkanoates (PHAs) are polymers that accumulate within microorganisms in nutrient-limited environments, and they have been studied as promising alternatives to conventional petroleum-based plastics because of their biodegradability and bio-based characteristics [1]. Moreover, the excellent biodegradability of PHAs, which decompose not only in soil but also in marine environments, makes them attractive candidates for use as bio-plastics [2]. PHAs are composed of various hydroxyalkanoate monomers that are present in microorganisms [3]. The most common PHA is poly(3-hydroxybutyrate) (P(3HB)), which consists of a 3HB monomer. P(3HB) had the advantage of being naturally produced in various microorganisms without the need for engineering [4]. However, its rigid and brittle nature, along with its high melting point, makes industrial utilization of PHB difficult [5]. These issues have been addressed by integrating other monomers into the polymer chains to synthesize copolymers, terpolymers, and tetrapolymers [6].

Poly(3-hydroxybutyrate-*co*-5-hydroxyvalerate) (PHB5HV or P(3HB-*co*-5HV)) is an actively researched PHA copolymer. 5-hydroxyvalerate (5HV) contributes to the flexibility of PHA polymers and decreases their melting points [7]. Additionally, the incorporation of 5HV into a polymer enhances its degradation rate by lipases; P(3HB-*co*-3HP-*co*-5HV) has been reported to exhibit low cytotoxicity and support cell proliferation [8]. Poly(3HB-*co*-5HV) can be synthesized by *Aneurinibacillus thermoaerophilus* and *Methylocystis parvus* [9,10]. One of the most studied PHA-producing strains, wild-type *Cupriavidus necator*, was also reported to be able to synthesize PHA containing 5HV; however, the mole fraction of 5HV in PHA produced by *C. necator* was very low [11]. Nonetheless, PHB5HV with a high mole fraction of 5HV could be synthesized by simply introducing PHA synthase with broad specificity into *C. necator*. Recently, in our lab, we successfully produced poly(3HB-*co*-5HV) with the mole fraction of 5HV reaching 70% using an engineered strain of *C. necator*. The PHB5HV produced by this strain had an increased elongation at a break of up to 1400% [12]. However, despite the favorable properties of PHAs containing 5HV monomers, their production is still relatively low, and further research on large-scale production is needed.

C. necator, also known as *Ralstonia eutropha*, can utilize various carbon sources, such as fructose, CO_2, and plant oils, to produce PHA [13]. To date, various studies have been conducted to synthesize different types of PHA from fructose by constructing pathways for monomer synthesis through genetic engineering in *Cupriavidus necator* or by supplying precursors alongside fructose [14–17]. However, producing various PHAs from fructose through genetic modification significantly reduces the yield of PHA, and in subsequent scale-up processes, it may not be possible to produce PHA with the desired mole fraction of monomers [18,19] (Table 1). Moreover, considering that the highest reported PHB yield from fructose fermentation is 18.46 g/L from 350 g/L of fructose, the method of supplying precursors along with fructose is also not suitable for the mass production of various PHAs.

Table 1. Production of co- or terpolymers with various monomers using *Cupriavidus necator* strains.

	C-Source	PHA Type	DCW (g/L)	PHA (g/L)	Content (%)	Ref.
Copolymers	fructose	P (3HB-*co*-2.1 mol% 3HP)	2.5	-	31	[18]
	fructose	P (3HB-*co*-37.7 mol% 3HHx)	1.42 ± 0.05	-	41.1 ± 3.0	[19]
	fructose	P (3HB-*co*-64.9 mol% 3HV)	1.5 ± 0.1	-	42.5 ± 3.8	[20]
	fructose + ε-CL	P (3HB-*co*-4HB)	7.5	-	-	[14]
	waste rapeseed oil + propanol	P (3HB-*co*-3HV)	14.7 ± 0.3	11.7 ± 0.7	80	[21]
		P (3HB-*co*-3HV) *	138	105	76	
	fructose + coconut oil	P (3HB-*co*-3HHx)	19	15	-	[22]
	palm kernel oil + butyrate	P (3HB-*co*-3HHx) *	153–175	113–138	73.3–78.6	[23]

Table 1. *Cont.*

	C-Source	PHA Type	DCW (g/L)	PHA (g/L)	Content (%)	Ref.
Terpolymers	fructose + 4HVA + 5HVA	P (3HB-*co*-3HV-*co*-4HV-*co*-5HV)	8.7 ± 0.1	6.3 ± 0.1	72	[12]
	fructose + GVL	P (3HB-*co*-3HV-*co*-4HV)	8.2 ± 0.2	-	80 ± 2	[14]
	fructose	P (3HB-*co*-3HV-*co*-3H4MV-*co*-3H2MP)	1.67 ± 0.03	0.93 ± 0.03	55.9 ± 1.8	[16]
	tung oil	P (3HB-*co*-3HV-*co*-3HHx)	1.65	0.68	41.2	[24]
	fructose + bean oil + DVL	P (3HB-*co*-3HHx-*co*-5HV) *	66	49	73	this work
	fructose + bean oil + DVL	P (3HB-*co*-3HHx-*co*-5HV) *	90	69	77	this work

Abbreviation: ε-CL, ε-caprolactone; GVL, γ-valerolactone; 4HVA, 4-hydroxyvaleric acid; 5HVA, 5-hydroxyvaleric acid; DVL, δ-valerolactone; 3HB, 3-hydroxybutyrate; 3HV, 3-hydroxyvalerate; 3H4MV, 3-hydroxy-4-methylvalerate; 3H2MP, 3-hydroxy-2-methylpropionate; 3HP, 3-hydroxypropionate; 3HHx, 3-hydroxyhexanoate; 4HB, 4-hydroxybutyrate; 4HV, 4-hydroxyvalerate; 5HV, 5-hydroxyvalerate. PHA with the mark * was produced through fed-batch fermentation.

Plant oils, such as soybean oil, palm oil, rapeseed oil, and jatropha oil, have been widely used as inexpensive carbon sources for PHA production in *C. necator*, aiming to reduce the cost of PHA [25]. Furthermore, plant oils yield higher amounts of PHA than sugars, and with simple genetic modifications to *C. necator*, plant oils can be used to produce poly (3-hydroxybutyrate-*co*-3-hydroxyhexanoate) (PHBHHx or poly(3HB-*co*-3HHx)) with properties superior to those of PHB [26,27]. Engineered *C. necator* H16 produced 125.9 g/L of PHBHHx from 164.7 g/L of dry cell weight using palm kernel oil. Another study reported the production of 86% PHBHHx from 124 g/L biomass of *C. necator* Re2058/pCB113 using fructose and rapeseed oil [28,29]. Therefore, using plant oil instead of fructose for the synthesis of various PHAs in *C. necator* seems more advantageous in terms of yield and cost-effectiveness.

While plant oils have been widely utilized to produce poly(3HB-*co*-3HHx), research on their use for producing PHAs with other monomers remains limited. Therefore, this study aims to assess the feasibility of using vegetable oil and precursors to enable large-scale production of various PHAs, particularly focusing on 5HV monomers by co-supplying DVL with vegetable oil. Furthermore, by analyzing the thermal and physical properties of the resulting polymer, this research provides a deeper understanding of the role of 5HV within PHA.

2. Materials and Methods

2.1. Microorganisms and Culture Conditions

The *C. necator* PHB^{-4} wild-type strain and PHB^{-4} strains harboring various PHA synthase genes (*phaC*$_{\text{BP-M-CPF4}}$, *phaC*$_{Ac}$, and *phaC*$_{Ra}$) were cultured in 5 mL of tryptic soy broth (TSB) at 30 °C for 24 h to obtain the seed culture (Figure 1).

The culture was supplemented with 50 μg/mL kanamycin for plasmid activity. For PHA production, 5× *Ralstonia eutropha* minimal medium (5× ReMM; 20 g/L NaH_2PO_4, 23 g/L Na_2HPO_4, 2.25 g/L K_2SO_4), 100× $MgSO_4$ (39 g/L $MgSO_4$), 100× $CaCl_2$ (6.2 g/L $CaCl_2$), and 1000× trace element (15 g/L $FeSO_4{\cdot}7H_2O$, 2.4 g/L $MnSO_4{\cdot}H_2O$, 2.4 g/L $ZnSO_4{\cdot}7H_2O$, 0.48 g/L $CuSO_4{\cdot}5H_2O$ dissolved in 0.1 M hydrochloric acid) solutions were used. The 5× ReMM was sterilized at 121 °C for 15 min using an autoclave, and the 100× $MgSO_4$, 100× $CaCl_2$, and 1000× trace element solutions were filtered using a 28-mm syringe filter with a 0.22-μm polyethersulfone (PES) membrane (Sartorious, Goettingen, Germany). A fructose solution (200 g/L) was used as the carbon source, and 50 g/L of urea solution was used as the nitrogen source. The production test of PHA with 3HHx and 5HV units was carried out in a 5 mL culture in a 14 mL round test tube, and the main culture conditions were as follows: 4 g/L NaH_2PO_4, 4.6 g/L Na_2HPO_4, 0.45 g/L K_2SO_4; 0.39 g/L $MgSO_4$; 0.062 g/L $CaCl_2$; 0.0015 g/L $FeSO_4{\cdot}7H_2O$, 0.0024 g/L $MnSO_4{\cdot}H_2O$, 0.0024 g/L $ZnSO_4{\cdot}7H_2O$, 0.00048 g/L $CuSO_4{\cdot}5H_2O$; 10 g/L fructose; 1 g/L urea. Additionally, various concentrations of 5-hydroxyvaleric acid (5HVA), δ-valerolactone (DVL), and bean oil were added, and 50 μg/mL kanamycin was added for plasmid activity. Unless otherwise specified, medium components were purchased from Sigma-Aldrich (St. Louis, MO, USA).

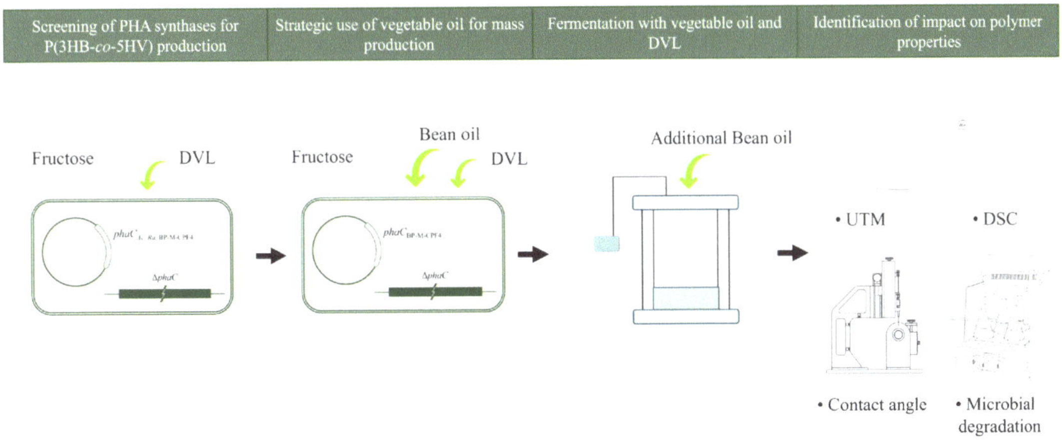

Figure 1. The overall methodological framework of the experiment. The experiment proceeded in the following order: establishing a strain capable of producing poly(3HB-*co*-5HV), confirming the increase in PHA production containing 5HV through vegetable oil feeding, verifying the feasibility of mass-producing PHA containing 5HV through vegetable oil-based fermentation, and finally, analyzing the physical properties of the produced PHA.

2.2. Analytical Methods

For PHA analysis, the culture broth was centrifuged, and the resulting pellet was washed twice with 1 mL distilled water (DW) and 1 mL hexane. Hexane was used to remove the residual bean oil and was not used when bean oil was not added to the culture. The cell pellet was washed once again with 1 mL of distilled water. The washed cells were then transferred into a glass vial for lyophilization, and the dry cell weight was measured. Next, 1 mL of chloroform and 1 mL of 15% (*v*/*v*) H_2SO_4/85% methanol solution were added to a glass vial, and methanolysis was performed at 100 °C for 2 h, followed by cooling to room temperature. Then, 1 mL of DW was added to the methyl ester solution, and the mixture was vortexed twice for 5 s each. The bottom of the chloroform layer was transferred to a microtube containing anhydrous Na_2SO_4 to remove residual water. The filtered 1 mL sample was analyzed using GC-FID (Young In Chromass 6500, Seoul, Republic of Korea) equipped with a fused silica capillary column (DB-FFAP, 30-mm length, 0.320-mm internal diameter, and 0.25 film, Agilent, Santa Clara, CA, USA). The injection volume was 1 μL, and the split ratio was 1/10. Helium was used as the carrier gas at a flow rate of 3.0 mL/min. The oven program for PHA analysis was as follows: 80 °C for 5 min, then increased from 80 °C to 220 °C at a rate of 20 °C/min, and held at 220 °C for 5 min. During the analysis, the injector temperature was maintained at 210 °C, and the FID temperature was maintained at 230 °C.

2.3. Gel Permeation Chromatography (GPC)

GPC was used to determine the number-average molar mass (Mn), weight-average molar mass (MW), and dispersity. GPC was performed using an HPLC system (Young In Chromass, Seoul, Republic of Korea) comprising a loop injector (Rheodyne 7725i), dual-headed isocratic pump system (YL9112), column oven (YL9131) with three columns (K-G 4A, guard column; K-804 8.0 × I.D. × 300 mm; K-805, 8.0 × 300 mm; Sho-dex), and a refractive index detector (YL9170). Chloroform was used as the mobile phase at a flow rate of 1 mL/min at 40 °C. The injection volume of the prepared samples was 20 µL. Polystyrene standards ranging from 5000 to 2,000,000 Da were used to calculate the MW and construct the calibration curve.

2.4. Analysis of the Physical and Thermal Characteristics

A Universal Testing Machine (UTM) Model was used to measure the tensile strength, Young's modulus, and elongation at break of the samples. The samples were cut into 10 × 60-mm pieces, and the gauge length was defined accordingly. The tests were conducted at a crosshead speed of 10 mm/min. The elongation at break was calculated using Equation (1):

$$EL = (d_{after} - d_{before})/d_{before} \times 100 \quad (1)$$

where d represents the distance between the grips holding the sample before and after the sample break.

Differential scanning calorimetry (DSC) was performed using a NEXTA DSC 200 instrument (Hitachi high-tech, Hitachi, Japan) to analyze the thermal properties of the PHAs. Approximately 5 mg of the PHA film containing 3HHx and 5HV units was measured in an aluminum pan for DSC. The experiment was conducted under a N_2 atmosphere. The heating and cooling rates were all 10 °C/min. The temperature program is as follows: 30 °C, 7 min → −60 °C, 10 min → 190 °C, 10 min (first heating) → −60 °C, 2 min → 30 °C, 10 min → −60 °C, 10 min → 190 °C, 10 min (second heating) → −60 °C, 0 min. The crystallization temperature (Tc), glass transition temperature (Tg), and melting temperature (Tm) of the polymer were determined via second heating.

2.5. Culture Conditions for a 5-L Fermenter

Precultures for the fermenter were prepared at a volume of 50 mL (TSB) in two 250-mL baffled flasks at 30 °C for 24 h. Each culture was centrifuged at 3511× *g* at 4 °C. The cell pellet was washed twice with 10 mL of DW. Each cell pellet was then suspended in 10 mL of DW to inoculate the fermenter with a total of 20 mL of cell culture. Fed-batch fermentation was conducted on a 2-L scale in a 5-L fermenter. The main culture conditions at the beginning of the culture were as follows: 4 g/L NaH_2PO_4, 4.6 g/L Na_2HPO_4, 0.45 g/L K_2SO_4; 0.39 g/L $MgSO_4$; 0.062 g/L $CaCl_2$; 0.0015 g/L $FeSO_4 \cdot 7H_2O$, 0.0024 g/L $MnSO_4 \cdot H_2O$, 0.0024 g/L $ZnSO_4 \cdot 7H_2O$, 0.00048 g/L $CuSO_4 \cdot 5H_2O$; 10 g/L fructose; 5 g/L Bean oil; 1 g/L NH_4NO_3. Additionally, 100 g/L of bean oil was gradually supplied for 10–20 h, which is the exponential phase, and 5 g/L of DVL was supplied at 48 h in one stroke. For pH control, phosphoric acid was used as the acid, and ammonia water as the base. The pH was adjusted to 6.8, and the dissolved oxygen (DO) level was set at 20%. The initial stirring rate was 200 rpm and was increased to 600 rpm to maintain the DO level during culture.

3. Results

3.1. PhaC Screening for Efficient 5HV Polymerization in PHA

Wild-type *C. necator* has been reported to synthesize PHB5HV [11]. However, because of the high specificity of *C. necator phaC* for 3HB, the mole fraction of 5HV was very low. Additionally, during PHA mass production, if a large amount of carbon source was supplied to produce 3HB monomers, it was expected that the mole fraction of 5HV would decrease further because of the accumulation of large amounts of 3HB in PHA. Therefore, for the efficient production of PHAs containing 5HV, screening for *phaC* with broad specificity not only toward 3HB but also toward 5HV was necessary. In addition, as 5HV is obtained by an additional saponification process requiring an increase in pH using DVL for ring opening and then neutralization of the pH for biological utilization, the ability of strains to use DVL as a precursor was also important [12,30].

To confirm the individual activity of *phaC*, a plasmid carrying three types of *phaC* (*phaC*$_{BP\text{-}M\text{-}CPF4}$, *phaC*$_{Ac}$, and *phaC*$_{At}$) was inserted into *C. necator* PHB^{-4}. *C. necator* PHB^{-4} is a strain that cannot produce PHB because of the abnormal expression of PHA synthase from *C. necator* H16, achieved through random mutagenesis techniques [31,32]. Next, we tested the ability of the constructed strains to synthesize poly (3HB-*co*-5HV) using delta-valerolactone as a precursor; along with wild-type *C. necator* H16; 1% fructose was used to support cell growth.

Wild-type *C. necator* H16 was able to synthesize poly (3HB-*co*-5HV) from DVL. The highest achieved mole fraction of 5HV in PHA was 1.7%, which was noted at a DVL concentration of 0.1%; however, it decreased to 0.7% when the DVL concentration was 0.2% (Figure 2a). As the DVL concentration increased, both dry cell weight (DCW) and PHB also increased. This indicates that DVL contributes to 3HB accumulation in *C. necator*. However, when the concentration of DVL exceeded 0.5%, *C. necator* H16 wild type did not grow.

phaC$_{Ac}$ and *phaC*$_{Ra}$ from *Aeromonas caviae* and *Rhodococcus aetherivorans*, respectively, have been widely used to synthesize poly(3HB-*co*-3HHx) in *C. necator* because of their broad specificity [33–36]. *phaC*$_{Ac}$ and *phaC*$_{Ra}$ were codon-optimized and inserted into the pBBR1MCS2 vector along with a high-expression ribosome binding site (RBS). As expected, *C. necator* PHB^{-4} with inserted *phaC*$_{Ac}$ produced PHA with a 5HV mole fraction of approximately 6% from DVL, with the highest 5HV mole fraction of 6.7% being achieved at a DVL concentration of 0.5% (Figure 2b). However, *C. necator* PHB^{-4} carrying *phaC*$_{Ra}$ was unable to synthesize poly(3HB-*co*-5HV). Additionally, as the concentration of 5HV increased, the PHA content decreased (Figure 2c).

phaC$_{BP\text{-}M\text{-}CPF4}$, discovered by a research team in Malaysia, is a PHA synthase derived from uncultured bacteria found through the metagenomic analysis of mangrove soil. *phaC*$_{BP\text{-}M\text{-}CPF4}$ has been reported to possess broad substrate specificity and is capable of polymerizing 3HB, but also 3HV, 4HB, 4HV, 5HV, 3HHx, and others [37–39]. Furthermore, when codon-optimized *phaC*$_{BP\text{-}M\text{-}CPF4}$ is expressed in *C. necator* PHB^{-4} with a high-expression RBS, it can synthesize poly(3HB-*co*-5HV) with a mole fraction of 70% 5HV from 5-hydroxyvaleric acid (5-HVA) [12]. The PHA synthase with the highest 5HV polymerization ability was *phaC*$_{BP\text{-}M\text{-}CPF4}$. When we confirmed the production of poly (3HB-*co*-5HV) from DVL in *C. necator* PHB^{-4} with inserted *phaC*$_{BP\text{-}M\text{-}CPF4}$ (PHB^{-4}/BP), we observed that with increasing DVL concentration, both the production of PHA and the mole fraction of 5HV in PHA increased (Figure 2d). PHB^{-4}/BP produced poly(3HB-*co*-5HV) with a 5HV mole fraction of 72.7% at a concentration of 0.5% DVL, yielding 6.85 ± 0.20 g/L. This is comparable to a previous study in which the same strain produced PHA with a 5HV mole fraction of 72.2%, yielding 5.87 ± 0.10 g/L when 0.5% 5HVA was used as the precursor [12]. The similarity in the 5HV mole fraction and PHA production when DVL and its ring-opening form, 5HVA, were used suggests that *C. necator* PHB^{-4} expresses a lactonase capable of opening lactone rings, such as DVL, and its activity was already sufficient.

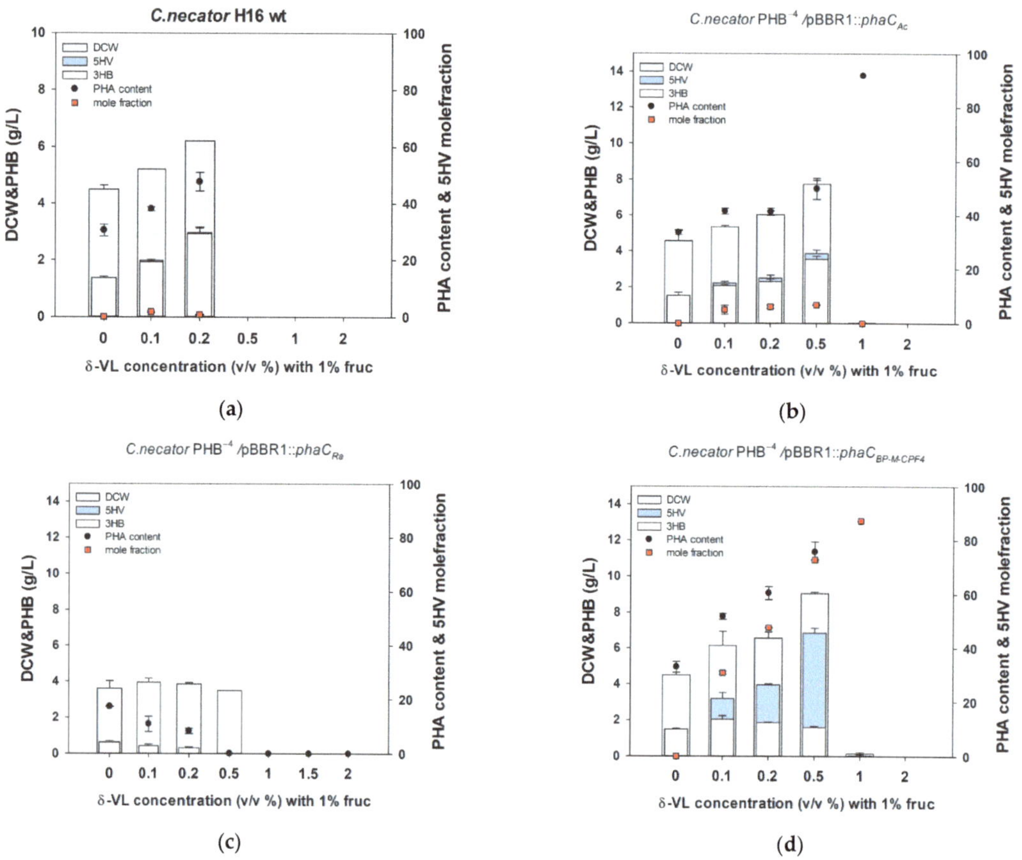

Figure 2. Validation of the 5HV polymerization ability of wild-type *Cupriavidus necator* H16 and *C. necator* PHB^{-4} harboring various PHA synthases. (**a**) Wild-type *C. necator* H16 (**b**) *C. necator* PHB^{-4} harboring *phaC*$_{Ac}$ (**c**) *C. necator* PHB^{-4} harboring *phaC*$_{Ra}$ (**d**) *C. necator* PHB^{-4} harboring *phaC*$_{\text{BP-M-CPF4}}$. Statistical analysis was performed by applying ANOVA with the level of significance at 5%.

3.2. Validation of Plant Oil for Increased Production of PHA Containing 5HV

Plant oil has been widely used to enhance the price competitiveness of PHA because of its cost-effectiveness and high conversion rate to PHA compared to sugar [40,41]. In addition, plant oils could help in the production of considerable amounts of cell mass and PHA [21,29,42–44]. Therefore, we aimed to investigate whether additional supplementation with plant oil could increase the production of poly(3HB-*co*-5HV) by the PHB^{-4}/BP strain to maximize the production of PHA containing 5HV. Additionally, in a previous study, considering the replacement of the native-*phaC* of *C. necator* H16 with the broad-specificity *phaC* from *Rhodococcus aetheriborans*, resulting in the production of PHA with a 1–1.5% mole fraction of 3HHx, we anticipated that feeding the PHB^{-4}/BP strain with an oil source would lead to the integration of additional 3HHx into PHA, thereby enhancing the flexibility of polymer [45].

In the previous experiment, we confirmed that the optimal concentration of DVL to produce poly(3HB-*co*-5HV) with a high 5HV mole fraction was 0.5% (Figure 2). Therefore, we used 1% fructose as the carbon source and 0.5% 5HV or DVL as the precursors, with bean oil supplemented at concentrations ranging from 0% to 2%. We observed that with the use of 5HV as the precursor, the addition of bean oil from 0% to 1.5% led to an increase in both PHA production and DCW (Figure 3a). Additionally, as expected, a small amount of 3HHx was polymerized, resulting in the formation of a poly(3HB-*co*-3HHx-*co*-5HV) ter-polymer. However, as the concentration of bean oil increased beyond 1.5%, the mole fraction of 5HV in PHA decreased from 64% to 10%.

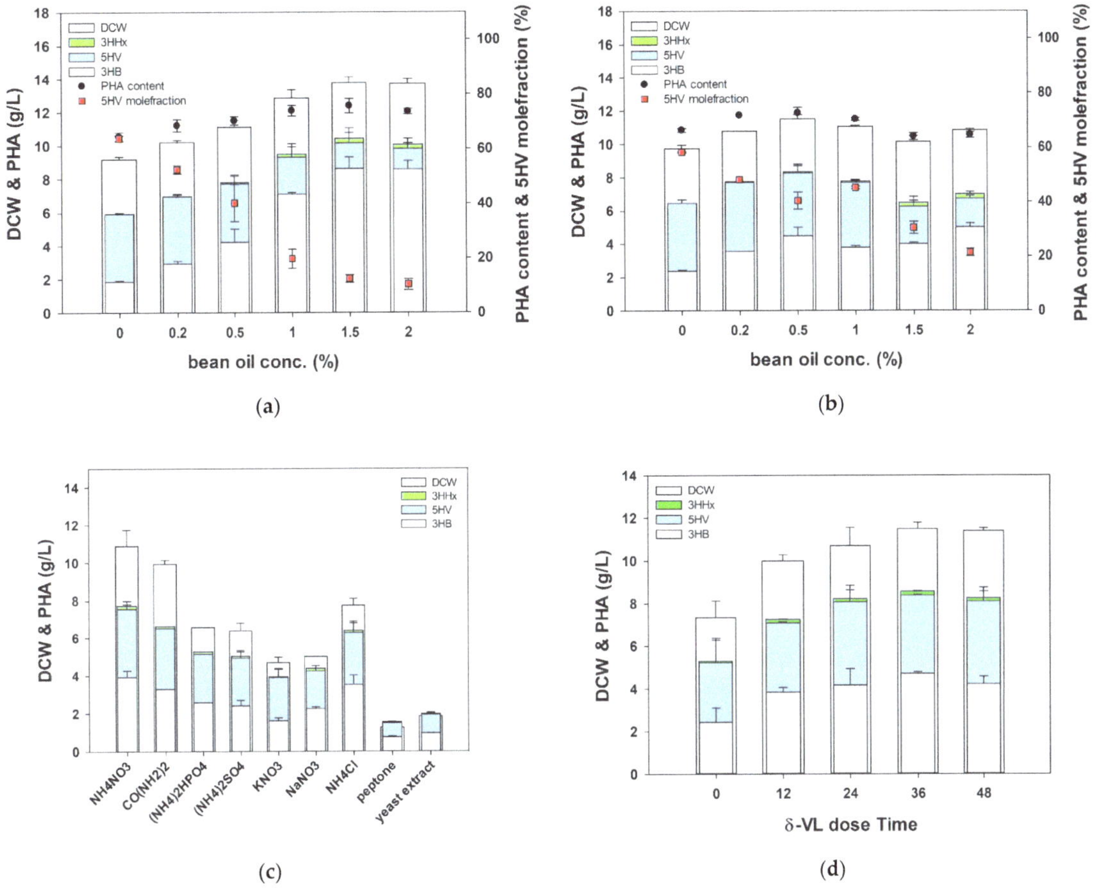

Figure 3. Confirmation of increased production of poly(3HB-*co*-3HHx-*co*-5HV) after supplying bean oil. (**a**) 5-hydroxyvaleric acid (5HVA) or (**b**) δ-valerolactone (DVL) was used as the 5HV precursor. (**c**) N-source optimization for poly(3HB-*co*-3HHx-*co*-5HV) production. (**d**) DVL feeding time optimization. Statistical analysis was performed by applying ANOVA with a level of significance at 5%.

On the contrary, when DVL was used as the precursor, increasing the concentration of bean oil from 0% to 0.5% resulted in an approximately 30% increase in PHA production (Figure 3b). However, when bean oil was added at concentrations of 1% or higher, both DCW and PHA decreased. This decrease might be attributed to the fact that 5HVA, as a free fatty acid, could directly participate in beta-oxidation along with bean oil, whereas DVL, as a lactone, could not directly participate in beta-oxidation. Therefore, the reduced compatibility between bean oil and DVL may be attributed to the requirement for lactonase, an enzyme essential for the ring-opening process of DVL. However, 5HVA is generally difficult to obtain, and an additional saponification process is required. Therefore, we focused on the 30% increase in PHA production when bean oil was added to the DVL precursor and designed further experiments.

The type of nitrogen source is crucial for PHA production from various substrates [46]. Furthermore, since the carbon source has changed, the corresponding control of the nitrogen source is also necessary. To enhance PHA production from bean oil and DVL, after including 1% fructose for initial cell growth, nine different nitrogen sources were administered at a concentration of 0.1%, and the production of poly(3HB-*co*-3HHx-*co*-5HV) was compared (Figure 3c). When urea and NH_4Cl were employed, the PHA production was similar, yielding 6.66 g/L and 6.41 ± 0.35 g/L, respectively. However, the PHA content when urea was used was 67.9%, whereas it was higher at 82.6 ± 1.85% when NH_4Cl was used. On the contrary, when NH_4NO_3 was used, PHB^{-4}/BP accumulated the highest PHA at 7.73 ± 0.19 g/L, and the DCW was also the highest at 10.9 ± 0.6 g/L. In our study, the type of nitrogen source used resulted in significant differences in PHA production from bean oil and DVL. Notably, the highest cell mass and PHA yield were achieved when using NH_4NO_3, a less commonly used nitrogen source, instead of the more commonly used sources such as urea and $(NH_4)_2SO_4$ in *Cupriavidus necator* fermentation. This highlights the importance of optimizing the nitrogen source as a key aspect of fermentation optimization.

The precursors used in the synthesis of PHA copolymers are known to inhibit cell growth [47]. Previous experiments revealed that DVL, the precursor of 5HV, also inhibited cell growth at concentrations of 0.5–1%. The toxicity of such precursors could be mitigated by supplying precursors after the cells had grown to a certain extent following the initiation of the culture. To confirm that the delayed supply of the DVL precursor could overcome its toxicity, DVL was supplied at 0, 12, 24, 36, and 48 h after the start of the culture, and PHA production was subsequently observed. It was observed that the delayed supply of DVL resulted in higher PHA production, with the production increasing with the extent of delay. When DVL was supplied at 0 h, the DCW and PHA production were 7.35 ± 0.55 g/L and 5.31 ± 0.75 g/L, respectively (Figure 3d). When DVL supply was delayed, both DCW and PHA production increased, reaching a maximum of 11.5 ± 0.2 g/L DCW and 8.56 ± 0.02 PHA, respectively.

3.3. Fed-Batch Production of Poly(3HB-co-3HHx-co-5HV)

Based on the previously identified conditions, the feasibility of mass-producing poly(3HB-*co*-3HHx-*co*-5HV) using PHB^{-4}/BP was tested on a 5-L scale. The working volume in the 5-L fermenter was 2 L. Initially, 1% fructose, 0.5% bean oil, and 0.1% NH_4NO_3 were fed during cultivation, with 0.5% DVL added after 48 h of fermentation, and terpolymer production and cell growth of the PHB^{-4}/BP strain were observed. At 42 h, i.e., 6 h before DVL feeding, a PHA production of 0.57 ± 0.01 g/L was observed, with a PHA content of 9.18 ± 0.34%. The low PHA production and PHA content in the fermenter were attributed to the use of ammonia water as a pH regulator. Owing to the continuous decrease in pH with the growth of PHB^{-4}/BP, ammonia water was continuously supplied to maintain the pH, leading to excessive N-source feeding and a subsequent decrease in PHA content. However, after DVL feeding at 48 h, the 5HV molar fraction continued to increase until 84 h, resulting in 3.17 ± 0.06 g/L of poly(3HB-*co*-2 mol% 3HHx-*co*-76 mol% 5HV) at 84 h.

To increase PHA production and content, an additional 100 g/L of bean oil was supplied from 10 h to 20 h with all other conditions remaining unchanged. Consequently, both DCW and PHA increased until 72 h, accumulating 48.63 ± 4.53 g/L of poly(3HB-*co*-3.7 mol% 3HHx-*co*-5.3 mol% 5HV) with 66.25 ± 2.95 g/L of DCW (Figure 4). Subsequently, when the supply of bean oil was increased to 200 g/L from 10 h to 20 h, 68.6 ± 7.8 g/L of poly(3HB-*co*- 5.6 mol% 3HHx-*co*- 2.7 mol% 5HV) was accumulated at 144 h with 90.3 ± 0.9 g/L of DCW (Supplementary Figure S1). However, the growth rate decreased, possibly because of the reduced oxygen supply resulting from the rapid supply of oil over a short period. Through this experiment, we confirmed the potential for the mass production of PHA-containing 5HV monomers using the PHB^{-4}/BP strain with bean oil supplementation.

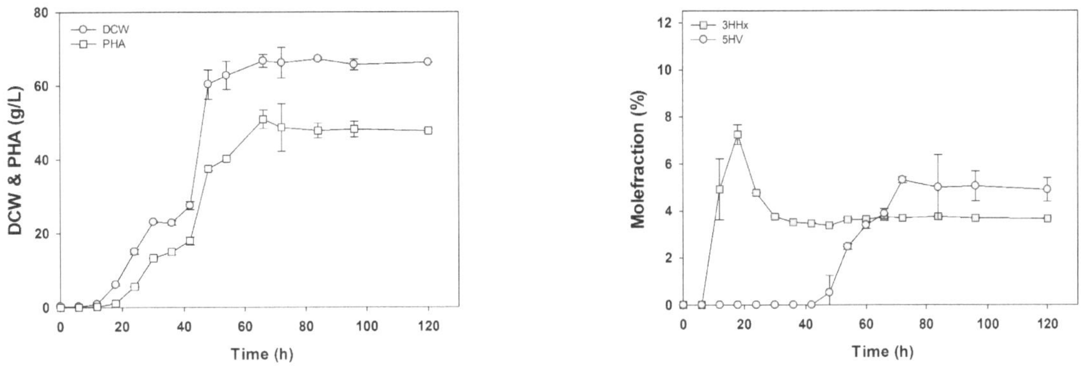

(a) (b)

Figure 4. Fed-batch Poly(3HB-*co*-3HHx-*co*-5HV) production by *C. necator* PHB^{-4} harboring *phaC*$_{BP\text{-}M\text{-}CPF4}$ in a 5-L jar fermenter. The initial culture conditions included 1% fructose, 0.5% bean oil, and 0.1% NH_4NO_3. From 10 h to 20 h of culture, 100 g/L of bean oil was supplied, and 5 g/L of DVL was added after 48 h. The changes in (**a**) DCW (Dry Cell Weight) and PHA, as well as (**b**) the molar fractions of 3HHx and 5HV, were monitored over the cultivation period. Statistical analysis was performed by applying ANOVA with the level of significance at 5%.

3.4. Physical and Mechanical Properties of Poly(3HB-co-3HHx-co-5HV)

To characterize the produced Poly (3HB-*co*-3.6 mol% 3HHx-*co*-4.9 mol% 5HV), the cell pellet was concentrated using a continuous centrifuge. PHA was then extracted using chloroform and cast into films. The thermal and physical properties of the poly(3HB-*co*-3HHx-*co*-5HV) film were analyzed using a UTM, DSC, and GPC and compared with those of the poly(3HB-*co*-7.1 mol% 3HHx) film.

In a previous study, a poly(3HB-*co*-5HV) film with a 70% 5HV mole fraction was reported to exhibit an elongation at a break of 1400% [12]. However, the increase in elongation at break owing to the 5HV content in PHA was found to be minimal at low 5HV mole fractions. Upon confirming the elongation at break using UTM for both films, the elongation at break of poly(3HB-*co*-7.1 mol% 3HHx) was 176.6 ± 18.2%, whereas that of Poly(3HB-*co*- 3.6 mol% 3HHx-*co*- 4.9 mol% 5HV) was 16.6 ± 1.4% (Table 2). This indicates that at low mole fractions, 3HHx has a more significant effect on elongation at break than 5HV.

Table 2. Physical and mechanical properties of poly(3HB-*co*-3.6 mol% 3HHx-*co*-4.9 mol% 5HV) produced by *C. necator* PHB^{-4} harboring *phaC*$_{BP\text{-}M\text{-}CPF4}$ and poly(3HB-*co*-7.1 mol% 3HHx).

PHA Composition			Properties					
3HB	3HHx	5HV	Tensile Strength (MPa)	Elongation at Break (%)	Young's Modulus (MPa)	Mn (10^3)	Mw (10^3)	Dispersity
91.5	3.6	4.9	9.03 ± 0.18	16.64 ± 1.36	112.14 ± 4.65	60 ± 10	306 ± 19	5.2 ± 0.5
92.9	7.1	-	6.81 ± 0.42	176.6 ± 18.24	81.71 ± 5.25	53 ± 3	354 ± 11	6.7 ± 0.2

The high melting point of PHB hinders its industrial application because it requires more energy for plastic molding. Additionally, the degradation of PHA at high temperatures is unavoidable, which can lead to changes in the polymer's properties. Therefore, to facilitate the molding of PHA to increase its industrial applicability and prevent degradation during the molding process, it is necessary to decrease its melting point. Analysis of the thermal behavior of the two films using DSC revealed that the Tm of Poly(3HB-*co*-

3HHx-*co*-5HV) was 151.5 °C, and no Tg and Tc were observed (Table 3). In contrast, for poly(3HB-*co*-3HHx), the Tm was 175.5 °C, with Tg and Tc values of 2.5 °C and 48.3 °C, respectively. Considering the significantly lower Tm of Poly(3HB-*co*-3HHx-*co*-5HV) than that of poly(3HB-*co*-3HHx), it can be inferred that the 5HV monomer in PHA contributes more to the decrease in Tm than 3HHx.

Table 3. Thermal properties of poly(3HB-*co*-3.6 mol% 3HHx-*co*-4.9 mol% 5HV) produced by *C. necator* PHB^{-4} harboring *phaC*$_{BP\text{-}M\text{-}CPF4}$ and poly(3HB-*co*-7.1 mol% 3HHx).

	Tg (°C)	Tc (°C)	Tm (°C)	ΔH (mJ/mg)
Poly(3HB-*co*-3HHx-*co*-5HV)	n.a.	n.a.	151.5 ± 0.1	16.6 ± 0.3
Poly(3HB-*co*-3HHx)	2.6 ± 0.2	48.3 ± 0.3	175.5 ± 0.0	19.2 ± 0.2

Abbreviations: Tg, glass transition temperature; Tc, crystallization temperature, Tm, melting temperature.

The roles of 3HHx and 5HV as PHA monomers have also been confirmed in two previous studies that produced poly(3HB-*co*-33 mol% 5HV) and poly(3HB-*co*-31.1 mol% 3HHx) using engineered *C. necator* [12,22]. The elongation at break for poly(3HB-*co*-31.1 mol% 3HHx) produced from lauric acid was 243.4 ± 36.4%, whereas poly(3HB-*co*-33 mol% 5HV) exhibited an elongation at break of 157.7%. Furthermore, PHBHHx achieved an elongation at a break of 132.2 ± 17.3%, similar to that of poly(3HB-*co*-33 mol% 5HV), with just the incorporation of 9.4% of 3HHx. Regarding thermal properties, poly(3HB-*co*-33 mol% 5HV) exhibited melting points of 31.5 °C and 153.3 °C, whereas poly(3HB-*co*-31.1 mol% 3HHx) showed a relatively higher melting point of 166.4 °C. A relatively lower melting point was observed in the presence of 5HV, even at low mole fractions. Poly(3HB-*co*-12.0 mol% 3HHx) produced from *Cupriavidus eutrophus* B10646 had a melting point of 170 °C, whereas poly(3HB-*co*-5HV) produced from *Escherichia coli* showed a melting point of 159.2 °C with only 4.7 mol% polymerization of 5HV [48,49].

For commercialization and improved processability of PHA, enhancement of flexibility and reduction of melting point should be achieved simultaneously. Typically, increasing the content of monomers other than 3HB in PHA copolymers improves the flexibility of the polymer and reduces its melting point. However, the mole fraction of the monomers required to achieve ideal mechanical properties and melting points may vary. For instance, in PHBHHx, a mole fraction of approximately 10% 3HHx may be optimal because higher 3HHx mole fractions result in amorphous polymers with reduced tensile strength and Young's modulus. However, a higher molar fraction of 3HHx may be required to achieve a lower melting point. In such cases, the production of PHA terpolymers by incorporating additional monomers with different properties is a good solution.

This study confirmed that 3HHx is effective in increasing elongation even at low molar fractions, whereas 5HV is effective in reducing the melting point at low molar fractions. Therefore, it is anticipated that optimizing the molar fraction of 3HHx to achieve the desired mechanical properties, followed by the addition of 5HV to achieve a lower melting point, can be an effective approach.

3.5. Comparison of PHBHHx5HV and Conventional PHBHHx Degradation and Contact Angle

Previous reports have identified the role of 5HV units in facilitating polymer degradation of PHA by pig pancreatic lipase [50]. However, the role of 5HV in PHA degradation by microorganisms has not been reported. *Microbulbifer* sp. Sol66, a PHB-degrading strain isolated from the coastal area of Korea, demonstrated a rapid degradation rate, reaching up to 98% in just 4 days of cultivation [51]. To verify the role of 5HV in PHA degradation by microorganisms, the degradation rate of Poly(3HB-*co*-3.6 mol% 3HHx-*co*-4.9 mol% 5HV) by Sol66 was compared to that of the Poly (3HB-*co*-7.1 mol% 3HHx) film.

The Poly(3HB-*co*-3.6 mol% 3HHx-*co*-4.9 mol% 5HV) film exhibited a lower degradation rate than the Poly(3HB-*co*-7.1 mol% 3HHx) film. The Poly(3HB-*co*-7.1 mol% 3HHx) film degraded by 76.4% in just 3 days, whereas the Poly(3HB-*co*-3.6 mol% 3HHx-*co*-4.9 mol%

5HV) film degraded by only 30.4% (Figure 5). Additionally, while the PHBHHx film was completely degraded by the fifth day, the Poly(3HB-*co*-3.6 mol% 3HHx-*co*-4.9 mol% 5HV) film degraded by only 82.3% by the seventh day. Throughout the 7-day degradation period, the molar fraction of 5HV remained between 4 and 5%, and the molar fraction of 3HHx remained between 3 and 4%. This indicates that the degradation of PHA terpolymers by Sol66 did not specifically involve certain monomers.

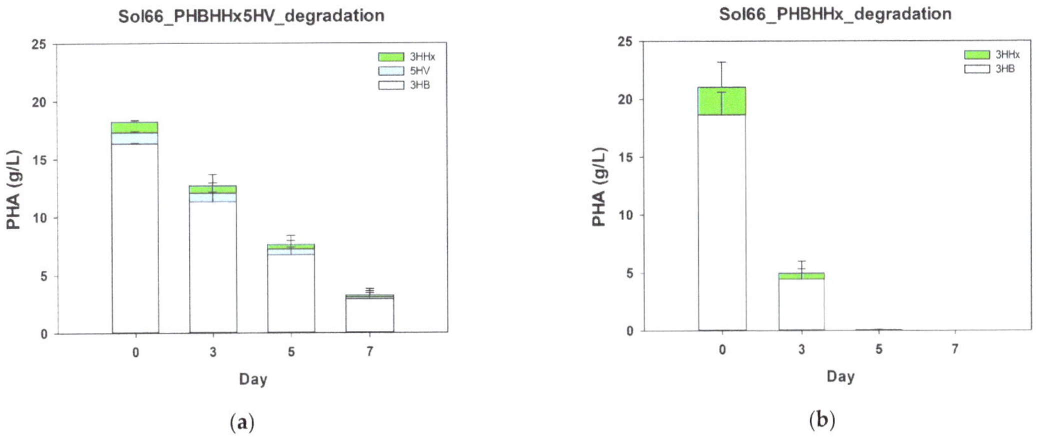

Figure 5. Time-dependent degradation rate of (**a**) poly(3HB-*co*-3.7 mol% 3HHx-*co*-5.3 mol% 5HV) and (**b**) poly(3HB-*co*-7.1 mol% 3HHx) films by *Microbulbifer* sp. Sol66. Statistical analysis was performed by applying ANOVA with the level of significance at 5%.

The interaction of the two PHA films with liquids was analyzed using a contact angle meter. For poly(3HB-*co*-3HHx-*co*-5HV), the initial contact angle was 109.23°, which is higher than that of 98.08° for poly(3HB-*co*-3HHx) (Figure 6). This indicated that the poly(3HB-*co*-3HHx-*co*-5HV) film had a more hydrophobic surface. Furthermore, when observing the change in contact angle over time, Poly(3HB-*co*-3HHx-*co*-5HV) maintained a contact angle of approximately 100° for about 3 min. In contrast, the contact angle of the Poly(3HB-*co*-3HHx) film decreased to 76.92° after 46 s of measurement, and after 3 min, the water droplet was completely absorbed by the polymer. The PHA films showed differences in the contact angle depending on the presence of 5HV monomers. However, because the films may contain components derived from plant oils or microbial sources that can be extracted with chloroform and may influence the contact angle, it may be difficult to attribute these changes solely to the presence of 5HV.

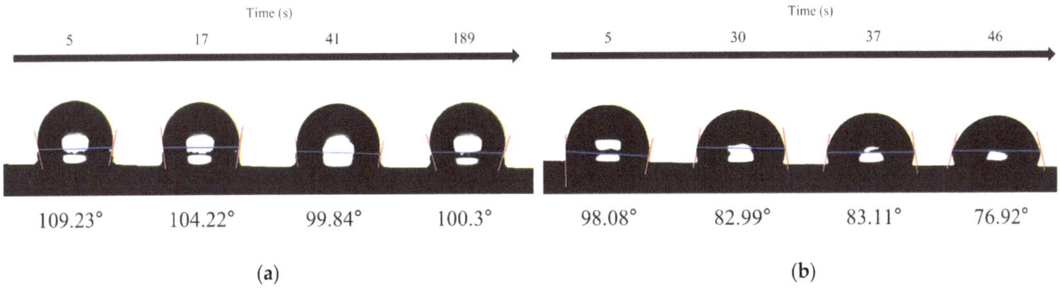

Figure 6. Analysis of changes in contact angle over time for extracted (**a**) poly(3HB-*co*-3.7 mol% 3HHx-*co*-5.3 mol% 5HV) and (**b**) poly(3HB-*co*-7.1 mol% 3HHx) films. Statistical analysis was performed by applying ANOVA with the level of significance at 5%.

4. Conclusions

Previous efforts to produce PHA copolymers or terpolymers through genetic modification of organisms aimed at synthesizing precursors from fructose or adding precursors alongside fructose have been limited by low production yields. Therefore, in this study, we aimed to explore the feasibility of mass-producing PHA containing 5HV by using vegetable oil together with DVL, a precursor of 5HV.

First, to efficiently incorporate 5HV into PHA, we tested the ability of three PHA synthases, known for their broad substrate specificity, to synthesize P(3HB-*co*-5HV). As a result, we developed a *Cupriavidus necator* PHB^{-4} strain harboring $phaC_{BP\text{-}M\text{-}CPF4}$, which exhibited a very high ability to synthesize P(3HB-*co*-5HV). Additionally, by feeding bean oil, we found that while the addition of vegetable oil leads to further synthesis of 3HHx in the PHA, the overall PHA production increased by about 30% when DVL was used.

In a 5-L jar fermenter, the potential for large-scale production of poly(3HB-*co*-3HHx-*co*-5HV) through the addition of plant oil was confirmed. Under fed-batch fermentation, PHB^{-4}/BP produced 48.63 ± 4.53 g/L of poly(3HB-*co*-3.7 mol% 3HHx-*co*-5.3 mol% 5HV) from 66.25 ± 2.95 g/L of DCW when 100 g/L of bean oil was additionally supplied. Furthermore, although the lag phase was long because the conditions were not thoroughly optimized, when 200 g/L of bean oil was supplied, 90.3 ± 0.9 g/L of DCW yielded 68.6 ± 7.8 g/L of poly(3HB-*co*-5.6 mol% 3HHx-*co*-2.7 mol% 5HV) at 144 h. The terpolymers were collected by continuous centrifugation and extracted with chloroform. By comparing the physical properties of the produced terpolymer and conventional PHBHHx, it was confirmed that, at lower mole fractions, the 3HHx monomer in PHA effectively increased the elongation at break, whereas the 5HV monomer effectively reduced the melting point.

The roles of 5HV and 3HHx in the produced PHA were elucidated through a comparison with conventional Poly(3HB-*co*-3HHx). Thermal and physical property analyses revealed that at low molar fractions, 3HHx effectively enhanced the elongation of PHA, while 5HV reduced its melting point. Additionally, PHA containing 5HV exhibited slower degradation by marine microorganisms compared to Poly(3HB-*co*-3HHx), with a higher contact angle, indicating improved water resistance.

Vegetable oil is an inexpensive feedstock known to produce significantly higher PHA yield and biomass in *Cupriavidus necator* fermentation compared to sugars. However, despite these advantages, the use of vegetable oil for PHA production has been primarily limited to P(3HB-*co*-3HHx), with few examples of its use for large-scale production of PHAs containing other monomers. In this study, we strategically utilized vegetable oil to enhance biomass production for the mass production of PHA containing 5HV. Furthermore, we propose using a combination of vegetable oil and precursors as an effective strategy for the large-scale production of PHAs with diverse monomers. The analysis of polymer properties suggests that incorporating 5HV can lower the melting point, regulate degradation rates in marine environments, and enhance water resistance. Given the demand for large quantities of samples in polymer applications, this oil-based fermentation strategy facilitates greater PHA production, offering expanded opportunities to investigate the properties of novel PHAs.

Supplementary Materials: The following supporting information can be downloaded at: https://www.mdpi.com/article/10.3390/polym16192773/s1, Figure S1: Fed-batch Poly(3HB-*co*-3HHx-*co*-5HV) production by *Cupriavidus necator* PHB-4 harboring $phaC_{BP\text{-}M\text{-}CPF4}$ in a 5-L jar fermenter. The initial culture conditions included 1% fructose, 0.5% bean oil, and 0.1% NH4NO3. From 10 h to 20 h of culture, 200 g/L of bean oil was supplied, and 5 g/L of DVL was added after 48 h. The changes in (A) DCW (Dry Cell Weight) and PHA, as well as (B) the molar fractions of 3HHx and 5HV were monitored over the cultivation period.

Author Contributions: S.-J.O.: Writing—Original Draft, Conceptualization, Investigation, Y.S.: Visualization, J.O.: Methodology, S.K.: Formal analysis, Y.L.: Formal analysis, S.C.: Investigation, G.L.: Investigation, J.-C.J.: Validation, J.-M.J.: Validation, J.-J.Y.: Writing—Reviewing and Editing, S.K.B.: Writing—Reviewing and Editing, J.A.: conceptualization. H.-T.K.: supervision, funding, Y.-H.Y.:

Supervision, Writing—Review and Editing. All authors have read and agreed to the published version of the manuscript.

Funding: This research was supported by the R&D Program of KEIT (20009508, 20018072, 20025698), the Cooperative Research Program for Agriculture Science and Technology Development (PJ01708201) through the Rural Development Administration and the support of 'The R&D Program for Forest Science Technology (Project No. "2023473E10-2325-EE02")' provided by Korea Forest Service (Korea Forestry Promotion Institute).

Institutional Review Board Statement: Not applicable.

Data Availability Statement: Data will be available on request.

Conflicts of Interest: The authors declare no conflicts of interest.

References

1. Rueda, E.; Gonzalez-Flo, E.; Mondal, S.; Forchhammer, K.; Arias, D.M.; Ludwig, K.; Drosg, B.; Fritz, I.; Gonzalez-Esquer, C.R.; Pacheco, S.; et al. Challenges, Progress, and Future Perspectives for Cyanobacterial Polyhydroxyalkanoate Production. *Rev. Environ. Sci. Biotechnol.* **2024**, *23*, 321–350. [CrossRef]
2. Meereboer, K.W.; Misra, M.; Mohanty, A.K. Review of Recent Advances in the Biodegradability of Polyhydroxyalkanoate (PHA) Bioplastics and Their Composites. *Green Chem.* **2020**, *22*, 5519–5558. [CrossRef]
3. Prados, E.; Maicas, S. Bacterial Production of Hydroxyalkanoates (PHA). *Univers. J. Microbiol. Res.* **2016**, *4*, 23–30. [CrossRef]
4. Alves, M.I.; Macagnan, K.L.; Rodrigues, A.A.; De Assis, D.A.; Torres, M.M.; De Oliveira, P.D.; Furlan, L.; Vendruscolo, C.T.; Moreira, A.D.S. Poly(3-Hydroxybutyrate)-P(3HB): Review of Production Process Technology. *Indus. Biotechnol.* **2017**, *13*, 192–208. [CrossRef]
5. Briassoulis, D.; Tserotas, P.; Athanasoulia, I.G. Alternative Optimization Routes for Improving the Performance of Poly(3-Hydroxybutyrate) (PHB) Based Plastics. *J. Clean. Prod.* **2021**, *318*, 128555. [CrossRef]
6. Volova, T.; Kiselev, E.; Nemtsev, I.; Lukyanenko, A.; Sukovatyi, A.; Kuzmin, A.; Ryltseva, G.; Shishatskaya, E. Properties of Degradable Polyhydroxyalkanoates with Different Monomer Compositions. *Int. J. Biol. Macromol.* **2021**, *182*, 98–114. [CrossRef]
7. Yan, X.; Liu, X.; Yu, L.P.; Wu, F.; Jiang, X.R.; Chen, G.Q. Biosynthesis of Diverse α,ω-Diol-Derived Polyhydroxyalkanoates by Engineered Halomonas Bluephagenesis. *Metab. Eng.* **2022**, *72*, 275–288. [CrossRef]
8. Lakshmanan, M.; Foong, C.P.; Abe, H.; Sudesh, K. Biosynthesis and Characterization of Co and Ter-Polyesters of Polyhydroxyalkanoates Containing High Monomeric Fractions of 4-Hydroxybutyrate and 5-Hydroxyvalerate via a Novel PHA Synthase. *Polym. Degrad. Stab.* **2019**, *163*, 122–135. [CrossRef]
9. Rehakova, V.; Pernicova, I.; Kourilova, X.; Sedlacek, P.; Musilova, J.; Sedlar, K.; Koller, M.; Kalina, M.; Obruca, S. Biosynthesis of Versatile PHA Copolymers by Thermophilic Members of the Genus Aneurinibacillus. *Int. J. Biol. Macromol.* **2023**, *225*, 1588–1598. [CrossRef]
10. Myung, J.; Flanagan, J.C.A.; Waymouth, R.M.; Criddle, C.S. Expanding the Range of Polyhydroxyalkanoates Synthesized by Methanotrophic Bacteria through the Utilization of Omega-Hydroxyalkanoate Co-Substrates. *AMB Express* **2017**, *7*, 118. [CrossRef]
11. Doi, Y.; Tamaki, A.; Kunioka, M.; Soga, K. Biosynthesis of Terpolyesters of 3-Hydroxybutyrate, 3-Hydroxyvalerate, and 5-Hydroxyvalerate in *Alcaligenes eutrophus* from 5-Chloropentanoic and Pentanoic Acids. *Die Makromol. Chem. Rapid Commun.* **1987**, *8*, 631–635. [CrossRef]
12. Oh, S.J.; Kim, S.; Lee, Y.; Shin, Y.; Choi, S.; Oh, J.; Bhatia, S.K.; Joo, J.C.; Yang, Y.H. Controlled Production of a Polyhydroxyalkanoate (PHA) Tetramer Containing Different Mole Fraction of 3-Hydroxybutyrate (3HB), 3-Hydroxyvalerate (3 HV), 4 HV and 5 HV Units by Engineered *Cupriavidus necator*. *Int. J. Biol. Macromol.* **2024**, *266*, 131332. [CrossRef] [PubMed]
13. Sohn, Y.J.; Son, J.; Jo, S.Y.; Park, S.Y.; Yoo, J.I.; Baritugo, K.A.; Na, J.G.; Choi, J.-i.; Kim, H.T.; Joo, J.C.; et al. Chemoautotroph *Cupriavidus necator* as a Potential Game-Changer for Global Warming and Plastic Waste Problem: A Review. *Bioresour. Technol.* **2021**, *340*, 125693. [CrossRef]
14. Zhila, N.O.; Sapozhnikova, K.Y.; Kiselev, E.G.; Shishatskaya, E.I.; Volova, T.G. Biosynthesis of Poly(3-Hydroxybutyrate-Co-4-Hydroxybutyrate) from Different 4-Hydroxybutyrate Precursors by New Wild-Type Strain *Cupriavidus necator* IBP/SFU-1. *Processes* **2023**, *11*, 1423. [CrossRef]
15. Duvigneau, S.; Dürr, R.; Behrens, J.; Kienle, A. Advanced Kinetic Modeling of Bio-Co-Polymer Poly(3-Hydroxybutyrate-Co-3-Hydroxyvalerate) Production Using Fructose and Propionate as Carbon Sources. *Processes* **2021**, *9*, 1260. [CrossRef]
16. Wang, C.-T.; Sivashankari, R.M.; Miyahara, Y.; Tsuge, T. Polyhydroxyalkanoate Copolymer Production by Recombinant Ralstonia Eutropha Strain 1F2 from Fructose or Carbon Dioxide as Sole Carbon Source. *Bioengineering* **2024**, *11*, 455. [CrossRef]
17. Park, S.; Roh, S.; Yoo, J.; Ahn, J.H.; Gong, G.; Lee, S.M.; Um, Y.; Han, S.O.; Ko, J.K. Tailored Polyhydroxyalkanoate Production from Renewable Non-Fatty Acid Carbon Sources Using Engineered *Cupriavidus necator* H16. *Int. J. Biol. Macromol.* **2024**, *263*, 130360. [CrossRef]
18. Fukui, T.; Suzuki, M.; Tsuge, T.; Nakamura, S. Microbial Synthesis of Poly((R)-3-Hydroxybutyrate-Co- 3-Hydroxypropionate) from Unrelated Carbon Sources by Engineered *Cupriavidus necator*. *Biomacromolecules* **2009**, *10*, 700–706. [CrossRef]

19. Insomphun, C.; Xie, H.; Mifune, J.; Kawashima, Y.; Orita, I.; Nakamura, S.; Fukui, T. Improved Artificial Pathway for Biosynthesis of Poly(3-Hydroxybutyrate-Co-3-Hydroxyhexanoate) with High C6-Monomer Composition from Fructose in *Ralstonia eutropha*. *Metab. Eng.* **2015**, *27*, 38–45. [CrossRef]
20. Jo, Y.Y.; Park, S.; Gong, G.; Roh, S.; Yoo, J.; Ahn, J.H.; Lee, S.M.; Um, Y.; Kim, K.H.; Ko, J.K. Enhanced Production of Poly(3-Hydroxybutyrate-Co-3-Hydroxyvalerate) with Modulated 3-Hydroxyvalerate Fraction by Overexpressing Acetolactate Synthase in *Cupriavidus necator* H16. *Int. J. Biol. Macromol.* **2023**, *242*, 125166. [CrossRef]
21. Obruca, S.; Marova, I.; Snajdar, O.; Mravcova, L.; Svoboda, Z. Production of Poly(3-Hydroxybutyrate-Co-3-Hydroxyvalerate) by *Cupriavidus necator* from Waste Rapeseed Oil Using Propanol as a Precursor of 3-Hydroxyvalerate. *Biotechnol. Lett.* **2010**, *32*, 1925–1932. [CrossRef] [PubMed]
22. Oh, S.J.; Choi, T.R.; Kim, H.J.; Shin, N.; Hwang, J.H.; Kim, H.J.; Bhatia, S.K.; Kim, W.; Yeon, Y.J.; Yang, Y.H. Maximization of 3-Hydroxyhexanoate Fraction in Poly(3-Hydroxybutyrate-Co-3-Hydroxyhexanoate) Using Lauric Acid with Engineered *Cupriavidus necator* H16. *Int. J. Biol. Macromol.* **2024**, *256*, 128376. [CrossRef] [PubMed]
23. Sato, S.; Maruyama, H.; Fujiki, T.; Matsumoto, K. Regulation of 3-Hydroxyhexanoate Composition in PHBH Synthesized by Recombinant *Cupriavidus necator* H16 from Plant Oil by Using Butyrate as a Co-Substrate. *J. Biosci. Bioeng.* **2015**, *120*, 246–251. [CrossRef] [PubMed]
24. Lee, H.S.; Lee, S.M.; Park, S.L.; Choi, T.R.; Song, H.S.; Kim, H.J.; Bhatia, S.K.; Gurav, R.; Kim, Y.G.; Kim, J.H.; et al. Tung Oil-Based Production of High 3-Hydroxyhexanoate-Containing Terpolymer Poly(3-Hydroxybutyrate-Co-3-Hydroxyvalerate-Co-3-Hydroxyhexanoate) Using Engineered *Ralstonia eutropha*. *Polymers* **2021**, *13*, 1084. [CrossRef]
25. Chien Bong, C.P.; Alam, M.N.H.Z.; Samsudin, S.A.; Jamaluddin, J.; Adrus, N.; Mohd Yusof, A.H.; Muis, Z.A.; Hashim, H.; Salleh, M.M.; Abdullah, A.R.; et al. A Review on the Potential of Polyhydroxyalkanoates Production from Oil-Based Substrates. *J. Environ. Manag.* **2021**, *298*, 113461. [CrossRef]
26. Tang, H.J.; Neoh, S.Z.; Sudesh, K. A Review on Poly(3-Hydroxybutyrate-Co-3-Hydroxyhexanoate) [P(3HB-Co-3HHx)] and Genetic Modifications That Affect Its Production. *Front. Bioeng. Biotechnol.* **2022**, *10*, 1057067. [CrossRef]
27. Ciesielski, S.; Mozejko, J.; Pisutpaisal, N. Plant Oils as Promising Substrates for Polyhydroxyalkanoates Production. *J. Clean. Prod.* **2015**, *106*, 408–421. [CrossRef]
28. Sato, S.; Fujiki, T.; Matsumoto, K. Construction of a Stable Plasmid Vector for Industrial Production of Poly(3-Hydroxybutyrate-Co-3-Hydroxyhexanoate) by a Recombinant *Cupriavidus necator* H16 Strain. *J. Biosci. Bioeng.* **2013**, *116*, 677–681. [CrossRef]
29. Santolin, L.; Waldburger, S.; Neubauer, P.; Riedel, S.L. Substrate-Flexible Two-Stage Fed-Batch Cultivations for the Production of the PHA Copolymer P(HB-Co-HHx) With *Cupriavidus necator* Re2058/PCB113. *Front. Bioeng. Biotechnol.* **2021**, *9*, 623890. [CrossRef]
30. Wang, J.; Li, C.; Zou, Y.; Yan, Y. Bacterial Synthesis of C3-C5 Diols via Extending Amino Acid Catabolism. *Korea Adv. Inst. Sci. Technol.* **2020**, *117*, 19159–19167. [CrossRef]
31. SelL, H.G. The Isolation of Mutants Not Accumulating Poly-β-Hydroxybutyric Acid. *Arch. Mikrobiol.* **1970**, *71*, 283–294.
32. Raberg, M.; Voigt, B.; Hecker, M.; Steinbüchel, A. A Closer Look on the Polyhydroxybutyrate- (PHB-) Negative Phenotype of *Ralstonia eutropha* PHB-4. *PLoS ONE* **2014**, *9*, e95907. [CrossRef] [PubMed]
33. Chek, M.F.; Hiroe, A.; Hakoshima, T.; Sudesh, K.; Taguchi, S. PHA Synthase (PhaC): Interpreting the Functions of Bioplastic-Producing Enzyme from a Structural Perspective. *Appl. Microbiol. Biotechnol.* **2019**, *103*, 1131–1141. [CrossRef] [PubMed]
34. Bhatia, S.K.; Kim, J.H.; Kim, M.S.; Kim, J.; Hong, J.W.; Hong, Y.G.; Kim, H.J.; Jeon, J.M.; Kim, S.H.; Ahn, J.; et al. Production of (3-Hydroxybutyrate-Co-3-Hydroxyhexanoate) Copolymer from Coffee Waste Oil Using Engineered *Ralstonia eutropha*. *Bioprocess Biosyst. Eng.* **2018**, *41*, 229–235. [CrossRef]
35. Wang, H.; Ye, J.W.; Chen, X.; Yuan, Y.; Shi, J.; Liu, X.; Yang, F.; Ma, Y.; Chen, J.; Wu, F.; et al. Production of PHA Copolymers Consisting of 3-Hydroxybutyrate and 3-Hydroxyhexanoate (PHBHHx) by Recombinant *Halomonas bluephagenesis*. *Chem. Eng. J.* **2023**, *466*, 143261. [CrossRef]
36. Harada, K.; Kobayashi, S.; Oshima, K.; Yoshida, S.; Tsuge, T.; Sato, S. Engineering of Aeromonas Caviae Polyhydroxyalkanoate Synthase Through Site-Directed Mutagenesis for Enhanced Polymerization of the 3-Hydroxyhexanoate Unit. *Front. Bioeng. Biotechnol.* **2021**, *9*, 627082. [CrossRef]
37. Foong, C.P.; Lakshmanan, M.; Abe, H.; Taylor, T.D.; Foong, S.Y.; Sudesh, K. A Novel and Wide Substrate Specific Polyhydroxyalkanoate (PHA) Synthase from Unculturable Bacteria Found in Mangrove Soil. *J. Polym. Res.* **2018**, *25*, 23. [CrossRef]
38. Tan, H.T.; Chek, M.F.; Lakshmanan, M.; Foong, C.P.; Hakoshima, T.; Sudesh, K. Evaluation of BP-M-CPF4 Polyhydroxyalkanoate (PHA) Synthase on the Production of Poly(3-Hydroxybutyrate-Co-3-Hydroxyhexanoate) from Plant Oil Using *Cupriavidus necator* Transformants. *Int. J. Biol. Macromol.* **2020**, *159*, 250–257. [CrossRef]
39. Tan, H.T.; Chek, M.F.; Miyahara, Y.; Kim, S.Y.; Tsuge, T.; Hakoshima, T.; Sudesh, K. Characterization of an (R)-Specific Enoyl-CoA Hydratase from Streptomyces Sp. Strain CFMR 7: A Metabolic Tool for Enhancing the Production of Poly(3-Hydroxybutyrate-Co-3-Hydroxyhexanoate). *J. Biosci. Bioeng.* **2022**, *134*, 288–294. [CrossRef]
40. Lim, S.W.; Kansedo, J.; Tan, I.S.; Tan, Y.H.; Nandong, J.; Lam, M.K.; Ongkudon, C.M. Microbial Valorization of Oil-Based Substrates for Polyhydroxyalkanoates (PHA) Production—Current Strategies, Status, and Perspectives. *Process Biochem.* **2023**, *130*, 715–733. [CrossRef]
41. Du, C.; Sabirova, J.; Soetaert, W.; Ki, S.; Lin, C. Polyhydroxyalkanoates Production From Low-Cost Sustainable Raw Materials. *Curr. Chem. Biol.* **2012**, *6*, 14–25.

42. Surendran, A.; Lakshmanan, M.; Chee, J.Y.; Sulaiman, A.M.; Van Thuoc, D.; Sudesh, K. Can Polyhydroxyalkanoates Be Produced Efficiently from Waste Plant and Animal Oils? *Front. Bioeng. Biotechnol.* **2020**, *8*, 169. [CrossRef] [PubMed]
43. Fadzil, F.I.B.M.; Tsuge, T. Bioproduction of Polyhydroxyalkanoate from Plant Oils. In *Microbial Applications*; Springer International Publishing: Cham, Switzerland, 2017; Volume 2, pp. 231–260. ISBN 9783319526690.
44. Ng, K.S.; Ooi, W.Y.; Goh, L.K.; Shenbagarathai, R.; Sudesh, K. Evaluation of Jatropha Oil to Produce Poly(3-Hydroxybutyrate) by *Cupriavidus necator* H16. *Polym. Degrad. Stab.* **2010**, *95*, 1365–1369. [CrossRef]
45. Budde, C.F.; Riedel, S.L.; Willis, L.B.; Rha, C.K.; Sinskey, A.J. Production of Poly(3-Hydroxybutyrate-Co-3-Hydroxyhexanoate) from Plant Oil by Engineered Ralstonia Eutropha Strains. *Appl. Environ. Microbiol.* **2011**, *77*, 2847–2854. [CrossRef]
46. Shenbagarathai, R.; Saranya, V. Effect of Nitrogen and Calcium Sources on Growth and Production of PHA of Pseudomonas Sp. LDC-5 and Its Mutant. *Curr. Res. J. Biol. Sci.* **2010**, *2*, 164–167.
47. Gumel, A.M.; Annuar, M.S.M.; Chisti, Y. Recent Advances in the Production, Recovery and Applications of Polyhydroxyalkanoates. *J. Polym. Environ.* **2013**, *21*, 580–605. [CrossRef]
48. Satoh, K.; Kawakami, T.; Isobe, N.; Pasquier, L.; Tomita, H.; Zinn, M.; Matsumoto, K. Versatile Aliphatic Polyester Biosynthesis System for Producing Random and Block Copolymers Composed of 2-, 3-, 4-, 5-, and 6-Hydroxyalkanoates Using the Sequence-Regulating Polyhydroxyalkanoate Synthase PhaCAR. *Microb. Cell. Fact.* **2022**, *21*, 84. [CrossRef]
49. Volova, T.G.; Syrvacheva, D.A.; Zhila, N.O.; Sukovatiy, A.G. Synthesis of P(3HB-Co-3HHx) Copolymers Containing High Molar Fraction of 3-Hydroxyhexanoate Monomer by Cupriavidus Eutrophus B10646. *J. Chem. Technol. Biotechnol.* **2016**, *91*, 416–425. [CrossRef]
50. Chuah, J.A.; Yamada, M.; Taguchi, S.; Sudesh, K.; Doi, Y.; Numata, K. Biosynthesis and Characterization of Polyhydroxyalkanoate Containing 5-Hydroxyvalerate Units: Effects of 5HV Units on Biodegradability, Cytotoxicity, Mechanical and Thermal Properties. *Polym. Degrad. Stab.* **2013**, *98*, 331–338. [CrossRef]
51. Park, S.L.; Cho, J.Y.; Kim, S.H.; Bhatia, S.K.; Gurav, R.; Park, S.H.; Park, K.; Yang, Y.H. Isolation of *Microbulbifer* Sp. Sol66 with High Polyhydroxyalkanoate-Degrading Activity from the Marine Environment. *Polymers* **2021**, *13*, 4257. [CrossRef]

polymers

MDPI

Review

Application of Chitosan-Based Hydrogel in Promoting Wound Healing: A Review

Xueyan Che [1,†], Ting Zhao [1,†], Jing Hu [1], Kaicheng Yang [1], Nan Ma [1], Anning Li [2], Qi Sun [3], Chuanbo Ding [1,*] and Qiteng Ding [4,*]

1 College of Traditional Chinese Medicine, Jilin Agriculture Science and Technology University, Jilin City 132101, China; cxy17376253128@163.com (X.C.); lyguiwandingding@163.com (T.Z.); 17657368028@163.com (J.H.); 15633806068@163.com (K.Y.); 18325968349@163.com (N.M.)
2 Jilin Aodong Yanbian Pharmaceutical Co., Ltd., Dunhua 133000, China; jladcpxstgb@163.com
3 Jilin Zhengrong Pharmaceutical Development Co., Ltd., Dunhua 133700, China; sunqi001428@163.com
4 College of Traditional Chinese Medicine, Jilin Agricultural University, Changchun 130118, China
* Correspondence: chuanboding0506@163.com (C.D.); ding152778@163.com (Q.D.); Tel.: +86-138-0446-0499 (C.D.); +86-155-6581-5336 (Q.D.)
† These authors have contributed equally to this work.

Abstract: Chitosan is a linear polyelectrolyte with active hydroxyl and amino groups that can be made into chitosan-based hydrogels by different cross-linking methods. Chitosan-based hydrogels also have a three-dimensional network of hydrogels, which can accommodate a large number of aqueous solvents and biofluids. CS, as an ideal drug-carrying material, can effectively encapsulate and protect drugs and has the advantages of being nontoxic, biocompatible, and biodegradable. These advantages make it an ideal material for the preparation of functional hydrogels that can act as wound dressings for skin injuries. This review reports the role of chitosan-based hydrogels in promoting skin repair in the context of the mechanisms involved in skin injury repair. Chitosan-based hydrogels were found to promote skin repair at different process stages. Various functional chitosan-based hydrogels are also discussed.

Keywords: chitosan; hydrogel; wound healing

Citation: Che, X.; Zhao, T.; Hu, J.; Yang, K.; Ma, N.; Li, A.; Sun, Q.; Ding, C.; Ding, Q. Application of Chitosan-Based Hydrogel in Promoting Wound Healing: A Review. *Polymers* **2024**, *16*, 344. https://doi.org/10.3390/polym16030344

Academic Editor: Masoud Ghaani

Received: 10 December 2023
Revised: 14 January 2024
Accepted: 24 January 2024
Published: 26 January 2024

1. Introduction

Hydrogels possess network structures in three dimensions (3D) composed of cross-linked polymer chains, enabling them to absorb substantial amounts of wound exudate. These hydrogels are distinguished by their elevated water content, soft composition, and porosity [1,2], exhibiting considerable potential as dressings for skin repair in recent years. Chitosan is the product of the chemical treatment of the deacetylation of chitin. It offers a variety of physiological functions, including biodegradability, biocompatibility, non-toxicity, bacteriostatic properties, anticancer potential, lipid-lowering capabilities, immune-enhancement effects, etc. Additionally, chitosan's exceptional properties, such as its anticoagulant features and wound healing promotion, have led to its extensive employment in medical dressings [3,4]. The carboxyl and amino groups in chitosan can be utilized to create chitosan-based hydrogels with diverse properties. These hydrogels are considered excellent materials for wound dressing due to their biodegradable, biocompatible, and antimicrobial properties. Moreover, chitosan-based hydrogels hold great potential for clinical application in the treatment of skin wounds [5]. The properties of these hydrogels can be customized by incorporating different natural or synthetic polymers, resulting in functional chitosan-based hydrogels [6].

The process of skin repair Involves several stages: hemostasis, inflammation, proliferation, and tissue remodeling. During this process, various types of cells, such as epithelial cells, immune cells, endothelial cells, and fibroblasts, work together. Additionally, skin repair is influenced by environmental factors, but it is important because the skin serves as the body's first line of defense against UV rays, microbial invasion, and other forms of aggression. As a result, prompt repair after a skin injury is of utmost importance [7].

Conventional wound dressings are currently unable to effectively address the requirements for treating chronic wound healing. However, novel smart chitosan-based hydrogel scaffolds have emerged as a potential solution. These hydrogels can be infused with active factors or drugs to not only inhibit inflammatory responses, oxidative stress, and microbial proliferation in chronic wounds but also promote the proliferation of skin cells. As a result, this review aims to summarize the recent applications of chitosan-based hydrogels in wound repair and explore the utilization of different drugs within these hydrogels. The findings of this review can serve as a valuable theoretical foundation for the future development and utilization of chitosan-based hydrogels.

2. Methods for Preparing Chitosan-Based Hydrogel

Chitosan is widely distributed in nature and is a renewable resource. People can extract chitosan from the shells of marine arthropods such as shrimps and crabs, mollusks and insects, and the cell walls of higher plants, and then prepare it by deacetylation [8]. Chitosan can be chemically and physically cross-linked to prepare a variety of chitosan-based hydrogels for wound repair (Table 1).

Table 1. Preparation of chitosan-based hydrogels.

Cross-Linking Type	Method	Mechanism	Refs.
Physical cross-linking	Electrostatic interaction	Electrostatic interactions occur through an interaction between anionic molecules and the amino groups of chitosan to gelate them.	[9]
	Metal ion coordination	Metal ion coordination is used to synthesize gels through intermolecular coordination bonds, forming more stable hydrogels.	[10]
	Hydrophobic interaction	Chitosan can undergo gelation through hydrophobic interactions, and this interacting biopolymer system can advantageously avoid the potential side effects of in situ polymerization associated with monomer or initiator toxicity.	[11]
Chemical cross-linking	Cross-linking agent initiated cross-linking method	Initiators are substances that can cause monomers to undergo polymerization, which can cause chitosan and its derivative molecules to combine through covalent bonds to form a reticulated structure and improve the strength, elasticity, and other properties of the material.	[12]
Radiation cross-linking		The radiation cross-linking method uses the action of a radiation source with a substance to ionize and excite the generation of activated atoms and molecules, causing cross-linking between the substances.	[13]

The main methods of physical cross-linking can be categorized into three ways. The first method involves electrostatic interaction, where the interaction between anionic molecules and the amino groups of chitosan leads to gelation. The second method is metal-ion coordination, which synthesizes gels through intermolecular coordination bonds, resulting in more stable hydrogels [14]. Yang J et al. utilized the α-hydroxyl in situ photoreduction and semi-soluble sol-gel transition (SD-A-SGT) method of silver ions and CMCTS to construct a multifunctional chitosan/carboxymethyl chitosan/silver nanoparticles (CTS/CMCTS/AgNPs) polyelectrolyte composite physical cross-linking hydrogel. The X-ray diffraction (XRD), transmission electron microscopy (TEM), and IR characterization revealed the highly uniform structure and good mechanical strength of the composite hydrogel. Additionally, the incorporation of AgNPs provided the hydrogel with excellent antimicrobial activity while also exhibiting good self-healing ability [10]. The third method involves hydrophobic interaction, where chitosan can undergo gelation through hydrophobic interactions. This biopolymer system allows for advantageous avoidance of potential side effects associated with in situ polymerization, such as monomer or initiator toxicity [15,16].

There are two main methods of chemical cross-linking. The first method is initiator-initiated cross-linking, where an initiator substance causes the monomer to undergo polymerization. This process allows chitosan and its derivative molecules to form a reticulation structure through covalent bonding, enhancing the material's strength, elasticity, and other properties [17]. For example, Park et al. utilized ethylene glycol as a cross-linking agent and the blowing method to prepare a chitosan-based super porous hydrogel [18]. Another approach involves exposing chitosan matrices containing photoinitiators to high-energy UV radiation, which triggers interactions between molecular chains and promotes the formation of hydrogel structures. Drabczyk et al. synthesized chitosan hydrogels containing Aloe vera by UV radiation using a cross-linking agent (polyethylene glycol diacrylate), a photoinitiator (2-hydroxy-2-methylpropiophenone), chitosan, and Aloe vera juice (99.5%) [12]. The second method is the radiation cross-linking method, which utilizes radiation sources and substances to generate activated atoms and molecules, leading to cross-linking between substances. Fan et al. prepared chitosan (CS)/gelatin (Gel)/poly(vinyl alcohol) (PVA) hydrogels using the γ-irradiation method for wound dressing application. The physical properties and coagulation activity of these hydrogels were tested and yielded positive results. Additionally, the hydrogels exhibited good pH sensitivity, swelling ability, and water evaporation rate [19].

Physically cross-linked hydrogels are commonly made using biologically safe polymers as a matrix to avoid any potential adverse chemical reactions in the human body. However, these hydrogels often suffer from poor stability, weak tissue adhesion, and inadequate mechanical properties. To address these limitations, the incorporation of additional polymers is typically necessary during the preparation of physically cross-linked hydrogels. On the other hand, chemical hydrogels, which are formed through chemical bonding, necessitate comprehensive human hazard testing to ensure their biosafety. Furthermore, the production of chemically cross-linked hydrogels requires specialized techniques and specific conditions, which can be costly and have environmental implications. However, the advantage of chemically cross-linked hydrogels lies in their ability to be tailored for wound repair based on the specific wound type. Therefore, when developing chitosan-based hydrogels for wound repair, it is crucial to carefully consider both the wound characteristics and the biosafety of the hydrogel. This consideration is essential in selecting suitable hydrogel matrices and cross-linking methods.

3. Chitosan-Based Hydrogel for Skin Injury Repair

Skin wound healing can be further categorized into acute and chronic wounds, depending on the healing time. Acute wounds are usually caused by sudden abrasions, scratches, puncture wounds, etc. These wounds usually heal in 2–7 days for superficial wounds and in 2–4 weeks for deeper wounds. Even in the case of wounds with a small amount of bleeding and a small area, patients can treat them at home by themselves. The healing time of chronic wounds is usually more than four weeks [20], and the main common chronic wounds are burns [21], diabetic wounds [22–24], and decubitus ulcers [25,26]. Skin wounds are often accompanied by painful bleeding, and the long healing time means that they can cause prolonged suffering for the patient. Skin wound healing can be accelerated by the use of appropriate wound dressings during this process. The process of wound healing is complex and dynamic, involving various stages such as hemostasis, inflammation, proliferation, and tissue remodeling.

Ideal wound dressings are flexible, stable, biodegradable, and biocompatible, with the ability to maintain wound moistness, hemostasis, and the adsorption of exudate (refer to Figure 1), and chitosan-based hydrogels meet most of the conditions. Additionally, diverse variants of wound dressings composed of chitosan-based hydrogel possess the potential to enhance the healing process of wounds at varying time intervals and mitigate any detrimental factors impeding wound healing [27]. These formulations predominantly exert their functions during the initial three stages of wound healing. Primarily, they facilitate the cessation of bleeding by stimulating the aggregation of platelets and red blood cells while simultaneously impeding fibrinolysis during the hemostatic phase. Subsequently, they aid in the elimination of bacteria from the wound during the inflammatory phase. Lastly, they expedite the proliferation of skin cells by promoting the growth of granulation tissue, known as the proliferative phase. Consequently, the wound proceeds to recover, and the skin undergoes remodeling to ultimately accomplish the healing process.

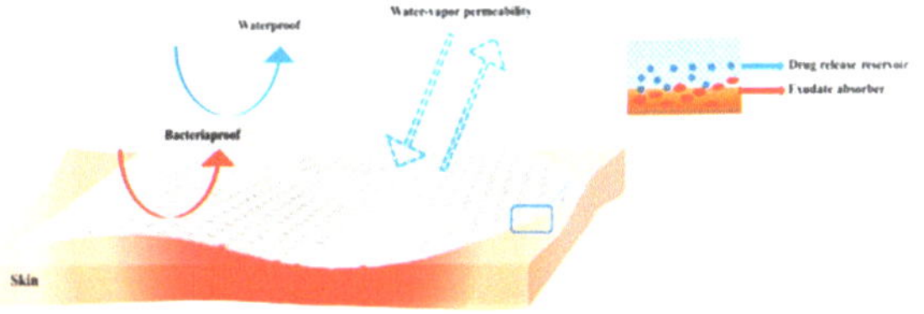

Figure 1. Ideal wound dressing [28], with permission from MDPI.

3.1. *Intelligent Chitosan-Based Hydrogel*

Chitosan as a carrier can be physically and chemically modified to prepare multifunctional chitosan hydrogels to form smart hydrogels for releasing restorative components. Smart hydrogels produce a sol-gel phase transition while subjected to environmental stimuli such as pH, temperature, and light. Because of this property, smart chitosan-based hydrogels have also become a focus of research in the past few years. Smart chitosan-based hydrogels are usually heat-sensitive hydrogels, photosensitive hydrogels, and pH-sensitive hydrogels [29,30], as presented in Table 2.

Table 2. Smart chitosan-based hydrogels.

Type	Added Ingredients	Action Mechanism	Refs.
pH - sensitive	Red cabbage extract (RCE)/ chitosan (cs)/Methylenebisacrylamide (MBAA)	The colorimetric properties of RCE-loaded chitosan hydrogels show that RCE undergoes visual color changes in both acidic and alkaline media. Monitoring wound pH changes can protect, heal, and indicate the healing process	[31]
	Polyacrylamide-quaternary ammonium/chitosan-carbon quantum dots (CQDs) phenol red hydrogel	Hybridization of CQD and pH indicator (phenol red) with the hydrogel resulted in a highly responsive, reversible, and accurate indication of pH variability to reflect dynamic wound states in both UV and visible backgrounds.	[32]

Table 2. *Cont.*

Type	Added Ingredients	Action Mechanism	Refs.
Heat - sensitive	Curcumin/Carboxymethyl/ cellulose	Cur-micellar-loaded hydrogels promote tissue regenerative capacity with enhanced fibroblasts, keratin-forming cells, and collagen deposition to stimulate epidermal junctions. Interestingly, chitosan-CMC-g-PF127 injectable hydrogel exhibited rapid wound repair potential by increasing cell migration and proliferation at the site of injury and providing a continuous drug delivery platform to the hydrophobic fraction.	[33]
	Synthesis by free radical-mediated polymerization of tannic acid-assisted New-gel hydrogels.	The new gel has excellent chemical/physical properties and can effectively load and release drugs and maintain drug activity. At the same time, New-gel has excellent oxygen loading capacity, which provides significant practical therapeutic benefits for diabetic wound repair.	[34]
Photo - responsive	Prussian blue nanoparticles (PBNPs)/Glycidyl trimethylammonium chloride (GTAC)/ Glycidyl methacrylate (GMA)	The positively charged QC in the hydrogel can capture bacteria through electrostatic attraction, change the potential of the bacterial membrane, destroy the bacterial membrane, and ultimately reduce the activity of the bacteria or even kill them. At the same time, the heat therapy generated by PBNPs under near-infrared (NIR) light irradiation can effectively and quickly kill these weak bacteria at mild temperatures (<55 °C).	[35]
	Porphyrin photosensitizer/polylactic acid-glycolic acid (PLGA)/basic fibroblast growth factor (bFGF)/carboxymethyl chitosan (CMCS)	The hydrogel block helps with repeated photodynamic stimulation and inhibits bacterial growth, while the aFGF content promotes wound healing.	[36]
Magnetic field corresponding hydrogel	Magnetite precursor/chitosan	The direct remote control of drug release behavior using low-frequency alternating magnetic field (LAMF) also avoids potential adverse thermal effects.	[37]

Table 2. *Cont.*

Type	Added Ingredients	Action Mechanism	Refs.
Electric field responsive hydrogel	Chitosan (CS)/Hydroxyethyl cellulose oxide (OHEC)/Reduced graphene oxide (RGO)/Salicylide liposomes	CS/OHEC/rGO/asiaticoside liposome hydrogel was prepared by dispersing RGO and asiaticoside into the hydrogel. This hydrogel serves as a filler for hollow nerve conduits, leveraging the benefits of OHEC and CS to enhance the mechanical and degradation properties of CS. Moreover, the hydrogel incorporates conductive rGO, which facilitates electric stimulation and scar inhibition, thereby promoting peripheral nerve regeneration.	[38]
Plasma-activated hydrogel	Gallic acid-modified chitosan-based (CS-GA)	Through self-cross-linking reaction exposed to oxygen, CS-GA solution may become bioadhesive hydrogel with high biocompatibility and blood compatibility, which is helpful for wound healing and hemostasis.	[39]

Thermosensitive hydrogels are a type of hydrogels that exhibit temperature-sensitive behavior. These hydrogels remain in a sol state at low temperatures but undergo a phase change to a gel state when exposed to body temperature. This unique property makes them suitable for use as an alternative to conventional wound dressings. Chitosan temperature-sensitive hydrogel can be produced by incorporating specific materials such as β-glycerol phosphate, HPMC, and poloxamer [40]. Odinokov et al. synthesized pH- and thermosensitive hydrogels by cross-linking chitosan with Ter phthaloyl diazide [41]. Bhattarai et al. developed a hydrogel using a chitosan solution neutralized with a polyol counter-ionic single-head salt. This hydrogel remains in a liquid state at low temperatures and forms a gel at body temperature. These thermosensitive chitosan-based hydrogels show promise as therapeutic drug delivery systems for promoting skin tissue repair and regeneration. They achieve this through the controlled and sustained release of loaded drugs. Additionally, the injectable nature of these hydrogels allows for better adaptation to various irregular skin wounds. If used as a new type of dressing, they can closely conform to wounds with greater flexibility compared to traditional wound dressings [42].

The photosensitizer can be crosslinked with chitosan hydrogel to prepare a photoresponsive intelligent hydrogel, thereby expanding its application range. This hydrogel shows potential in drug delivery and can effectively inhibit bacterial growth through repeated photodynamic stimulation [43–47]. The combination of the photothermal effect and drug release, with the control of spatial and temporal factors, provides a synergistic therapy. This approach effectively prevents the disadvantages associated with separate treatment methods. The outstanding sterilizing effect of this therapy was demonstrated through in vitro antimicrobial assays and experiments on mouse wounds infected with *S. aureus*. Moreover, the hydrogel used in this treatment exhibits strong antioxidant properties. These properties are instrumental in eliminating the inflammatory response caused by the bacteria's remnants at the site of infection. Consequently, this prevents any further harm to the tissue of the wound and facilitates its healing process [48].

The special feature of pH-sensitive hydrogels is their ability to change dimensions based on the pH of the environment [49]. During wound healing, the pH of the injured area undergoes dynamic changes. Normal skin typically has a pH below 5 [50]. However, when the skin surface is damaged, the underlying tissue with a pH of 7.4 becomes exposed. Chitosan, which has a pH of approximately 6.5, is responsive to these changes in environmental pH. In the early stages of wound healing, chitosan hydrogels expand in the acidic environment, which can enhance cellular infiltration, proliferation, and oxygen permeation. This hydrogel has potential applications in releasing anti-inflammatory agents during the initial wound-healing phase. Such release would reduce inflammation during the inflammatory phase and prevent excessive fibroblast growth during the proliferative phase [51].

Multiple studies have demonstrated that intelligent chitosan-based hydrogels exhibit superior adaptability to meet specific wound management requirements. These hydrogels offer promising solutions for treating special wounds and effectively leveraging the benefits of chitosan-based hydrogels in wound repair. However, it is important to note that these hydrogel products are still in the research phase, and further enhancements are required to improve their mechanical properties. To fully utilize them for wound treatment, extensive and thorough research is necessary.

3.2. Self-Healing Chitosan-Based Hydrogel

It is widely recognized that hydrogels can be injected to conform to different wound shapes. Injectable hydrogel wound dressings effectively fill and adhere to the wound site, protecting external factors. However, injectable hydrogels with weak elasticity may experience deformations or damage when exposed to mechanical forces after application. These damages can reduce the lifespan of the hydrogel, increase the risk of infections, and trigger inflammatory reactions [52–54]. The development of a hydrogel material with self-repairing ability has the potential to extend the lifespan of the hydrogel and minimize the occurrence of wound infections. In their study, Chen et al. successfully created a self-healing hydrogel using dynamic covalent bonding. This hydrogel exhibited remarkable self-healing properties, as it was able to regenerate a complete hydrogel structure even after being injected or fragmented due to injury [55]. Li et al. developed a biobased hydrogel by combining quaternized chitosan with bis-formaldehyde bacterial cellulose. The hydrogel exhibited rapid self-healing and injectable behavior, which was attributed to dynamic Schiff base interactions. Furthermore, it demonstrated remarkable antibacterial activity against both *Escherichia coli* (Gram negative) and Staphylococcus aureus (Gram positive). The hydrogels also showed suitable compression properties and an optimal water retention capacity [56].

Self-healing water gel offers self-healing capabilities for hydrogels due to the chemical bond linkage within the hydrogel matrix. With its unique properties, it is well-suited for wound dressings, effectively preventing damage to the hydrogel during daily activities. Furthermore, it can prolong the lifespan of the hydrogel, thereby reducing the discomfort experienced by patients during dressing changes and minimizing resource wastage.

3.3. Drug-Loaded Chitosan-Based Hydrogels

Chitosan-based hydrogels have emerged as an ideal material for the development of new-generation dressings, owing to their exceptional biocompatibility and wetting ability. Extensive research has demonstrated their efficacy in promoting the healing of acute wounds. However, the healing methods for chronic wounds differ. The selection of wound dressings for chronic wounds is contingent upon the specific conditions of the wound. Consequently, the treatment of chronic wounds necessitates specialized wound dressings. The limited effectiveness of chitosan-based hydrogels in promoting the healing of chronic wounds hampers their widespread application as wound dressings. To overcome this deficiency, researchers have explored the use of therapeutic drugs in conjunction with chitosan-based hydrogels to enhance their efficacy.

Drug loading in hydrogel scaffolds is a crucial strategy for enhancing the skin repair ability of wound dressings. By utilizing the pharmacological activity of drugs, they can be effectively employed for the treatment of chronic wounds through controlled release within hydrogel scaffolds. However, further experimental data are required to ascertain the biocompatibility of the loaded drug dosage.

3.3.1. Loading Metal Ion Chitosan Hydrogel

From a clinical perspective, the presence of infectious bacteria can result in the suppuration of wounds, thereby prolonging the healing process [57]. This can further lead to the formation of wounds that are difficult to treat. In severe cases, it may even result in sepsis, posing a significant threat to life and overall well-being [58]. Consequently, the utilization of wound dressings with antibacterial and antimicrobial properties becomes crucial. Hydrogel dressings with antimicrobial capabilities can impede the growth of detrimental microorganisms, safeguard against microbial infections, and effectively minimize the risk of wound infection in patients. Extensive research has consistently demonstrated the remarkable bactericidal activity possessed by a variety of nanoparticles, including silver [59], copper [60,61], gold [62], and platinum [63,64]. The incorporation of these particles can significantly improve the antimicrobial activity of hydrogels. For example, Nešović et al. added silver nanoparticles (AgNPs), which are a broad-spectrum antimicrobial agent, to a hydrogel. The hydrogel showed strong antimicrobial activity in subsequent experiments [65].

3.3.2. Chitosan-Based Hydrogel Loaded with Flavonoids

In recent years, flavonoids have garnered increased attention due to their significant antioxidant effects and their additional anti-inflammatory and antimicrobial properties. As a result, scholars are actively conducting in-depth research on flavonoids to develop new wound dressings. Apigenin, also known as apigenin and psoralen, is a flavonoid compound that is widely distributed in nature. It is primarily found in plants of the Rafflesiaceae, Verbenaceae, and Curculionaceae families, as well as in various vegetables and fruits in temperate zones. Celery, in particular, has a high content of apigenin. Numerous studies conducted both domestically and internationally have demonstrated the antitumor, cardiovascular, and cerebrovascular protective, antiviral, antibacterial, and other biological activities of apigenin. Building upon these properties, Shukla et al. proposed the development of a new wound dressing using a chitosan-based hydrogel loaded with apigenin. This dressing offers unique benefits such as biocompatibility, biodegradability, moisturization, and antioxidant efficacy, all of which effectively promote the healing of diabetic wounds [66].

Dihydroquercetin, also known as paclitaxel, is a potent flavonoid antioxidant. It is commonly found in onion, silymarin, French marine bark, and Douglas fir bark [67,68]. Ding et al. fabricated hydrogels sensitive to temperature by incorporating the bioactive compound taxifolin (TAX) into Poloxamer 407, chitosan (CS), and hyaluronic acid (HA). The findings demonstrated that the interconnected hydrogels showcased enhanced resistance to temperature changes, antioxidative properties, controlled drug release, and ensured safety. Moreover, the combination of paclitaxel and hydrogels expedited the healing process of skin injuries in mice [69].

3.3.3. Chitosan-Based Hydrogels Loaded with Phenolic Acids

Phenolic acids, also known as phenolic compounds, are a class of compounds that contain multiple phenolic hydroxyl groups on the same benzene ring. Some examples of phenolic acids include caffeic acid, chlorogenic acid, pentapic acid, gallic acid, protocatechuic acid, ferulic acid, and mesic acid. These compounds have a wide range of physiological activities, such as antioxidant properties, free radical scavenging, protection against ultraviolet radiation, bacteriostatic effects, and antiviral effects. Among these activities, antioxidant and bacteriostatic effects play an important role in promoting wound healing.

Protocatechuic acid (3,4-dihydroxybenzoic acid, PCA) is a simple phenolic acid that can be found in various edible plants. It possesses a variety of pharmacological activities, including anti-inflammatory, antioxidant, antihyperglycemic, antimicrobial, anticancer, anti-aging, antitumor, antiasthmatic, antiulcer, and anti-spasmodic properties [70]. Zhou et al. experimented with developing injectable hydrogels with antioxidant properties. They used protocatechuic acid (PA) and carboxymethyl chitosan (CCS), which are conjugated with oxidized hyaluronic acid (OHA). The resulting OHA/CCS-PA composite hydrogels demonstrated excellent chemical stability, physical characteristics, and remarkable antimicrobial efficacy, surpassing 99% against common pathogens. Furthermore, these hydrogels showed satisfactory antioxidant capabilities by scavenging more than 85% of excessive free radicals, ultimately reducing cellular damage caused by oxidative stress. In vitro testing revealed that the sustained release of phenolic hydroxyl elements contributed to the reduction in reactive oxygen species expression. Notably, an animal study confirmed that the OHA/CCS-PA hydrogel outperformed commercially available Tegaderm™ film and a control hydrogel containing physically mixed PA, as it significantly enhanced wound healing with a remarkable 86.29% increase in collagen content. This indicates that OHA/CCS-PA hydrogels hold promising potential for wound healing applications [71].

Gallic acid (GA, 3,4,5-trihydroxybenzoic acid) is a natural phenolic antioxidant that can be extracted from plants, especially green tea [72]. Sun et al. prepared chitosan-copper-gallic acid nanocomposites (CS-Cu-GA NCs) with bifunctional nano-enzymatic properties (oxidative and peroxidase-like functions) using a combination of ionic cross-linking, in-situ reduction, and microwave-assisted methods. CS-Cu-GA NCs integrated the inherent antimicrobial properties of chitosan, Cu NPs, and Cu. Animal experiments showed that the antimicrobial dressings doped with CS-Cu-GA NCs could effectively promote the healing of Staphylococcus aureus-infected wounds with no damage to normal tissues. Additionally, the antimicrobial dressing was formulated into a bandage possessing exceptional water-swelling antimicrobial characteristics. Furthermore, it was securely affixed to medical adhesive tape in order to fabricate a portable antimicrobial commodity that can be administered to the human skin's surface. This innovation showcases remarkable waterproof capabilities, thereby introducing novel perspectives regarding the development of antimicrobial products for clinical wound healing [73].

The aforementioned studies suggest that phenolic acid analogs have shown promising results in enhancing the antioxidant and antibacterial properties of chitosan-based hydrogels, thereby facilitating wound healing. These findings indicate that further research on the utilization of phenolic acid analogs in chitosan-based hydrogels can lead to the development of more efficient wound dressings.

3.3.4. Chitosan-Based Hydrogel Carrying Plant Essential Oil

Plant essential oils are versatile substances that are extracted from various parts of plants, such as flowers, seeds, leaves, fruits, and roots. These oils are characterized by their hydrophobic, aromatic, and volatile nature, and they find widespread use in the food industry, perfumery, and aromatherapy. Notably, essential oils contain numerous bioactive components that possess antibacterial, antifungal, antiviral, insecticidal, and antioxidant properties, thereby making them highly suitable for diverse medical applications. In recent years, there has been research and application of essential oils to promote skin wound healing. Essential oils can effectively inhibit the growth of bacteria. The loading of essential oils on chitosan-based hydrogels is currently more intensively studied. Wang et al. investigated the antimicrobial activity of several plant essential oils, namely grape seed, rose, bergamot, lemon, chamomile, lavender, tea tree, ginger, cumin, and eucalyptus. The results showed that among them, EEO (eucalyptus essential oil), GEO (ginger essential oil), and CEO (cumin essential oil) exhibited the highest antimicrobial activity. Furthermore, due to their volatile nature, these essential oils are continuously released during the interaction with the hydrogel matrix. It was observed that when the added EEO, GEO, or CEO phases were added, they diffused on the surface of the hydrogel network structure, roughening the hydrogel surface and promoting cell adhesion, which is more favorable for wound repair [74]. Muscimol (2-isopropyl-5-methylphenol), a natural phenolic monoterpene found in essential oils mainly from thyme, oregano, and cow daylily, is one of the natural compounds with therapeutic abilities. Its various therapeutic properties, such as antioxidant, anti-inflammatory, local anesthetic, anti-injury sensory, distemper acid, antiseptic, and especially antibacterial and antifungal, have brought it to the forefront of interest in wound repair applications. Koosehgol et al. prepared blended films of chitosan, poly(ethylene glycol fumarate), and muscimol by solvent casting method. The films with a certain concentration of muscimol exhibited improved water vapor permeability, water vapor absorption, equilibrium water absorption, air permeability, swelling behavior, and antimicrobial activity against both Gram-negative and Gram-positive bacteria [75].

These studies demonstrate that the use of a plant essential oil in combination with a chitosan-based hydrogel has a positive effect on wound repair. However, it is widely recognized that plant essential oils are volatile, and further research is needed to investigate the stability of essential oil volatilization on the antibacterial properties of the gel and its impact on wound repair activity.

3.3.5. Chitosan-Based Hydrogel Carrying Polypeptide

Peptide is a kind of amphoteric compound that is dehydrated from amino acids and contains carboxyl and amino groups. These compounds are easy to absorb, require low energy consumption to produce, and exhibit high affinity, specificity, and low toxicity. Existing studies have shown that peptides have the potential to stimulate tissue repair and wound healing [76–78]. Using chitosan-based hydrogels to carry peptides, stable delivery at the wound is more conducive to the peptide's ability to promote skin wound healing. Ouyang et al. loaded tilapia peptide on chitosan-based hydrogel to prepare a wound dressing and explored the effectiveness of the dressing in treating burn wounds. The results showed that the chitosan-based hydrogel loaded with the peptide not only had good antimicrobial activity against *Escherichia coli* (*E. coli*) and *Staphylococcus aureus* (*S. aureus*) but also excellent antimicrobial activity. It also promotes skin regeneration [79,80].

Collagen is an extracellular protein known for its triple-helical structure. Its exceptional biodegradability, biocompatibility, antioxidant properties, and minimal immunogenic response make it a favorable option for various restorative applications. Collagen peptide (COP), a functional ingredient derived from collagen hydrolysis, possesses a lower molecular weight and is readily absorbed by the body. It offers advantages for bone health, Achilles tendon strength, and skin rejuvenation. Furthermore, COP facilitates cell attachment and proliferation [81]. Hu et al. grafted collagen peptide (COP) molecules onto the amino group of carboxymethyl chitosan sulfate (CMCS) using microbial glutamine transferase (MTGase) as a catalyst to enhance the antioxidant ability of hydrogels. The degree of substitution (DS) of CMCS-COP could be controlled experimentally by adjusting the reaction conditions. The ability of each sample to scavenge and reduce tends to increase significantly with increasing concentration. Meanwhile, no relevant cytotoxicity of the copolymer was found in NIH-3T3 mouse fibroblasts. These results indicate the promising potential of CMCS-COP as a novel wound dressing [82]. It is not difficult to see the role of chitosan-based hydrogels loaded with animal peptides in promoting wound repair. However, the use of animal peptides still carries some risks. Several factors limit the application of collagen from animal tissues. For example, their quality and purity can affect performance; non-human protein composition can lead to immune responses in susceptible patients; and there is a risk of contamination with pathogenic substances.

Recombinant human collagen peptides (RHCs) are considered reliable, predictable, and chemically defined as non-allergenic alternative biomaterials. Furthermore, studies have demonstrated that these RHCs are non-cytotoxic and can be effectively utilized in tissue engineering. Deng et al. discovered that by combining RHCs with chitosan, a thermosensitive hydrogel with improved mechanical properties was created, addressing the limitations of traditional thermosensitive hydrogels. This modified hydrogel was found to be more suitable for cell encapsulation and wound repair. The test results of the RHC chitosan hydrogel revealed that cells cultured with the modified hydrogel exhibited excellent vitality, in contrast to the hydrogel without RHC contact. Additionally, when injected into second-degree burned rats, the RHC chitosan hydrogel promoted cell infiltration, angiogenesis, and wound healing [83]. From the above, it is easy to see that chitosan-based hydrogels loaded with peptides have great potential as novel dressings for skin repair in the future.

3.3.6. Chitosan-Based Hydrogel Carrying Other Therapeutic Components

In addition to the above types, Chitosan-based hydrogels can also be loaded with other therapeutic ingredients, such as bacteriostatic agents, antibiotics, phages, etc., which can inhibit bacteria, reduce wound infection, and promote chronic wound healing.

Antibiotics are chemical substances produced by microorganisms, plants, and animals that have properties that can potentially combat pathogens or disrupt the growth of other cells. There are several ways in which antibiotics can work against bacteria, including inhibiting cell wall synthesis, increasing cell membrane permeability, disrupting protein synthesis, and preventing nucleic acid replication and transcription. These mechanisms are the primary means by which antibiotics can destroy bacteria.

Amiri and colleagues investigated the incorporation of ticlopidine into chitosan-PEO nanofibers to enhance their antibacterial activity by 1.5 to 2 times. The results demonstrated that the ticlopidine-loaded nanofibers exhibited no cytotoxic effects on human fibroblasts. Additionally, in vivo experiments conducted using a rat total wound model confirmed the safety and effectiveness of utilizing ticlopidine-loaded nanofibers, which greatly accelerated wound closure [84]. Moghaddam et al. conducted a study in which they synthesized two O-carboxymethyl chitosan hydrogels combined with caffeic acid using electron beam irradiation. The researchers investigated the release of doxycycline from the chitosan-based hydrogel and found that it followed a non-Fick diffusion mechanism and exhibited different behavior in different media. In vitro release tests demonstrated that the composite hydrogel released more doxycycline compared to other matrices. The cytotoxicity study confirmed that the hydrogel dressings were non-toxic. Additionally, the chitosan hydrogel loaded with doxycycline showed inhibitory effects on the growth of Staphylococcus aureus and *Escherichia coli*. Therefore, the synthesized hydrogel holds the potential for the development of new wound dressings with antibacterial properties [85]. The application of antibiotics is more common in treating wounds and promoting wound healing. The loading of appropriate doses of antibiotics on chitosan-based hydrogels can be effective in antimicrobial treatment, reducing the risk of wound infection and promoting wound healing.

Antibiotic resistance has emerged as a significant concern within the medical community, particularly in the context of chronic wound infections caused by antibiotic-resistant bacteria. This global health issue arises due to the ability of bacteria to form biofilms on wound surfaces, facilitating their continued growth. As a consequence, the healing process becomes challenging and, in severe cases, can even result in death. To address this pressing issue, Fasiku et al. devised a hydrogel that employs a chitosan-based carrier infused with the antibacterial agent hydrogen peroxide (HP) and antibacterial peptides (Ps). This hydrogel can be directly applied to wounds, effectively combating biofilm-related bacteria, biofilms, and wound infections. The study findings demonstrate the hydrogel's inhibitory activity against methicillin-resistant Staphylococcus aureus (MRSA) bacteria [86]. Ilomuanya et al., on the other hand, developed an encapsulation of a mixture of Acinetobacter baumannii phage by chitosan microparticles in a hydrogel matrix for the treatment of multidrug-resistant chronic wound infections. In vivo results showed a significant reduction in wound size [87]. These two studies mentioned above, in response to the problem of antibiotic misuse, have opened up the idea of applying antibiotics in correspondence to the challenges of wound repair. It brings hope for safer and more efficient wound repair in the future.

4. Process of Skin Repair

Skin wounds are common occurrences resulting from surgery, burns, chronic ulcers, and various traumatic injuries. Nevertheless, the process of wound healing is an intricate physiological phenomenon that is impacted by numerous elements [88]. Ordinarily, the complete process of wound healing encompasses four phases: hemostasis, inflammation, proliferation, and tissue remodeling (refer to Figure 2).

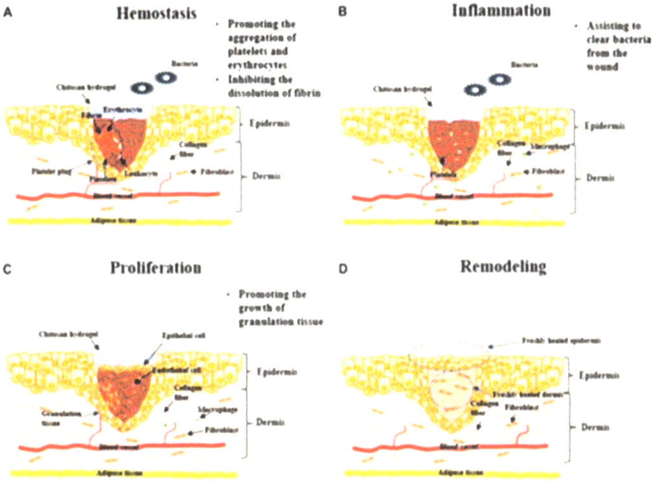

Figure 2. Four stages of skin repair [89], with permission from Frontiers. (**A**) Platelets, white blood cells, insoluble fibrin, and red blood cells combine to form a platelet thrombus, which effectively prevents bleeding during the hemostatic stage. Chitosan hydrogel plays a crucial role in hemostasis by enhancing the aggregation of platelets and red blood cells while inhibiting fibrin dissolution. (**B**) During the "inflammation" stage, chitosan hydrogel supports the activity of inflammatory cells, notably macrophages, aiding in the elimination of bacteria and necrotic tissue from the wound site. (**C**) The "proliferation" stage involves the proliferation and migration of epithelial cells to generate epithelial tissue, which acts as a protective layer for the wound. Chitosan hydrogel facilitates the growth of granulation tissue, effectively filling any tissue gaps that may exist. (**D**) The concluding stage encompasses reshaping, which signifies the completion of the entire skin repair process. Although chitosan-based hydrogels primarily exert their effects during the first three stages, their critical involvement in promoting healing is indisputable.

4.1. Hemostasis Stage

The creation of skin wounds inevitably leads to bleeding, and the process of repairing skin wounds begins with hemostasis. As soon as a skin wound starts bleeding, the organism's spontaneous hemostatic response is triggered immediately. During the hemostatic phase, the coagulation system is activated after vasoconstriction and platelet aggregation. Fibrinogen is converted into insoluble fibrin, which forms a clot to stop the bleeding. However, the body's spontaneous hemostatic function may not be sufficient for wound hemostasis in cases of large wound areas or other factors. Uncontrolled bleeding can result in various complications, including hypothermia, decreased blood pressure, bacterial infection, and even shock. If left untreated, it can lead to difficulties in wound healing and a significant increase in morbidity and mortality [90]. Bandages and gauze dressings are commonly used as traditional materials for wound hemostasis, relying on direct pressure to control bleeding. These materials have the advantages of being easily manufactured, cost-effective, and reusable. Nevertheless, their susceptibility to bacterial infection becomes evident in the presence of blood or tissue fluid [91]. In addition, dressing tears may lead to pain and increased wound healing time, plus they cannot accommodate irregular, deep, and narrow wound shapes. In addition to hemostasis by compression, there are other ways to stop bleeding. Topical hemostats, adhesives, and closures have been developed over the past few decades with good hemostatic results in surgical and emergency settings. Hemostatic agents enhance the blood-clotting cascade to achieve hemostasis, while adhesives hold various tissues and blood vessels together. However, both agents have limitations. For example, fibrin sealants based on fibrinogen and thrombin lack good adhesion properties and tend to shift during blood flushing. They may fail to stop bleeding and cause infection. Strong hemostatic adhesives, cyanoacrylates, have been developed to solve these problems, but they can cause allergic reactions [92]. In addition, cyanoacrylates rapidly generate heat during curing, and the degradation products may be toxic and have adverse effects on the human body [93]. Therefore, finding a safe, fast, and efficient hemostatic material is an urgent requirement.

Chitosan has been found to have the ability to induce platelet and plasma protein aggregation, promoting coagulation and vasoconstriction at the site of injury [94,95]. It can also enhance the function of polymorphonuclear leukocytes, macrophages, and fibroblasts, leading to faster wound healing. Chitosan can be transformed into chitosan-based hydrogels that can be customized to fit irregular wound shapes (refer to Figure 3). Xia et al. developed a degradable chitosan-based hydrogel by cross-linking carboxymethyl chitosan with oxidized hyaluronic acid as the base material. This hydrogel exhibited significant hemostatic properties, as well as favorable rheology and cytocompatibility [96]. Similarly, Zhao et al. synthesized an antimicrobial antioxidant electroactive injectable hydrogel using quaternate chitosan-g-polyaniline (QCSP) and benzaldehyde-based functionalized poly (ethylene glycol)-copolymerization (sebacic acid glycerol) (PEGS-FA) as the main raw materials. The hemostatic properties of the hydrogel QCSP3/PEGS-FA were evaluated in a hemorrhagic liver mouse model. The application of the hydrogel at the bleeding site resulted In excellent hemostasis, with minimal blood stains on the filter paper, while the control group exhibited significant blood spots. Quantitative analysis of blood loss aligned with the macroscopic findings, showing a total blood loss of 214.7 ± 65.1 mg in the hydrogel adhesive group compared to 2025.9 ± 507.9 mg in the control group, indicating a significant difference ($p < 0.01$). The results showed that QCSP3/PEGSFA1.5 hydrogel has fast in situ gel properties, good adhesive properties, and excellent hemostatic properties, which makes it an effective anti-bleeding hydrogel barrier for practical applications [97].

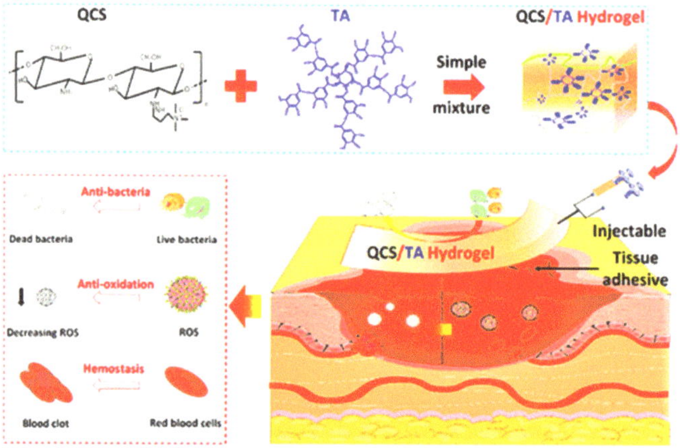

Figure 3. Chitosan-based hydrogels in the hemostatic phase [98]. Quaternary chitosan (QCS) Tannic acid (TA), with permission from ACS.

The utilization of chitosan-based hydrogels during the hemostatic phase of wound healing has the potential to prevent life-threatening hemorrhage. Previous studies have demonstrated that chitosan, which carries a positive charge, interacts with activated platelets that carry a negative charge, thereby facilitating platelet adhesion and aggregation. Consequently, chitosan-based hydrogels can effectively contribute to the cessation of bleeding.

4.2. Inflammatory Stage

The second stage of skin repair is the inflammatory stage. Inflammation is an important response of the body's immune system to ensure the survival of the body during infection and tissue damage. The inflammatory response is necessary to maintain normal tissue homeostasis. During the inflammatory phase, inflammatory cells remove bacteria and necrotic tissue. However, excessive and prolonged inflammatory reactions can be harmful instead of beneficial [99]. For the repair of damaged skin, a wound dressing with a good anti-inflammatory effect is required. Liu et al. developed alginate microspheres (Ms) loaded with basic fibroblast growth factor (bFGF) and incorporated them into carboxymethyl chitosan (CMCS)-poly (vinyl alcohol) (PVA) to create a composite chitosan-based hydrogel. The experimentally proven bFGF/Ms-CMCS-PVA composite hydrogel effectively affected inflammatory factors, inhibited inflammation, and successfully treated full-thickness skin burns on the backs of rats. Experiments have demonstrated that the bFGF/Ms-CMCS-PVA composite hydrogel has significant potential for rapidly restoring the structural and functional properties of damaged skin in burn patients [100]. Furthermore, Gull et al. also developed a hydrogel for the targeted release of diclofenac sodium with anti-inflammatory properties, using chitosan and poly (vinylidene beryllophthalide) as base polymers crosslinked with epichlorohydrin [101]. The prepared hydrogel exhibited good biodegradability, excellent antimicrobial properties, and promising cytotoxicity, as confirmed by in vitro studies. The drug release profile of the hydrogel showed 56.130% release of diclofenac sodium over 87 min, with a drug encapsulation efficiency of 84%.

The anti-inflammatory capacity of chitosan-based hydrogels alone may not be enough to inhibit the inflammatory response at the wound site and promote wound healing. However, researchers are exploring the incorporation of active ingredients with significant anti-inflammatory effects into chitosan-based hydrogel scaffolds. This area of research is considered a key focus in the development of tissue repair engineering scaffolds.

4.3. Proliferation Stage

The third stage of skin repair is proliferation, which involves the regeneration of tissues and the formation of granulation tissue. During this stage, epithelial cells proliferate and migrate to cover the wounds, while inflammatory cells, fibroblasts, and new capillaries form the granulation tissue. Research suggests that chitosan can accelerate skin wound repair by promoting the growth of inflammatory cells, fibroblasts, and capillaries. Chitosan stimulates the secretion of cytokines such as transforming growth factor-β (TGF-β), platelet-derived growth factor (PDGF), and interleukin-1 (IL-1) by macrophages. These cytokines play a crucial role in promoting migration, proliferation, collagen synthesis, and angiogenesis. Chitosan also increases the secretion of IL-8 by fibroblasts, further enhancing the inflammatory process and stimulating angiogenesis [102]. Chitosan hydrogels possess significant antibacterial properties and biocompatibility, making them highly promising in the fields of tissue engineering and regenerative medicine. However, the mechanical properties of pure chitosan hydrogels may be limited. In a study by Liu et al., a multifunctional chitosan hydrogel was developed using chitosan methacrylate (CTSMA) and vulcanized chitosan (CTSSH). The hydrogel was crosslinked through both free radical polymerization and mercaptan ene reaction. Comparatively, CTSMA/CTSSH (CMS) hydrogels demonstrated superior tissue adhesion and mechanical properties when compared to pure CTSMA hydrogels. Furthermore, this chitosan-based hydrogel showed potential in wound healing through its ability to promote angiogenesis, dermal repair, and epidermal regeneration [103].

4.4. Remodeling Phase

In the final remodeling phase, platelets in the blood clot secrete growth factors, including platelet-derived factor (PDGF), transforming growth factor (TGF)-α, and TGF-β, to promote wound healing. PDGF repairs connective tissue by attracting fibroblasts and promoting collagen deposition, thereby specifically promoting angiogenesis. Fresh epidermis and dermis will regenerate to complete the skin repair process [104]. Growth factors in skin wounds play a crucial role in promoting the healing process, specifically by regulating the proliferation, epithelialization, extracellular matrix remodeling, and angiogenesis of keratinocytes and fibroblasts. Chen et al. conducted a study to investigate the impact of chitosan hydrogel modified with SIKVAV (Ser Ile Lys Val Ala Val) peptide on skin wound healing. The experimental results demonstrate that the application of chitosan matrix hydrogel promotes the remodeling of skin wounds, thereby facilitating the wound healing process [105].

Based on the aforementioned research, chemical modification can be employed to enhance the limitations of chitosan-based hydrogels by combining them with other hydrogel matrices. Nevertheless, it is crucial to conduct comprehensive biocompatibility tests to ensure the safety of modified chitosan in hydrogel applications.

5. Conclusions and Prospects

In recent years, researchers have conducted studies on smart chitosan-based hydrogels. One notable example is the addition of silver and other metal nanoparticles with antimicrobial activity to chitosan-based pH-responsive hydrogels. This addition has resulted in hydrogels with high antimicrobial activity, which have been successfully used in experimental treatments for full-layer skin burns. These treatments have shown excellent therapeutic efficacy and effectively prevent wound infections. Additionally, there are various types of smart chitosan-based hydrogels, including thermosensitive, photosensitive, and pH-sensitive hydrogels. Thermosensitive hydrogels exhibit reversible sol-gel properties, while photosensitive hydrogels utilize light stimulation to enhance bacterial inhibition. pH-sensitive hydrogels respond to the different pH environments of wounds and promote wound healing. Another area of research focuses on self-repairing chitosan-based hydrogels, which have the potential to increase the lifespan of hydrogels. Furthermore, pharmaceutical hydrogels can be loaded with potent drugs to promote wound repair. These advancements represent significant progress in the field of chitosan-based hydrogels.

There are currently several types of chitosan-based hydrogel wound dressings available in the market. However, these products do not fully meet the requirements for clinical applications. While the extraction process of chitosan is well established, the technology for industrialized production of hydrogels is still in its early stages. Further research is needed to improve the preparation methods of hydrogels, enabling their stable use in industrial production. Collaboration between researchers and enterprises is crucial to promoting the application of chitosan-based hydrogels in wound repair. Additionally, there are some challenges that need to be addressed, such as the poor adhesion of chitosan-based hydrogels and the precise control of drug release from drug-carrying hydrogels. Biosafety concerns also need to be fully considered during the preparation of chitosan-based hydrogels for wound repair. In conclusion, chitosan-based hydrogels have great potential for wound repair applications. However, it is important to thoroughly evaluate the stability, safety, and reliability of the preparation method, and the feasibility of industrialized production of hydrogels during their development.

Author Contributions: Writing—Original draft preparation, Conceptualization, X.C.; Writing—Original draft preparation, Financial, T.Z.; Conceptualization, J.H. and K.Y.; Investigation N.M.; Supervision, Investigation, A.L. and Q.S.; Writing—Reviewing and Editing, Funding acquisition, C.D. and Q.D. All authors have read and agreed to the published version of the manuscript.

Funding: This work was supported by the Jilin Provincial Department of Education Science and Technology Research Project (JJKH20240513KJ).

Data Availability Statement: No data was used for the research described in the article.

Conflicts of Interest: Author Anning Li was employed by the company Jilin Aodong Yanbian Pharmaceutical Co., Author Qi Sun was employed by the company Jilin Zhengrong Pharmaceutical Development Co. Ltd., The remaining authors declare that the research was conducted in the absence of any commercial or financial relationships that could be construed as a potential conflict of interest.

References

1. Ho, T.C.; Chang, C.C.; Chan, H.P.; Chung, T.W.; Shu, C.W.; Chuang, K.P.; Duh, T.H.; Yang, M.H.; Tyan, Y.C. Hydrogels: Properties and applications in biomedicine. *Molecules* **2022**, *27*, 2902. [CrossRef] [PubMed]
2. Brumberg, V.; Astrelina, T.; Malivanova, T.; Samoilov, A. Modern wound dressings: Hydrogel dressings. *Biomedicines* **2021**, *9*, 1235. [CrossRef] [PubMed]
3. Ogawa, K.; Yui, T.; Okuyama, K. Three d structures of chitosan. *Int. J. Biol. Macromol.* **2004**, *34*, 1–8. [CrossRef] [PubMed]
4. Hamedi, H.; Moradi, S.; Hudson, S.M.; Tonelli, A.E. Chitosan based hydrogels and their applications for drug delivery in wound dressings: A review. *Carbohydr. Polym.* **2018**, *199*, 445–460. [CrossRef] [PubMed]
5. Desai, N.; Rana, D.; Salave, S.; Gupta, R.; Patel, P.; Karunakaran, B.; Sharma, A.; Giri, J.; Benival, D.; Kommineni, N. Chitosan: A potential biopolymer in drug delivery and biomedical applications. *Pharmaceutics* **2023**, *15*, 1313. [CrossRef] [PubMed]
6. Pella, M.; Lima-Tenorio, M.K.; Tenorio-Neto, E.T.; Guilherme, M.R.; Muniz, E.C.; Rubira, A.F. Chitosan-based hydrogels: From preparation to biomedical applications. *Carbohydr. Polym.* **2018**, *196*, 233–245. [CrossRef] [PubMed]
7. Broughton, G.N.; Janis, J.E.; Attinger, C.E. Wound healing: An overview. *Plast. Reconstr. Surg.* **2006**, *117*, 1e–32e. [CrossRef] [PubMed]
8. Tian, B.; Hua, S.; Tian, Y.; Liu, J. Chemical and physical chitosan hydrogels as prospective carriers for drug delivery: A review. *J. Mat. Chem. B* **2020**, *8*, 10050–10064. [CrossRef]
9. Kim, G.O.; Kim, N.; Kim, D.Y.; Kwon, J.S.; Min, B.H. An electrostatically crosslinked chitosan hydrogel as a drug carrier. *Molecules* **2012**, *17*, 13704–13711. [CrossRef]
10. Yang, J.; Chen, Y.; Zhao, L.; Feng, Z.; Peng, K.; Wei, A.; Wang, Y.; Tong, Z.; Cheng, B. Preparation of a chitosan/carboxymethyl chitosan/agnps polyelectrolyte composite physical hydrogel with self-healing ability, antibacterial properties, and good biosafety simultaneously, and its application as a wound dressing. *Compos. Pt. B Eng.* **2020**, *197*, 108139. [CrossRef]
11. Mirzaei, E.B.; Ramazani, A.S.A.; Shafiee, M.; Danaei, M. Studies on glutaraldehyde crosslinked chitosan hydrogel properties for drug delivery systems. *Int. J. Polym. Mater. Polym. Biomat.* **2013**, *62*, 605–611. [CrossRef]
12. Drabczyk, A.; Kudlacik-Kramarczyk, S.; Glab, M.; Kedzierska, M.; Jaromin, A.; Mierzwinski, D.; Tyliszczak, B. Physicochemical investigations of chitosan-based hydrogels containing aloe vera designed for biomedical use. *Materials* **2020**, *13*, 3073. [CrossRef] [PubMed]
13. Nguyen, N.T.; Liu, J.H. Fabrication and characterization of poly (vinyl alcohol)/chitosan hydrogel thin films via uv irradiation. *Eur. Polym. J.* **2013**, *49*, 4201–4211. [CrossRef]
14. Pita-Lopez, M.L.; Fletes-Vargas, G.; Espinosa-Andrews, H.; Rodriguez-Rodriguez, R. Physically cross-linked chitosan-based hydrogels for tissue engineering applications: A state-of-the-art review. *Eur. Polym. J.* **2021**, *145*, 110176. [CrossRef]
15. Schuetz, Y.B.; Gurny, R.; Jordan, O. A novel thermoresponsive hydrogel based on chitosan. *Eur. J. Pharm. Biopharm.* **2008**, *68*, 19–25. [CrossRef] [PubMed]
16. Xu, Y.; Yuan, S.; Han, J.; Lin, H.; Zhang, X. Design and fabrication of a chitosan hydrogel with gradient structures via a step-by-step cross-linking process. *Carbohydr. Polym.* **2017**, *176*, 195–202. [CrossRef] [PubMed]
17. Berger, J.; Reist, M.; Mayer, J.M.; Felt, O.; Peppas, N.A.; Gurny, R. Structure and interactions in covalently and ionically crosslinked chitosan hydrogels for biomedical applications. *Eur. J. Pharm. Biopharm.* **2004**, *57*, 19–34. [CrossRef]
18. Park, H.; Park, K.; Kim, D. Preparation and swelling behavior of chitosan-based superporous hydrogels for gastric retention application. *J. Biomed. Mater. Res. Part A* **2006**, *76*, 144–150. [CrossRef]
19. Fan, L.; Yang, H.; Yang, J.; Peng, M.; Hu, J. Preparation and characterization of chitosan/gelatin/pva hydrogel for wound dressings. *Carbohydr. Polym.* **2016**, *146*, 427–434. [CrossRef]
20. Siddiqui, A.R.; Bernstein, J.M. Chronic wound infection: Facts and controversies. *Clin. Dermatol.* **2010**, *28*, 519–526. [CrossRef]
21. Burn injury. *Nat. Rev. Dis. Primers* **2020**, *6*, 12. [CrossRef]
22. Falanga, V. Wound healing and its impairment in the diabetic foot. *Lancet* **2005**, *366*, 1736–1743. [CrossRef]
23. Peppa, M.; Stavroulakis, P.; Raptis, S.A. Advanced glycoxidation products and impaired diabetic wound healing. *Wound Repair Regen.* **2009**, *17*, 461–472. [CrossRef] [PubMed]
24. Burgess, J.L.; Wyant, W.A.; Abdo, A.B.; Kirsner, R.S.; Jozic, I. Diabetic wound-healing science. *Medicina* **2021**, *57*, 1072. [CrossRef] [PubMed]
25. Kottner, J.; Black, J.; Call, E.; Gefen, A.; Santamaria, N. Microclimate: A critical review in the context of pressure ulcer prevention. *Clin. Biomech.* **2018**, *59*, 62–70. [CrossRef] [PubMed]
26. Vangilder, C.; Lachenbruch, C.; Algrim-Boyle, C.; Meyer, S. The international pressure ulcer prevalence survey: 2006–2015: A 10-year pressure injury prevalence and demographic trend analysis by care setting. *J. Wound Ostomy Cont. Nurs.* **2017**, *44*, 20–28. [CrossRef] [PubMed]

27. Liu, H.; Wang, C.; Li, C.; Qin, Y.; Wang, Z.; Yang, F.; Li, Z.; Wang, J. A functional chitosan-based hydrogel as a wound dressing and drug delivery system in the treatment of wound healing. *RSC Adv.* **2018**, *8*, 7533–7549. [CrossRef]
28. Negut, I.; Grumezescu, V.; Grumezescu, A.M. Treatment strategies for infected wounds. *Molecules* **2018**, *23*, 2392. [CrossRef] [PubMed]
29. Mu, M.; Li, X.; Tong, A.; Guo, G. Multi-functional chitosan-based smart hydrogels mediated biomedical application. *Expert Opin. Drug Deliv.* **2019**, *16*, 239–250. [CrossRef] [PubMed]
30. Taokaew, S.; Kaewkong, W.; Kriangkrai, W. Recent development of functional chitosan-based hydrogels for pharmaceutical and biomedical applications. *Gels* **2023**, *9*, 277. [CrossRef]
31. Arafa, A.A.; Nada, A.A.; Ibrahim, A.Y.; Sajkiewicz, P.; Zahran, M.K.; Hakeim, O.A. Preparation and characterization of smart therapeutic ph-sensitive wound dressing from red cabbage extract and chitosan hydrogel. *Int. J. Biol. Macromol.* **2021**, *182*, 1820–1831. [CrossRef] [PubMed]
32. Zheng, K.; Tong, Y.; Zhang, S.; He, R.; Xiao, L.; Iqbal, Z.; Zhang, Y.; Gao, J.; Zhang, L.; Jiang, L.; et al. Flexible bicolorimetric polyacrylamide/chitosan hydrogels for smart real—Time monitoring and promotion of wound healing. *Adv. Funct. Mater.* **2021**, *31*, 2102599. [CrossRef]
33. Shah, S.A.; Sohail, M.; Karperien, M.; Johnbosco, C.; Mahmood, A.; Kousar, M. Chitosan and carboxymethyl cellulose-based 3D multifunctional bioactive hydrogels loaded with nano-curcumin for synergistic diabetic wound repair. *Int. J. Biol. Macromol.* **2023**, *227*, 1203–1220. [CrossRef] [PubMed]
34. Cai, Y.; Fu, X.; Zhou, Y.; Lei, L.; Wang, J.; Zeng, W.; Yang, Z. A hydrogel system for drug loading toward the synergistic application of reductive/heat-sensitive drugs. *J. Control. Release* **2023**, *362*, 409–424. [CrossRef]
35. Han, D.; Li, Y.; Liu, X.; Li, B.; Han, Y.; Zheng, Y.; Yeung, K.W.K.; Li, C.; Cui, Z.; Liang, Y.; et al. Rapid bacteria trapping and killing of metal-organic frameworks strengthened photo-responsive hydrogel for rapid tissue repair of bacterial infected wounds. *Chem. Eng. J.* **2020**, *396*, 125194. [CrossRef]
36. Mai, B.; Jia, M.; Liu, S.; Sheng, Z.; Li, M.; Gao, Y.; Wang, X.; Liu, Q.; Wang, P. Smart hydrogel-based dvdms/bfgf nanohybrids for antibacterial phototherapy with multiple damaging sites and accelerated wound healing. *ACS Appl. Mater. Interfaces* **2020**, *12*, 10156–10169. [CrossRef] [PubMed]
37. Wang, Y.; Li, B.; Xu, F.; Han, Z.; Wei, D.; Jia, D.; Zhou, Y. Tough magnetic chitosan hydrogel nanocomposites for remotely stimulated drug release. *Biomacromolecules* **2018**, *19*, 3351–3360. [CrossRef]
38. Zheng, F.; Li, R.; He, Q.; Koral, K.; Tao, J.; Fan, L.; Xiang, R.; Ma, J.; Wang, N.; Yin, Y.; et al. The electrostimulation and scar inhibition effect of chitosan/oxidized hydroxyethyl cellulose/reduced graphene oxide/asiaticoside liposome based hydrogel on peripheral nerve regeneration in vitro. *Mater. Sci. Eng. C* **2020**, *109*, 110560. [CrossRef]
39. Sun, C.; Zeng, X.; Zheng, S.; Wang, Y.; Li, Z.; Zhang, H.; Nie, L.; Zhang, Y.; Zhao, Y.; Yang, X. Bio-adhesive catechol-modified chitosan wound healing hydrogel dressings through glow discharge plasma technique. *Chem. Eng. J.* **2022**, *427*, 130843. [CrossRef]
40. Blacklow, S.O.; Li, J.; Freedman, B.R.; Zeidi, M.; Chen, C.; Mooney, D.J. Bioinspired mechanically active adhesive dressings to accelerate wound closure. *Sci. Adv.* **2019**, *5*, eaaw3963. [CrossRef]
41. Odinokov, A.V.; Dzhons, D.Y.; Budruev, A.V.; Mochalova, A.E.; Smirnova, L.A. Chitosan modified with terephthaloyl diazide as a drug delivery system. *Russ. Chem. Bull.* **2016**, *65*, 1122–1130. [CrossRef]
42. Bhattarai, N.; Ramay, H.R.; Gunn, J.; Matsen, F.A.; Zhang, M. Peg-grafted chitosan as an injectable thermosensitive hydrogel for sustained protein release. *J. Control. Release* **2005**, *103*, 609–624. [CrossRef] [PubMed]
43. He, M.; Han, B.; Jiang, Z.; Yang, Y.; Peng, Y.; Liu, W. Synthesis of a chitosan-based photo-sensitive hydrogel and its biocompatibility and biodegradability. *Carbohydr. Polym.* **2017**, *166*, 228–235. [CrossRef] [PubMed]
44. Chen, X.; Li, H.; Lam, K.Y. A multiphysics model of photo-sensitive hydrogels in response to light-thermo-pH-salt coupled stimuli for biomedical applications. *Bioelectrochemistry* **2020**, *135*, 107584. [CrossRef] [PubMed]
45. Tomatsu, I.; Peng, K.; Kros, A. Photoresponsive hydrogels for biomedical applications. *Adv. Drug Deliv. Rev.* **2011**, *63*, 1257–1266. [CrossRef] [PubMed]
46. Li, L.; Scheiger, J.M.; Levkin, P.A. Design and applications of photoresponsive hydrogels. *Adv. Mater.* **2019**, *31*, 1807333. [CrossRef] [PubMed]
47. Liu, J.; Xiao, Y.; Wang, X.; Huang, L.; Chen, Y.; Bao, C. Glucose-sensitive delivery of metronidazole by using a photo-crosslinked chitosan hydrogel film to inhibit porphyromonas gingivalis proliferation. *Int. J. Biol. Macromol.* **2019**, *122*, 19–28. [CrossRef]
48. Yang, N.; Zhu, M.; Xu, G.; Liu, N.; Yu, C. A near-infrared light-responsive multifunctional nanocomposite hydrogel for efficient and synergistic antibacterial wound therapy and healing promotion. *J. Mat. Chem. B* **2020**, *8*, 3908–3917. [CrossRef]
49. Yan, H.; Jin, B. Equilibrium swelling of a polyampholytic pH-sensitive hydrogel. *Eur. Phys. J. E* **2013**, *36*, 27. [CrossRef]
50. Lambers, H.; Piessens, S.; Bloem, A.; Pronk, H.; Finkel, P. Natural skin surface pH is on average below 5, which is beneficial for its resident flora. *Int. J. Cosmet. Sci.* **2006**, *28*, 359–370. [CrossRef]
51. Wu, J.; Su, Z.G.; Ma, G.H. A thermo- and pH-sensitive hydrogel composed of quaternized chitosan/glycerophosphate. *Int. J. Pharm.* **2006**, *315*, 1–11. [CrossRef] [PubMed]

52. Ren, Z.; Ke, T.; Ling, Q.; Zhao, L.; Gu, H. Rapid self-healing and self-adhesive chitosan-based hydrogels by host-guest interaction and dynamic covalent bond as flexible sensor. *Carbohydr. Polym.* **2021**, *273*, 118533. [CrossRef] [PubMed]
53. Ou, Y.; Tian, M. Advances in multifunctional chitosan-based self-healing hydrogels for biomedical applications. *J. Mat. Chem. B* **2021**, *9*, 7955–7971. [CrossRef] [PubMed]
54. Li, L.; Yan, B.; Yang, J.; Chen, L.; Zeng, H. Novel mussel-inspired injectable self-healing hydrogel with anti-biofouling property. *Adv. Mater.* **2015**, *27*, 1294–1299. [CrossRef] [PubMed]
55. Chen, M.; Tian, J.; Liu, Y.; Cao, H.; Li, R.; Wang, J.; Wu, J.; Zhang, Q. Dynamic covalent constructed self—Healing hydrogel for sequential delivery of antibacterial agent and growth factor in wound healing. *Chem. Eng. J.* **2019**, *373*, 413–424. [CrossRef]
56. Deng, L.; Wang, B.; Li, W.; Han, Z.; Chen, S.; Wang, H. Bacterial cellulose reinforced chitosan-based hydrogel with highly efficient self-healing and enhanced antibacterial activity for wound healing. *Int. J. Biol. Macromol.* **2022**, *217*, 77–87. [CrossRef] [PubMed]
57. Yu, Q.; Yan, Y.; Huang, J.; Liang, Q.; Li, J.; Wang, B.; Ma, B.; Bianco, A.; Ge, S.; Shao, J. A multifunctional chitosan-based hydrogel with self-healing, antibacterial, and immunomodulatory effects as wound dressing. *Int. J. Biol. Macromol.* **2023**, *231*, 123149. [CrossRef]
58. Robson, M.C. Wound infection. A failure of wound healing caused by an imbalance of bacteria. *Surg. Clin. North Am.* **1997**, *77*, 637–650. [CrossRef]
59. Wang, C.; Huang, X.; Deng, W.; Chang, C.; Hang, R.; Tang, B. A nano-silver composite based on the ion-exchange response for the intelligent antibacterial applications. *Mater. Sci. Eng. C-Mater. Biol. Appl.* **2014**, *41*, 134–141. [CrossRef]
60. Chatterjee, A.K.; Chakraborty, R.; Basu, T. Mechanism of antibacterial activity of copper nanoparticles. *Nanotechnology* **2014**, *25*, 135101. [CrossRef]
61. Tamayo, L.; Azocar, M.; Kogan, M.; Riveros, A.; Paez, M. Copper-polymer nanocomposites: An excellent and cost-effective biocide for use on antibacterial surfaces. *Mater. Sci. Eng. C-Mater. Biol. Appl.* **2016**, *69*, 1391–1409. [CrossRef] [PubMed]
62. Ramamurthy, C.H.; Padma, M.; Samadanam, I.D.; Mareeswaran, R.; Suyavaran, A.; Kumar, M.S.; Premkumar, K.; Thirunavukkarasu, C. The extra cellular synthesis of gold and silver nanoparticles and their free radical scavenging and antibacterial properties. *Colloid Surf. B Biointerfaces* **2013**, *102*, 808–815. [CrossRef] [PubMed]
63. Odularu, A.T.; Ajibade, P.A.; Mbese, J.Z.; Oyedeji, O.O. Developments in platinum-group metals as dual antibacterial and anticancer agents. *J. Chem.* **2019**, *2019*, 5459461. [CrossRef]
64. Liu, H.; Du, Y.; Wang, X.; Sun, L. Chitosan kills bacteria through cell membrane damage. *Int. J. Food Microbiol.* **2004**, *95*, 147–155. [CrossRef]
65. Nesovic, K.; Jankovic, A.; Radetic, T.; Vukasinovic-Sekulic, M.; Kojic, V.; Zivkovic, L.; Peric-Grujic, A.; Rhee, K.Y.; Miskovic-Stankovic, V. Chitosan-based hydrogel wound dressings with electrochemically incorporated silver nanoparticles—In vitro study. *Eur. Polym. J.* **2019**, *121*, 109257. [CrossRef]
66. Shukla, R.; Kashaw, S.K.; Jain, A.P.; Lodhi, S. Fabrication of apigenin loaded gellan gum-chitosan hydrogels (ggch-hgs) for effective diabetic wound healing. *Int. J. Biol. Macromol.* **2016**, *91*, 1110–1119. [CrossRef] [PubMed]
67. Ding, C.; Zhao, Y.; Chen, X.; Zheng, Y.; Liu, W.; Liu, X. Taxifolin, a novel food, attenuates acute alcohol-induced liver injury in mice through regulating the nf-κb-mediated inflammation and pi3k/akt signalling pathways. *Pharm. Biol.* **2021**, *59*, 866–877. [CrossRef]
68. Ding, Q.; Ding, C.; Liu, X.; Zheng, Y.; Zhao, Y.; Zhang, S.; Sun, S.; Peng, Z.; Liu, W. Preparation of nanocomposite membranes loaded with taxifolin liposome and its mechanism of wound healing in diabetic mice. *Int. J. Biol. Macromol.* **2023**, *241*, 124537. [CrossRef]
69. Ding, C.; Liu, Z.; Zhao, T.; Sun, S.; Liu, X.; Zhang, J.; Ma, L.; Yang, M. A temperature-sensitive hydrogel loaded with taxifolin promotes skin repair by modulating mapk-mediated autophagic pathway. *J. Mater. Sci.* **2023**, *58*, 14831–14845. [CrossRef]
70. Khan, A.K.; Rashid, R.; Fatima, N.; Mahmood, S.; Mir, S.; Khan, S.; Jabeen, N.; Murtaza, G. Pharmacological activities of protocatechuic acid. *Acta Pol. Pharm.* **2015**, *72*, 643–650.
71. Zhou, C.; Xu, R.; Han, X.; Tong, L.; Xiong, L.; Liang, J.; Sun, Y.; Zhang, X.; Fan, Y. Protocatechuic acid-mediated injectable antioxidant hydrogels facilitate wound healing. *Compos. Pt. B Eng.* **2023**, *250*, 110451. [CrossRef]
72. Pasanphan, W.; Chirachanchai, S. Conjugation of gallic acid onto chitosan: An approach for green and water-based antioxidant. *Carbohydr. Polym.* **2008**, *72*, 169–177. [CrossRef]
73. Sun, X.; Dong, M.; Guo, Z.; Zhang, H.; Wang, J.; Jia, P.; Bu, T.; Liu, Y.; Li, L.; Wang, L. Multifunctional chitosan-copper-gallic acid based antibacterial nanocomposite wound dressing. *Int. J. Biol. Macromol.* **2021**, *167*, 10–22. [CrossRef] [PubMed]
74. Wang, H.; Liu, Y.; Cai, K.; Zhang, B.; Tang, S.; Zhang, W.; Liu, W. Antibacterial polysaccharide-based hydrogel dressing containing plant essential oil for burn wound healing. *Burn. Trauma* **2021**, *9*, tkab41. [CrossRef] [PubMed]
75. Koosehgol, S.; Ebrahimian-Hosseinabadi, M.; Alizadeh, M.; Zamanian, A. Preparation and characterization of in situ chitosan/polyethylene glycol fumarate/thymol hydrogel as an effective wound dressing. *Mater. Sci. Eng. C-Mater. Biol. Appl.* **2017**, *79*, 66–75. [CrossRef] [PubMed]
76. Pickart, L. The human tri-peptide ghk and tissue remodeling. *J. Biomater. Sci.-Polym. Ed.* **2008**, *19*, 969–988. [CrossRef] [PubMed]
77. Pickart, L.; Vasquez-Soltero, J.M.; Margolina, A. Ghk peptide as a natural modulator of multiple cellular pathways in skin regeneration. *Biomed. Res. Int.* **2015**, *2015*, 648108. [CrossRef]
78. Wang, S.; Feng, C.; Yin, S.; Feng, Z.; Tang, J.; Liu, N.; Yang, F.; Yang, X.; Wang, Y. A novel peptide from the skin of amphibian rana limnocharis with potency to promote skin wound repair. *Nat. Prod. Res.* **2021**, *35*, 3514–3518. [CrossRef]

79. Ouyang, Q.Q.; Hu, Z.; Lin, Z.P.; Quan, W.Y.; Deng, Y.F.; Li, S.D.; Li, P.W.; Chen, Y. Chitosan hydrogel in combination with marine peptides from tilapia for burns healing. *Int. J. Biol. Macromol.* **2018**, *112*, 1191–1198. [CrossRef]
80. Qianqian, O.; Songzhi, K.; Yongmei, H.; Xianghong, J.; Sidong, L.; Puwang, L.; Hui, L. Preparation of nano-hydroxyapatite/chitosan/tilapia skin peptides hydrogels and its burn wound treatment. *Int. J. Biol. Macromol.* **2021**, *181*, 369–377. [CrossRef]
81. Xiao, Y.; Ge, H.; Zou, S.; Wen, H.; Li, Y.; Fan, L.; Xiao, L. Enzymatic synthesis of n-succinyl chitosan-collagen peptide copolymer and its characterization. *Carbohydr. Polym.* **2017**, *166*, 45–54. [CrossRef] [PubMed]
82. Hu, W.; Liu, M.; Yang, X.; Zhang, C.; Zhou, H.; Xie, W.; Fan, L.; Nie, M. Modification of chitosan grafted with collagen peptide by enzyme crosslinking. *Carbohydr. Polym.* **2019**, *206*, 468–475. [CrossRef] [PubMed]
83. Deng, A.; Yang, Y.; Du, S.; Yang, X.; Pang, S.; Wang, X.; Yang, S. Preparation of a recombinant collagen-peptide (rhc)-conjugated chitosan thermosensitive hydrogel for wound healing. *Mater. Sci. Eng. C-Mater. Biol. Appl.* **2021**, *119*, 111555. [CrossRef] [PubMed]
84. Amiri, N.; Ajami, S.; Shahroodi, A.; Jannatabadi, N.; Amiri, D.S.; Fazly, B.B.; Pishavar, E.; Kalalinia, F.; Movaffagh, J. Teicoplanin-loaded chitosan-peo nanofibers for local antibiotic delivery and wound healing. *Int. J. Biol. Macromol.* **2020**, *162*, 645–656. [CrossRef] [PubMed]
85. Hafezi, M.R.; Dadfarnia, S.; Shabani, A.; Amraei, R.; Hafezi, M.Z. Doxycycline drug delivery using hydrogels of o-carboxymethyl chitosan conjugated with caffeic acid and its composite with polyacrylamide synthesized by electron beam irradiation. *Int. J. Biol. Macromol.* **2020**, *154*, 962–973. [CrossRef] [PubMed]
86. Fasiku, V.O.; Omolo, C.A.; Devnarain, N.; Ibrahim, U.H.; Rambharose, S.; Faya, M.; Mocktar, C.; Singh, S.D.; Govender, T. Chitosan-based hydrogel for the dual delivery of antimicrobial agents against bacterial methicillin-resistant staphylococcus aureus biofilm-infected wounds. *ACS Omega* **2021**, *6*, 21994–22010. [CrossRef] [PubMed]
87. Ilomuanya, M.O.; Enwuru, N.V.; Adenokun, E.; Fatunmbi, A.; Adeluola, A.; Igwilo, C.I. Chitosan-based microparticle encapsulated acinetobacter baumannii phage cocktail in hydrogel matrix for the management of multidrug resistant chronic wound infection. *Turk. J. Pharm. Sci.* **2022**, *19*, 187–195. [CrossRef]
88. Guo, S.; Dipietro, L.A. Factors affecting wound healing. *J. Dent. Res.* **2010**, *89*, 219–229. [CrossRef]
89. Feng, P.; Luo, Y.; Ke, C.; Qiu, H.; Wang, W.; Zhu, Y.; Hou, R.; Xu, L.; Wu, S. Chitosan-based functional materials for skin wound repair: Mechanisms and applications. *Front. Bioeng. Biotechnol.* **2021**, *9*, 650598. [CrossRef]
90. Hoshi, R.; Murata, S.; Matsuo, R.; Myronovych, A.; Hashimoto, I.; Ikeda, H.; Ohkohchi, N. Freeze-dried platelets promote hepatocyte proliferation in mice. *Cryobiology* **2007**, *55*, 255–260. [CrossRef]
91. Khan, M.A.; Mujahid, M. A review on recent advances in chitosan based composite for hemostatic dressings. *Int. J. Biol. Macromol.* **2019**, *124*, 138–147. [CrossRef] [PubMed]
92. Leggat, P.A.; Smith, D.R.; Kedjarune, U. Surgical applications of cyanoacrylate adhesives: A review of toxicity. *ANZ J. Surg.* **2007**, *77*, 209–213. [CrossRef] [PubMed]
93. Fan, P.; Zeng, Y.; Zaldivar-Silva, D.; Aguero, L.; Wang, S. Chitosan-based hemostatic hydrogels: The concept, mechanism, application, and prospects. *Molecules* **2023**, *28*, 1473. [CrossRef] [PubMed]
94. Kozen, B.G.; Kircher, S.J.; Henao, J.; Godinez, F.S.; Johnson, A.S. An alternative hemostatic dressing: Comparison of celox, hemcon, and quikclot. *Acad. Emerg. Med.* **2008**, *15*, 74–81. [CrossRef]
95. Pusateri, A.E.; Mccarthy, S.J.; Gregory, K.W.; Harris, R.A.; Cardenas, L.; Mcmanus, A.T.; Goodwin, C.J. Effect of a chitosan-based hemostatic dressing on blood loss and survival in a model of severe venous hemorrhage and hepatic injury in swine. *J Trauma* **2003**, *54*, 177–182. [CrossRef] [PubMed]
96. Xia, L.; Wang, S.; Jiang, Z.; Chi, J.; Yu, S.; Li, H.; Zhang, Y.; Li, L.; Zhou, C.; Liu, W.; et al. Hemostatic performance of chitosan-based hydrogel and its study on biodistribution and biodegradability in rats. *Carbohydr. Polym.* **2021**, *264*, 117965. [CrossRef] [PubMed]
97. Zhao, X.; Wu, H.; Guo, B.; Dong, R.; Qiu, Y.; Ma, P.X. Antibacterial anti-oxidant electroactive injectable hydrogel as self-healing wound dressing with hemostasis and adhesiveness for cutaneous wound healing. *Biomaterials* **2017**, *122*, 34–47. [CrossRef]
98. Guo, S.; Ren, Y.; Chang, R.; He, Y.; Zhang, D.; Guan, F.; Yao, M. Injectable self-healing adhesive chitosan hydrogel with antioxidative, antibacterial, and hemostatic activities for rapid hemostasis and skin wound healing. *Acs Appl. Mater. Interfaces* **2022**, *14*, 34455–34469. [CrossRef]
99. Medzhitov, R. Origin and physiological roles of inflammation. *Nature* **2008**, *454*, 428–435. [CrossRef]
100. Liu, Q.; Huang, Y.; Lan, Y.; Zuo, Q.; Li, C.; Zhang, Y.; Guo, R.; Xue, W. Acceleration of skin regeneration in full-thickness burns by incorporation of bfgf-loaded alginate microspheres into a cmcs-pva hydrogel. *J. Tissue Eng. Regen. Med.* **2017**, *11*, 1562–1573. [CrossRef]
101. Gull, N.; Khan, S.M.; Butt, O.M.; Islam, A.; Shah, A.; Jabeen, S.; Khan, S.U.; Khan, A.; Khan, R.U.; Butt, M. Inflammation targeted chitosan-based hydrogel for controlled release of diclofenac sodium. *Int. J. Biol. Macromol.* **2020**, *162*, 175–187. [CrossRef]
102. Gonzalez, A.C.; Costa, T.F.; Andrade, Z.A.; Medrado, A.R. Wound healing—A literature review. *An. Brasil. Dermatol.* **2016**, *91*, 614–620. [CrossRef]
103. Liu, F.; Wang, L.; Zhai, X.; Ji, S.; Ye, J.; Zhu, Z.; Teng, C.; Dong, W.; Wei, W. A multi-functional double cross-linked chitosan hydrogel with tunable mechanical and antibacterial properties for skin wound dressing. *Carbohydr. Polym.* **2023**, *322*, 121344. [CrossRef]

104. Larouche, J.; Sheoran, S.; Maruyama, K.; Martino, M.M. Immune regulation of skin wound healing: Mechanisms and novel therapeutic targets. *Adv. Wound Care* **2018**, *7*, 209–231. [CrossRef]
105. Chen, X.; Cao, X.; Jiang, H.; Che, X.; Xu, X.; Ma, B.; Zhang, J.; Huang, T. Sikvav-modified chitosan hydrogel as a skinsubstitutes for wound closure in mice. *Molecules* **2018**, *23*, 2611. [CrossRef]

polymers

MDPI

Review

Embracing Sustainability: The World of Bio-Based Polymers in a Mini Review

Grazia Isa C. Righetti *, Filippo Faedi and Antonino Famulari *

Department of Chemistry, Materials and Chemical Engineering "Giulio Natta", Politecnico di Milano, Piazza Leonardo da Vinci 32, 20133 Milano, Italy
* Correspondence: graziaisacarla.righetti@polimi.it (G.I.C.R.); antonino.famulari@polimi.it (A.F.)

Abstract: The proliferation of polymer science and technology in recent decades has been remarkable, with synthetic polymers derived predominantly from petroleum-based sources dominating the market. However, concerns about their environmental impacts and the finite nature of fossil resources have sparked interest in sustainable alternatives. Bio-based polymers, derived from renewable sources such as plants and microbes, offer promise in addressing these challenges. This review provides an overview of bio-based polymers, discussing their production methods, properties, and potential applications. Specifically, it explores prominent examples including polylactic acid (PLA), polyhydroxyalkanoates (PHAs), and polyhydroxy polyamides (PHPAs). Despite their current limited market share, the growing awareness of environmental issues and advancements in technology are driving increased demand for bio-based polymers, positioning them as essential components in the transition towards a more sustainable future.

Keywords: bio-based polymers; polymers; sustainable materials; renewable sources; polymeric materials

Citation: Righetti, G.I.C.; Faedi, F.; Famulari, A. Embracing Sustainability: The World of Bio-Based Polymers in a Mini Review. *Polymers* **2024**, *16*, 950. https://doi.org/10.3390/polym16070950

Academic Editors: Stefano Farris and Masoud Ghaani

Received: 28 February 2024
Revised: 25 March 2024
Accepted: 28 March 2024
Published: 30 March 2024

1. Introduction

The interest in polymer science and technology has witnessed an unprecedented growth over the past few decades, leading to rapid advancements in the development of polymers and plastics. Throughout the 20th century, the polymer industry was heavily reliant on petroleum-related chemistry, refineries, and processes. The petrochemical boom in the latter half of the century played a pivotal role in propelling the progress of the chemical sciences, experiencing unparalleled growth in the organic chemical industry.

Despite ongoing debates about the detrimental environmental impact of petroleum-related processes, industries hesitated to alter their raw materials or methodologies until the situation reached a critical juncture. Notably, this era saw the discovery of various monomers for polymerizations, propelling synthetic polymers to a prominent position in the global market, as they could be utilized in the production of a wide array of everyday commodities.

This remarkable scientific and technological upswing led to the gradual displacement of naturally occurring polymers by their synthetic counterparts. This choice was primarily driven by economic considerations, as the cost of oil was significantly lower than the processing of natural fibers at that time. Although polymers derived from renewable sources continued to exist, they played a minor role due to limited investments compared to the substantial funds allocated to petrochemistry.

However, the polymer industry copes with two significant challenges: global warming and the depletion of fossil resources. A potential solution to address these issues involves exploring the use of renewable sources instead of fossil-based ones. Moreover, the decreased availability of fossil resources resulted in a surge in market prices, exacerbated by the fact that oil is not universally accessible and requires certain states to be subjected to others. In contrast, renewable resources are globally shared, making them highly valuable [1,2].

In the ever-evolving landscape of polymers for materials science, the quest for sustainable alternatives has become imperative. In light of the growing concerns and an increased awareness of the environmental impact of traditional plastics, researchers are turning their attention to bio-based polymers as a promising solution. The urgency to reduce the consumption of fossil fuels, together with the negative impact of plastic pollution on the environment, has exponentially increased the interest in these alternative materials. As we try to face the challenges of a global environmental crisis, the significance of bio-based polymers extends beyond their material properties. Their potential to mitigate carbon emissions, reduce plastic pollution, and contribute to a circular economy underscores their pivotal role in building a sustainable future.

Bio-based polymers are derived from renewable resources such as plants, microbes, and agricultural or forestry feedstock and represent a paradigm shift in the pursuit of eco-friendly materials. They can be produced from the raw sources through chemical, physical, or biochemical transformations, and due to the sustainable nature of the starting material, bio-based polymers offer several advantages such as a reduced carbon footprint and the potential for higher biodegradability [3,4]. Even if it was logical to assume that polymers crafted from renewable and frequently biodegradable sources would possess biodegradable properties as well, this feature is not always assured. Variations in functional groups, crosslink density, and the inclusion of non-biodegradable co-monomers through copolymerization may lead to materials that do not necessarily demonstrate substantial biodegradability [5,6]. As the terms "biodegradable" and "bio-based" are often mentioned in the literature, it is crucial to underline the key difference between these two types of polymers. A biodegradable polymer is defined as a material able to undergo degradation and deterioration. The biodegradation of a polymer (bio-based or synthetic) is defined as the chemical decomposition process of the substance into environmentally friendly compounds [7]. After a first fragmentation step of the high molecular mass (HMW) polymer to a lower molecular mass (LMW) group of chains [8,9], the final degradation is dependent on microorganisms which, based on the environment (aerobic or anaerobic), will convert the substances mainly into CO_2, H_2O, and CH_4 [10]. On the other hand, bio-based polymers are defined by IUPAC as "composed or derived in whole or in part of a biological product issued from biomass (including plant, animal, and marine or forestry materials)" [11].

The primary advantage of bio-based polymers relies on substituting non-renewable resources with renewable carbon sourced from biomass, a crucial step in achieving sustainability and promoting a climate-friendly plastics industry. Another key benefit is that biodegradable bio-based polymers constitute roughly a quarter of all bio-based polymers (depending on environmental conditions). These next-generation polymers might be the key to a lower reliance on fossil fuels, offering a potential solution to the problem of uncollected and unprocessed plastics in the environment [12,13].

Bio-based polymers still hold just a small fraction of the market, accounting for 0.5% of the overall global plastic production and 1.0% of the European market in 2023 (Figure 1) [14]. However, the global market request is estimated to expand at a compound annual growth rate of 18.8% from 2023 to 2030 [15].

Despite the currently low market fraction occupied by bio-based polymers, their crucial role in shifting plastic production towards a greener and more sustainable direction is underlined by the growing interest among researchers in both academia and industry. This is evident not only in the increasing number of scientific articles and citations dedicated to this subject (Figure 2) but also in the wide application of those polymers in different fields (Figure 3).

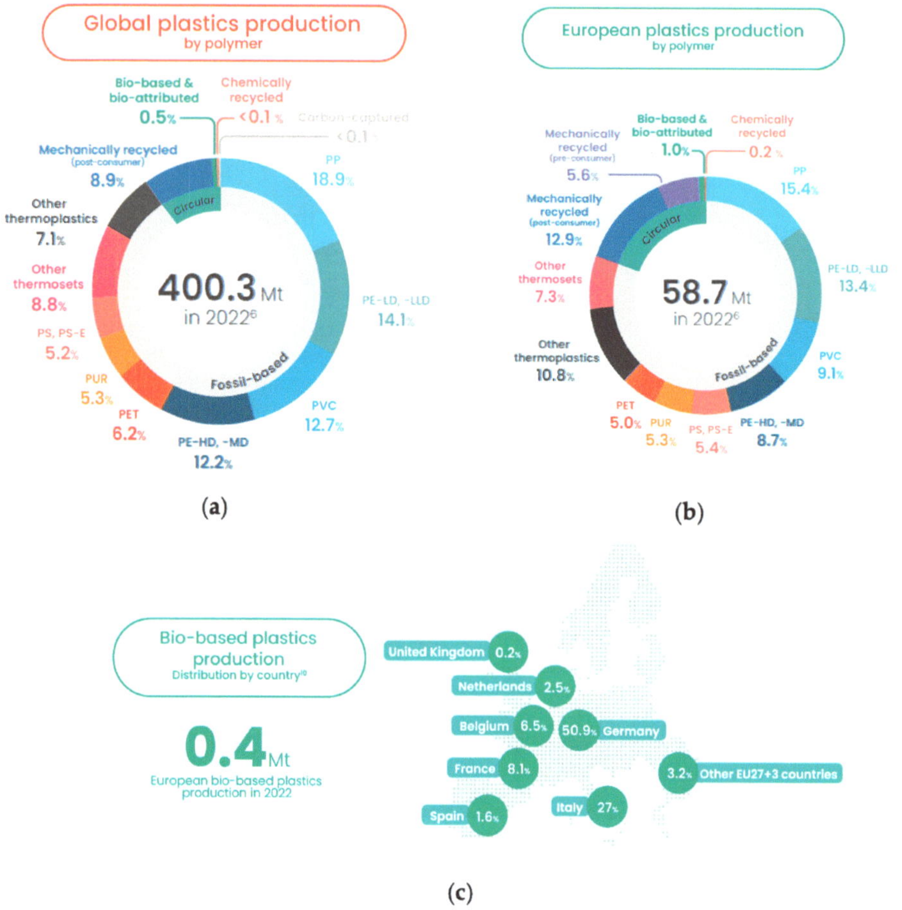

Figure 1. Plastic production in 2023: (**a**) global plastic production, (**b**) European plastic production, (**c**) bio-based plastic production per country in Europe. The credit for the graphs belongs to PlasticEurope and is attributed to the report *'Plastic Europe—The Fast Facts 2023'*. The graphs have been reproduced with the permission of PlasticEurope [12].

This mini-review aims to provide an overview on three important bio-based polymers: (a) polylactic acid (PLA), (b) polyhydroxy alkanoates (PHA), and (c) polyhydroxy polyamides (PHPAs).

These three polymers were chosen due to their importance among all the biobased polymers. In fact, polylactic acid (PLA) and polyhydroxy alkanoates (PHA) are the most-studied bio-based polymeric materials due to their potential to replace petroleum-based polymeric materials thanks to their peculiar properties. Their importance is also underlined by the fact that they have already found applications in many fields, ranging from packaging to biomedical fields [16]. On the other hand, polyhydroxy polyamides (PHPAs) have also gained increasing attention with the aim of synthetizing new, more hydrophilic and degradable polymeric materials [17]. What makes this class of polymers highly attractive is the nature of the dicarboxylic acid employed, which is a natural carbohydrate-based monomeric building block that displays peculiar structural features (e.g., stereochemistry) that can help tune the polyamide's final properties [18].

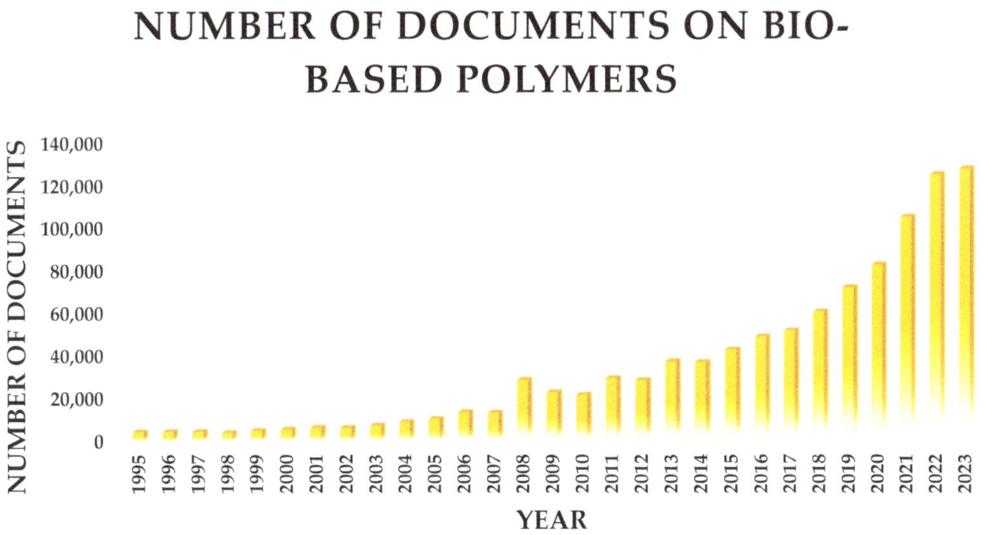

Figure 2. Total number of documents produced on bio-based polymers per year from 1995 to 2003, including both articles and patents. Dimension AI was used as a tool to retrieve the total number of documents per year on the given topic.

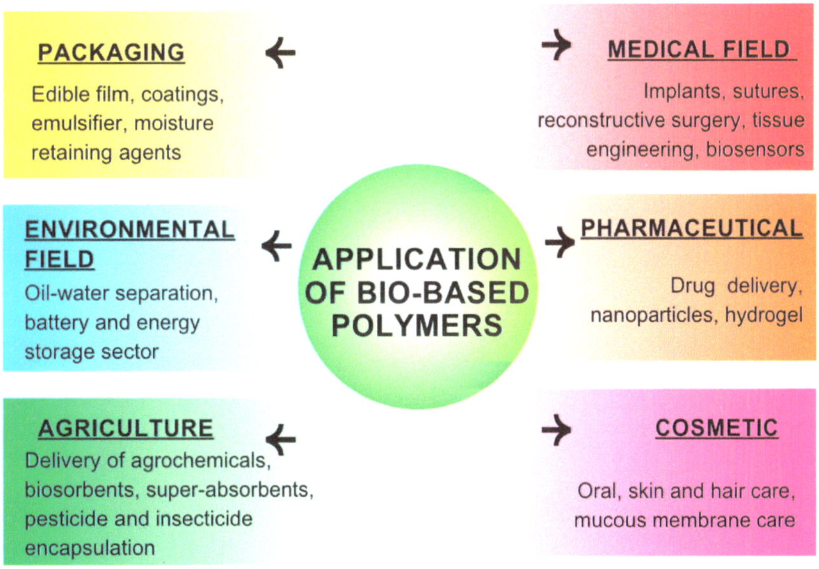

Figure 3. Examples of some applications of biopolymers and bio-based polymers.

2. Bio-Based Polymers

2.1. Polylactic Acid

Polylactic acid (PLA), a biodegradable aliphatic polyester, has been known since 1845 when Theophile-Jules Pelouze first synthesized it [19]. It has been extensively studied in the past 20–25 years due to its exceptional properties, such as its high biodegradability and biocompatibility.

PLA is composed of lactic acid as its fundamental building block. This monomer is a hydroxyl carboxylic acid that can be obtained from renewable sources like corn starch (via fermentation processes) or sugarcane. Even if various renewable resources can be employed, corn stands out, as it offers the advantage of yielding a high-quality feedstock for fermentation, resulting in the production of high-purity lactic acid. Depending on the microbial strain utilized in the fermentation process, both L-lactic acid and D-lactic can be obtained [20].

2.1.1. Polylactic Acid Synthesis

PLA can be synthetized via direct polymerization of lactic acid or via ring-opening polymerization (ROP) of the lactide monomer, where the lactide is the cyclic dimer of lactic acid (Scheme 1). Typically, the polycondensation process is associated with a low-molecular-weight polymeric product due to water formation associated with the lactic acid condensation process causing the degradation of the forming polymeric chain itself. Water removal systems and protocols utilizing high temperatures in the presence of catalysts have been proposed to attain higher degrees of polymerization in the production of polylactic acid (PLA) [21]. Nevertheless, these approaches still face limitations, primarily stemming from challenges in effectively removing water as the degree of polymerization (DP) of PLA increases, and from material degradation issues at elevated temperatures [22].

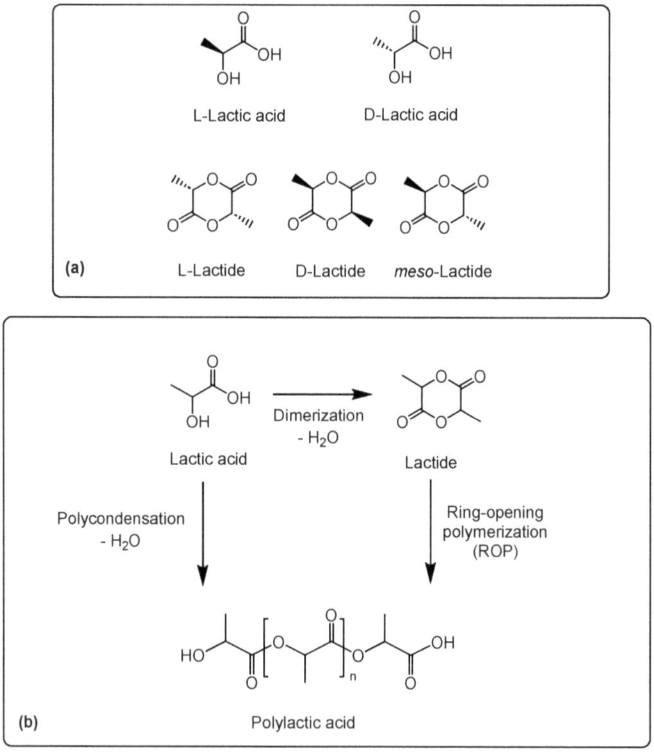

Scheme 1. (**a**) optical monomers of lactic acid and stereoisomers of lactide; (**b**) formation of polylactic acid via polycondensation and ring-opening polymerization.

The current preferred synthetic pathway for producing polylactic acid (PLA) is ring-opening polymerization. This method offers enhanced control over various reaction parameters such as polydispersity, molecular weight, and stereochemistry. Notably, ring-opening polymerization allows for precise control over the insertion of lactide monomers, and the stereochemistry of PLA can be easily manipulated by polymerizing d, l, or meso-lactide monomers [23]. The ability to control and modulate stereochemistry is pivotal, as it enables the production of polymers with markedly different characteristics. From this perspective, the investigation of metal complexes as catalysts in lactide polymerization via coordination–insertion mechanisms has experienced a surge in interest over the past decade. While tin (II) bis(2-ethylhexanoate), commonly referred to as $Sn(Oct)_2$, remains the most-utilized catalyst for industrial PLA synthesis, other examples such as aluminum, zinc, magnesium, and calcium have also gained widespread usage [24–27].

2.1.2. Polylactic Acid: Physical Properties

Extensive research has focused on understanding the properties of polylactic acid (PLA). These properties are influenced by factors such as stereochemistry, processing temperature, annealing time, and molecular weight (Mw). Specifically, the stereochemistry and thermal history of PLA have a direct impact on its crystallinity, which in turn affects its properties.

Crystallinity, which refers to the proportion of crystalline regions in the polymer compared to the amorphous content, plays a crucial role in determining various properties of PLA. These properties include hardness, modulus, tensile strength, stiffness, crease resistance, and melting points (the values of some of these parameters are reported in Table 1, with those of polypropylene and polyethylene provided for a comparison with synthetic non-bio-based polymers). The degree of crystallinity significantly influences the mechanical, thermal, and processing characteristics of PLA, making it a key parameter in understanding and optimizing PLA-based materials for diverse applications.

Table 1. Comparison between the physical properties of PLA [28–30], polypropylene (PP) [31–34], and high-density polyethylene (HDPE) [34–38].

Properties	Type of Polymer				
	PLA	**PLLA**	**PDLLA**	**PP Homopolymer**	**HDPE**
Polymer density—ρ (g/cm^3)	1.21–1.25	1.24–1.30	1.25–1.27	0.91–0.94	0.80–1.80
Tensil strength—σ (MPa)	21–60	15.5–150	27.6–50	22–34.4	11.6–40
Young's modulus—E (GPa)	0.35–3.5	2.7–4.14	1–3.45	0.545	0.30–1.30
Strain—ε (%)	2.5–6	3.0–10.0	2.0–10.0	//	20
Elongation at break (%)	10	//	//	3–700	3.5–800
Glass transition temperature—T_g (°C)	45–60	55–65	50–60	−25	−120
Melting temperature—T_m (°C)	150–162	170–220	n.a. [1]	160–166	127–255
Crystallinity (%)	42	//	//	30–60	70

[1] n.a.—not applicable: amorphous material does not have a well-defined melting point.

PLA is renowned for its exceptional optical properties and high tensile strength. However, its versatility and potential applications might be compromised by its rigidity and brittleness at room temperature; characteristics that stem from its glass transition temperature (T_g) of approximately 55 °C. PLA might meet the mechanical property requirements for most rigid objects, but it needs to be plasticized when it comes to its applications, especially as films in the packaging industry. Plasticizers play a vital role in enhancing the processability, flexibility, and ductility of polymers. In the case of semi-crystalline polymers like PLA, an effective plasticizer needs to not only reduce the glass transition temperature (T_g) but also depress the melting temperature (T_m) and overall crystallinity [30,39,40]. Different kinds of molecules are reported in the literature as plasticizing agents for PLA. Low-molecular-weight agents such as glycerol, sorbitol, bishydroxymethyl malonate (DBM), adipates and citrates [41], vegetable oils derivatives [42,43], and 10,16-dihydroxy hexadecenoic acid and its derivatives [29] have been tested, but these molecules showed a high mobility inside the PLA matrix due to their low molecular weight [44]. In order to reduce the plasticizer migration phenomena, blending PLA with higher-molecular-weight compounds such as polyethylene glycol (PEG) [45,46], poly(propylene) glycol (PPG) [47], or poly(diethylene adipate) (PDEA) [48] has emerged as an appealing solution. Some starch-blended PLA mixtures like Mater-Bi (Novamont, Novara, Italy) and Bioplast (Biotec, Emmerich am Rhein, Germany) [49] have also reached the market. Despite the wide range of plasticizer choices offered by the ever-growing scientific literature, this is often constrained by legislative or technical requirements that are specific to their application, especially in the medical and food industries [44,46,50]. This limitation makes the selection process more challenging. In fact, the plasticizer used for PLA must meet stringent criteria such as biodegradability, non-toxicity, and/or biocompatibility. As a result, the most common plasticizers used for PLA are low-molecular-weight polyethylene glycols (PEGs) [30,40,45,46].

2.1.3. Polylactic Acid Applications

Currently, PLA-based products have entered the market with various applications, ranging from drug carriers, temporary implants, and bone-fixing elements to degradable dishes and packaging. PLA, as a highly versatile and environmentally friendly polymer, possesses renewability, biodegradability, transparency, colorlessness, processability, and mechanical properties, making it increasingly interesting as a substitute polymer for those products where recycling, reuse, and product recovery of the materials are not feasible [51]. As the production costs for PLA manufacturing processes decrease, the potential for PLA to be utilized in a broad array of products continues to grow.

One of the largest potential application areas for PLA is in fibers, where the first commercial success as a fiber material was in the form of resorbable sutures [23]. Since then, PLA has been extensively studied, especially for its potential use in other medical applications due to its bioresorbability and biocompatible properties in the human body. In this sector, PLA is employed not only in wound management but also to produce resorbable stents [52], as encapsulation systems to enhance target-specific drug delivery [53–58], in the orthopedic field as fracture fixation systems and as temporary fillings in facial reconstructive surgery [59], and in tissue engineering and regenerative medicine fields [30,60,61].

The second-largest application for PLA is in the production of films, particularly in the food packaging industry. PLA films are prized for their transparency and excellent deadfold or twist retention properties, making them increasingly sought after as a green alternative in food packaging due to their biodegradability. In the packaging sector, PLA finds applications ranging from containers for fresh produce to disposable cutlery and drinking cups. While PLA presents several advantages, such as biodegradability, it also poses challenges that need to be addressed. These include inferior moisture barrier properties and brittleness [62]. To overcome these challenges, various approaches are being explored, including coatings, lamination, blends with other polymers, and chemical or physical modifications [63]. However, achieving a balance between stiffness and toughness while maintaining a high biobased content remains a significant challenge. Strategies such as plasticizing PLA with its own monomers or blending it with other polymers have been attempted, but they often encounter issues such as phase separation [24,64].

PLA is a good candidate material for 3D printing technology, and in the last three years, many investigations have been focused on tuning its properties for this application [65]. Melt extrusion (MEX) is the most-used 3D molding method, and it imposes some requirements to the molding material such as a low viscosity, sufficient adhesion, robustness, and stability. PLA as a 3D printing material partially meets these requirements, but it presents some defects such as a poor high-temperature resistance, brittleness, and low impact resistance. The performance of PLA's extruded filament can be improved by adding auxiliary material and changing the processing parameters. The research in this field has focused on changing the mechanical properties of PLA filaments and improving their printing performance. One promising root to make PLA suitable for 3D printing is to add natural fillers to the polymers, such as calcined shell particles (CSh), straw meal, and poplar fibers. The resulting materials are completely biodegradable and result in a higher tensile strength of the polymers [66–68]. Copper and lignin are used as fillers for impact resistance enhancements. The addition of lignin microspheres resulted in an increase in impact strength of 100% [69], and Cu-modified PLA achieved remarkable results in impact resistance tests as well [70]. Regarding the change in processing parameters, Fontana et al. analyzed the effects of layer height (LH) and fill rate on the ultimate tensile strength (UTS) values of 3D-printed tensile specimens of Makerbot PLA rigid materials [71,72]. The data analysis revealed that the LH had a more prominent effect on the mechanical strength than the fill rate.

The inherent ignitability properties of neat PLA restrict its application [71,72]. A way to improve PLA's fire safety is to add flame retardants to the neat polymers. Among them, the most successful have been intumescent flame retardants (IFR), P/N synergistic systems, metal oxides, hydroxides, and nanomaterials [73–75].

Intumescent flame retardants (IFR) proved to be very efficient, environmentally friendly, and also capable of increasing the mechanical properties of PLA [76]. Recently, a novel bio-based IFR with a nanosheet structure, PP-Fe, was fabricated using self-assembly technology. The new material presented an outstanding increase in fire safety and UV resistance. For fire safety, introduction of 20.0 wt% PP-Fe nanosheets to PLA resulted in significant reductions in the PHRRc (72.7%), THRc (41.6%), and TSP (64.7%) values in the CCT and achieved a UL-94 V-0 rating with an LOI of 31.2%. The UPF value of the PLA/20PP-Fe composite was 120%, categorized as excellent UV shielding [77].

Despite these challenges, PLA holds promise as a future material for eco-friendly food packaging, and ongoing research and innovation in this field are expected to lead to further improvements and applications.

2.2. Polyhydroxy Alcanoates

Poly(hydroxyalkanoates) (PHAs) represent a class of biopolymers prized for their biodegradability and their ability to be derived from natural resources. Falling under the umbrella of aliphatic polyesters, PHAs have gained significant attention from both the scientific and industrial sectors owing to their potential as biomaterials. In fact, due to the combination of their innate biodegradability, natural origin, functional versatility, and compatibility with biological systems and their mechanical and thermal properties, they are viewed as promising alternatives to petroleum-based plastics [78]. Notably, PHA stands out as an ideal candidate due to its ability to form small pores, facilitating efficient recycling and boasting a high volume-to-surface ratio [79].

PHA, comprised of hydroxyalkanoic acid monomers, owes its uniqueness primarily to its fascinating production process. This process is a captivating biological phenomenon orchestrated by a diverse array of microorganisms such as *Alcaligenes latus*, *Azobacter vinelandii*, *Bacillus megaterium*, *Cupriavidus necator*, *Pseudomonas oleovorans*, and *Escherichia coli* [80]. These microorganisms naturally accumulate PHAs, especially polyhydroxybutyrate (PHB), as their principal lipid reserve within cellular membranes in the form of granules under conditions of metabolic stress (Figure 4) [81].

Biopolymer biosynthesis occurs at the cytoplasmic level when there is an excess of carbon and a deficiency of other vital nutrients required for cell proliferation, including phosphorus, nitrogen, and magnesium. Under such circumstances, intracellular accumulation of the biopolymer in granular form serves as a reservoir of carbon and energy [82].

This phenomenon mirrors the industrial emulsion polymerization process with remarkable precision. It serves as a captivating and exemplary instance of how certain biological processes can be seamlessly integrated into industries, catalyzing a new revolution centered around the effective harnessing of biological mechanisms. These mechanisms are renowned for their energy efficiency, high selectivity, and environmental friendliness, making them ideal candidates for integration into traditional industrial processes [83,84].

From a structural point of view, polyhydroxy alcanoates are composed of hydroxylated fatty acid monomers that polymerize via ester bond formation with each monomeric unit bearing the alkylic chain as a side chain (Figure 5).

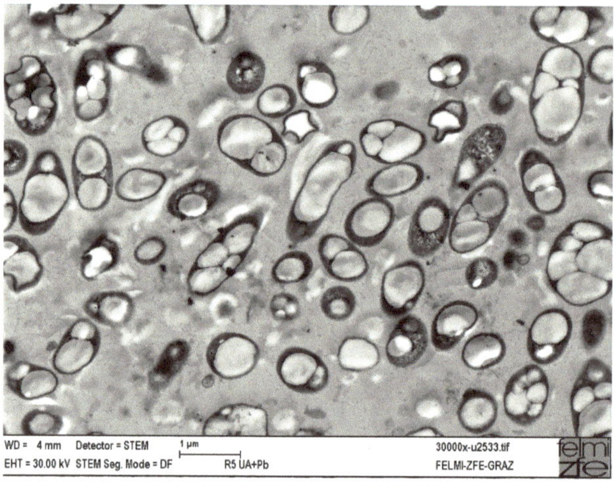

Figure 4. Cupriavidus necator DSM 545, a metalophilic strain, containing PHA granules (bright intracellular inclusions) cultivated in continuous mode on glucose as the carbon source, imaged using STEM, magnification ×30,000. Picture taken by E. Ingolić, FELMI-ZFE Graz, and provided by M. Koller, University of Graz, Austria [83].

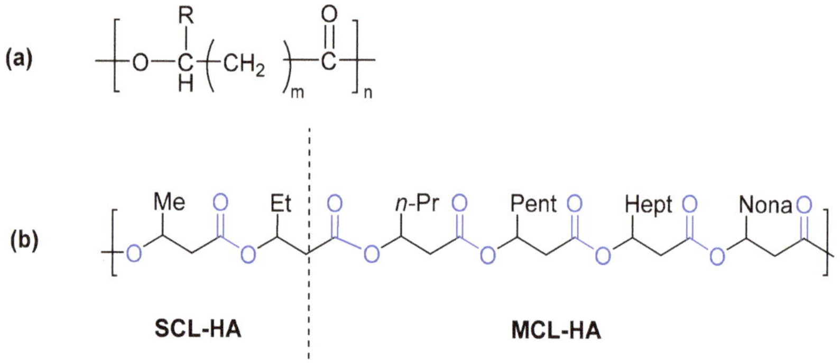

Figure 5. (**a**) General structure of polyhydroxy alkanoates (PHAs). Each monomer bears a side chain (-R). Usually, *m* can range from 1 to 8 carbon atoms, while *n* ranges from 100 to 3000 units. (**b**) Representation of some small-chain-length (SCL-HA) and medium-chain-length (MCL-HA) side groups. Adapted from reference [85].

2.2.1. Classification of Polyhydroxyoctanoate

PHA polymers are categorized into three groups based on their structure and molecular size: short chain length (SCL-PHA), medium chain length (MCL-PHA) and long chain length (LCL-HA) [84–86]. In terms of chain length, SCL-PHAs are the most commonly synthetized and studied [87].

Short-chain compounds consist of repeating 3 to 5 carbon building blocks (C_3-C_5). Some examples are poly (3-hydroxyvalerate) (PHV) and poly (3-hydroxybutyrate) (PHB), which is one of the most investigated PHA polymers. Medium-chain polymers (MCL-PHA, C_6-C_{14}) contain between 6 and 14 carbon monomers. A good example of such PHA is poly(3-hydroxyoctanoate) (PHO), which can be obtained from *Pseudomonas mendocina*. Lastly, long-chain PHAs encompass longer monomeric units (number of carbon atoms ≥ 15). They are the least common type of PHAs and are usually obtained from *Shewanella oneidensis* and *Aureispira marina* [84–86,88–91].

The differentiation in chain length/monomeric unit is achieved by selecting appropriate bacteria. Each microorganism has the ability to produce distinct types of monomeric units. This variance in chain size primarily stems from the substrate specificity of PHA synthases (PhaC): enzymes responsible for catalyzing hydroxy acids within a specific carbon length range and integrating them into the growing polymer chain [79,92].

For instance, *Cupriavidus necator* is renowned for synthesizing PHA plastics when provided with excess sugar substrate, with PHA accumulation levels reaching up to approximately 90% of the cell's dry weight [93]. The PHA produced by this bacterium typically consists of monomers with a small number of carbon atoms, such as poly(3-hydroxyvalerate) (PHV) and poly(3-hydroxybutyrate) (PHB) [94]. On the other hand, *Pseudomonas mendocina* produces PHA composed of medium-molecular-weight monomers, imparting elastomeric properties and a relatively low mechanical strength [95]. In contrast, *Pseudomonas aeruginosa* synthesizes PHA composed of high-molecular-weight monomers, exemplified by poly(3-hydroxypentadecanoate).

2.2.2. Biosynthesis of Polyhydroxy Alcanoates

The biosynthesis of PHAs is strictly linked to different kind of metabolic pathways. Among them, the most notorious is the one involved in the synthesis of PHB [96]. The method used to produce PHB involves starting from carbon sources such as glucose, sucrose, or lactose that are converted through different enzymatic reactions to acetyl-CoA. Under metabolic stress, acetyl-CoA enters the metabolic pathway of ketone bodies and is converted into PHB in three steps: 3-ketothiolase transform acetyl-CoA into acetoacetyl-CoA, which is then converted into (R)-3-hydroxybutyryl-CoA via the action of acetoacetyl-CoA reductase via a NADH-dependent mechanism of action. Finally, PHB synthase catalyze the polymer formation [81,97,98].

2.2.3. Chemical and Physical Properties of Polyhydroxy Alcanoates

The monomeric composition of microbial PHAs is influenced by many factors, such as the strain of bacteria used, the composition of the growth medium, the carbon source, and the overall cultivation conditions. Consequently, PHA polymers can display a wide range of chemical, thermal, and mechanical properties, varying from rigid and brittle with a high crystallinity percentage to elastomeric polymers, thereby enabling diverse industrial applications. It is worth remembering that the properties of the polymeric material are closely related to the structure of the monomer (e.g., length of the side chain, length of the monomeric backbone, etc.). Therefore, the properties of PHA can be modified physically, chemically, or biologically to tune them to different applications [86,99,100].

Despite their potential for commercial use, many PHA-based materials still present some challenges that researchers need to tackle. For instance, poly(3-hydroxybutyrate) (PHB), one of the most-studied and characterized PHA-based polymers, is highly hydrophobic and exhibits low heat distortion temperatures and poor gas barrier properties. PHB, with its high crystallinity percentage (up to 80%), suffers from poor mechanical properties, making it unsuitable for various applications including packaging, biomedical and pharmaceutical uses like heart valves, vascular applications, and controlled drug delivery systems [101]. Additionally, PHA production costs are higher compared to other bio-based plastics, limiting their applications due to their undesirable physical properties.

Various approaches have been devised to enhance the physical and chemical properties of PHAs, aiming to overcome these limitations by modifying the biopolymers physically, chemically, or biologically to broaden their applications across different fields (Figure 6) [102,103]. There are several methods to tune and improve PHA properties, the most important of which are physical modification via blending with other polymeric materials and chemical modification of the polymeric chain.

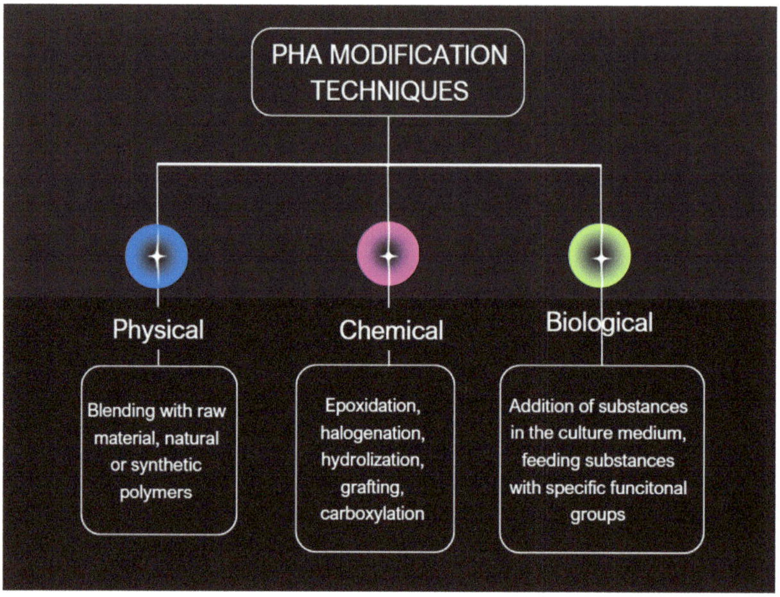

Figure 6. Graphical representation of the possible ways to achieve PHA modification [84,104].

Blending

With regard to physical modification, blending PHA with natural raw materials and with synthetic biodegradable polymers is a highly studied technique in the literature. A polymer blend has been defined by IUPAC as "a macroscopically homogeneous mixture of two or more different polymers" [105]. In polymer technology, blending is a captivating solution that, by mixing a defined polymeric matrix with other substances, can allow researchers to obtain new polymeric materials with improved properties.

One of the major advantages of this approach to polymer modification relies in the possibility of tuning the properties of the original polymer to make it compatible with a specific and desired application [106]. Researchers have explored numerous low-cost, high-value raw materials and bio-based polymers as additives to modify PHA properties and create materials with enhanced characteristics. Some of the most notable examples include blending with natural polymers such as cellulose, starch, and lignin.

- Cellulose derivatives have garnered significant attention as viable blending agents with PHA due to their compatibility and capacity to accelerate PHA degradation rates. The incorporation of cellulose derivatives like cellulose acetate butyrate (CAB) has demonstrated notable enhancements in the various physical and mechanical properties of PHA, including but not limited to a PHB concentration-dependent increase in the T_g and an increase in the elongation at break of up to 7%. The utilization of PHA blends with cellulose derivatives has found diverse applications in the pharmaceutical and medical domains. These blends serve as effective carriers for poorly soluble drugs, facilitate blood coagulation processes, and are employed in pharmaceutical tablet coating, among other uses [107–109].
- Starch, being among the most promising natural polymers, boasts inherent biodegradability and widespread availability. Poly-3-hydroxybutyrate (PHB) stands out as the predominant type of PHA utilized thus far in formulating blends with starch [110]. The blending of PHA with starch or its derivatives has shown promise in reducing production costs while enhancing mechanical properties. Notably, studies have demonstrated the viability of blending thermoplastic starch with PHB. Incorporating thermoplastic starch at levels of up to 30 wt% with PHB has yielded improvements in tensile strength, Young's modulus, and elongation at break when compared to pure

PHB, all while yielding a more cost-effective material. These blends have expanded the potential applications of PHB as a coating material for food packaging, particularly on paper or cardboard substrates. Despite these benefits, blending PHAs with starch still presents challenges to tackle due to their inherent incompatibility, leading, for example, to difficulties in the production of non-brittle films [110–112].

- Lignin is a natural amorphous biopolymer made of repeating units of phenylpropane, featuring both aliphatic and aromatic hydroxyl groups, along with carboxylic acid groups [113]. The presence of these functional groups makes lignin a valuable material for blending purposes with PHA. It was demonstrated that lignin has a good miscibility with PHB, and lignin's amorphous structure can decrease the formation of large spherulite crystals and secondary nucleation, which significantly affects the brittleness of PHB [114]. The studies of Kai et al. have shown that lignin and its butyrate derivative exhibit a high miscibility with PHB, and that it can affect PHB crystallization. In particular, it was demonstrated that lignin can effectively slow down PHB crystallization. These findings suggest that lignin fine powder could serve as a novel type of nucleating agent to modulate PHB crystallization [115]. The thermo-physical properties and rheology of PHB/lignin blends were deeply studied by Mousavioun et al. Their findings, based on TGA, DSC, and SEM analyses of the PHB/soda lignin blends, suggest that intermolecular interactions between PHB and soda lignin are favored at a soda lignin content of up to 40 wt%. These interactions are attributed to the formation of hydrogen bonds between the reactive functional groups of lignin and the carbonyl groups of PHB. While soda lignin enhances the overall thermal stability of PHB, it also reduces the initial decomposition temperature of PHB [116].

Chemical Modification

Chemical modification represents another promising solution for enhancing PHAs' properties by introducing additional chemical groups into the PHA structure. Polyhydroxyalkanoates (PHAs) can undergo modifications through the incorporation of various chemical groups, leading to the development of chemically modified PHAs with enhanced functionalities. These modified PHAs hold potential for utilization as multifunctional materials. The most common chemical modifications involve carboxylation, halogenation, hydroxylation, epoxidation, and grafting processes on the PHA polymeric matrix.

The carboxylation process consists of the creation of -COOH functional groups, often mediated by $KMnO_4$ as an oxidation agent [117]. This type of modification was demonstrated to be efficient in enhancing the hydrophilicity of the polymer [118]. The halogenation of PHA stands out as a remarkable method for enhancing the properties, functionalities, and applications of polymers.

The introduction of halogen atoms (such as chlorine, bromine, and fluorine) can be performed on both unsaturated and saturated PHA through addition or substitution reactions depending on the chemical characteristics of the substrate. This method offers a versatile means of broadening the functionalities and applications of the polymer. Depending on the quantity of halogen introduced, the resulting PHA demonstrates elevated melting and glass transition temperatures. However, this method also presents the drawback of reduced biocompatibility [119].

Hydroxylation presents a method to achieve a reduction in the molecular weight of the original polymer chain via hydrolysis reactions. This can proceed either through an acid- or base-catalyzed reaction and is usually carried out in the presence of low-molecular-weight mono or diol compounds. The most common methods employ either sodium hydroxide or para-toluene sulfonic acid (PTSA). PHA modified by these reactions might exhibit lower glass transition and melting temperatures [120].

Epoxidation represents a chemical modification method with the potential to enhance the thermal stability of PHA. This process involves converting double carbon–carbon bonds into epoxy groups. The high reactivity and facile conversion of epoxy groups into anionic and polar groups, even under mild conditions, underscore the significance of epoxidation as a pivotal strategy for PHA modification [121].

Grafting emerges as a highly reliable chemical modification method for PHA, wherein monomers are covalently bonded or "grafted" onto the polymer chain to achieve desired properties. This approach holds promise in preserving the inherent characteristics of PHA while introducing new ones. Various grafting methods are available, including radiation-based, enzyme-based, ionic, or radical grafting. Generally, this type of modification ensures minimal loss of original properties while imparting additional properties to the polymer [122,123].

2.2.4. Applications of Polyhydroxyalcanoates

In terms of applications, PHAs can be employed across various fields, such as medicine, agriculture, and packaging, owing to their biodegradability, biocompatibility, and non-toxic nature.

PHA materials are extensively utilized in the medical industry owing to their excellent biodegradability and biocompatibility. PHA particles exhibit non-toxicity and are devoid of any pyrogenic, allergenic, carcinogenic, inflammatory, or teratogenic properties [124]. The employment of PHA beads has proven to be useful for medical and industrial applications such as protein purification [125], diagnostics and imaging [126,127], tissue engineering, etc.

In the medical field, PHA serves as a valuable material for drug delivery systems including nanoparticles and encapsulation of active pharmaceutical ingredients [128,129]. The in vivo and in vitro cytotoxicity of PHA carriers have been studied. A recent paper by Pevic et al. proved how the use MCL-PHA as a drug delivery system can both improve therapeutic effects and reduce systemic drug toxicity [130]. Bokrova et al. developed novel combined PHB–liposome nanoparticles and assessed their cytotoxicity in mammalian cells in vitro. The newly combined PHB–liposome particles demonstrated no toxicity to HEK (human embryonic kidney) and HaCaT (human immortalized keratinocyte) cells [131]. These PHB nanoparticles (NPs) exhibit potential as active carriers for both hydrophobic and hydrophilic active ingredients, including antioxidants, anti-aging compounds, complex natural extracts, and antibacterial agents [124]. PHA particles offer versatility in modifications through various surface-binding proteins, thereby expanding their practical applications and reducing PHA nanoparticle cytotoxicity [132]. In a study by Fan et al., the natural PHA binding protein PhaP was effectively adsorbed onto PHB NPs through hydrophobic interactions and used as a drug delivery system for targeted accumulation in prostate tumors. This system allowed for a remarkably enhanced cellular uptake in the human prostate cancer cell line PC3 compared to non-functionalized NPs. This study demonstrates the potential of surface-modified PHA nanoparticles to enhance or target phagocytosis, thus paving the way for their potential clinical application in the future [133].

Spherical polyhydroxyalkanoates have also found many applications in the vaccine engineering field [134]. Given that PHB has received approval from the U.S. Food and Drug Administration (FDA) for clinical studies, and that its 3-hydroxybutyric acid component is a natural constituent of human blood, this type of PHA has been proposed as a safe option for use as an antigen carrier in vaccine formulations [135,136].

PHAs have been studied and used to produce PHA bead-based particles embodied with specific antigens as vaccines formulations against different biological agents such as *Pseudomonas aeruginosa* [137], *Mycobacterium tuberculosis* [138–141], *Streptococcus pneumoniae* [142–144], and *Neisseria meningitidis* [145]. Additionally, PHA is utilized in tissue engineering to fabricate scaffolds for regenerative medicine purposes [146–148]. PHA is also employed in the agricultural field, not only as an encapsulating agent for pesticides and to control their release kinetics but also as a seed-protecting agent [149]. The packaging and food service industries utilize PHA due to its biodegradability and permeability properties, making it an attractive alternative to petroleum-based plastics. PHA-based packaging films are utilized in the production of plastic bags, cups, and similar products [150].

2.3. Polyhydroxy Polyamides

A polyamide, by definition, is a condensation product in which the monomers are linked via an amide bond. Perhaps the most renowned polyamide is Nylon-6,6, which has led to polyamides being commonly referred to as "nylons" in a general sense. Linear polyamides are formed through a condensation reaction between bifunctional monomers. When amino acids or their lactam forms are used as monomers, the resulting polymer is classified as an AB type polymer, with A representing the amino group and B the carboxylic group. The resulting polyamide is denoted as nylon-n, with n representing the number of carbon atoms between the amino and carboxylic termini. Similarly, polymers resulting from the reaction between a diamine and a dicarboxylic acid are classified as AABB type, with two numbers (m, n) indicating the number of carbon atoms separating the amino and carboxylic functional groups, respectively (Scheme 2) [151].

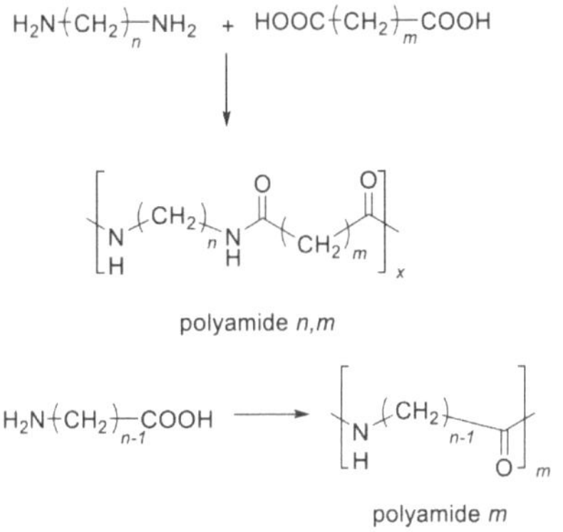

Scheme 2. Polyamide types.

The physical properties of nylons are closely tied to the distance between the functional groups and thus the length of the polyethylene chain. Consequently, common nylons exhibit very similar performances, displaying a "monotonous behavior", in contrast to natural polymers, which exhibit greater diversity in their physical, chemical, and biological functions [152]. This has led to the gradual replacement of these polymers with high-performance materials ad hoc designed to express improved biocompatibility, biodegradability, and other desired properties [153].

Efforts have already been made to synthesize biodegradable nylons, with findings indicating that polyamides containing methyl and hydroxyl groups demonstrate biodegradability. Furthermore, the presence of electron-withdrawing substituents near the carbonyl groups has been shown to enhance their hydrolysis rate [154]. Biodegradable synthetic polymers are typically designed to mimic natural peptides or proteins to enhance both chemical and enzymatic hydrolysis of the final product. From a synthetic perspective, introducing asymmetric carbons into the polyamide backbone allows for control over the physical properties of the polymer by modulating the tacticity of the macromolecule [155].

Carbohydrate-based synthetic polymers can be prepared through the polymerization reaction of appropriate activated monomers. Overall, the interest in polyaldaramides has increased in the past decade. Polyaldaramides, also known as polyhydroxypolyamides (PHPAs) or hydroxylated nylons, are hydroxylated, linear polyamides of the AABB type, where the diacid monomer units of typical nylons are replaced by an aldaric acid. Aldaric acids are dicarboxylic acids derived from the oxidation of aldoses. Their chemical nature and renewable origin make them promising candidates as monomers or building blocks for the synthesis of biobased polymers, particularly hydroxylated polyesters and polyamides [156,157]. In terms of environmental impact, these compounds have garnered significant attention due to the renewable nature of the carbohydrate backbone of the monomer unit and their high degradability in soil, especially when compared with traditional nylons or poly(ethylene terephthalate) [158,159].

The regio- and stereochemical properties of the macromolecule depend on the configuration of the chosen monomers. Although derivatives of aldaric acids could have been employed in polymerization reactions, their application in this regard was not investigated until the 1950s. In fact, the first example of these polymers was reported by Wolform et al. in 1958 [159], and later on, more in-depth studies of these polymers were carried out by Ogata et al. in Japan in the 1970s. The pioneering work by Ogata et al. demonstrated that the condensation between a diethyl ester of an aldaric acid (such as tartaric or galactaric) and a diamine could proceed under mild conditions. It was found that the presence of heteroatomic groups (such as ether or hydroxyl functional groups) in the α- or β-position to the ester group greatly enhanced the reactivity of the diester monomer in polymerization reactions carried out in polar solvents [158,160,161]. The polymerization of diethyl galactarate with hexamethylenediamine yielded a hydroxylated analogue of nylon-6,6 that did not melt or decompose at 200 °C. These reactions were performed in solvents such as methanol, dimethyl sulfoxide, and N-methylpyrrolidone under mild conditions. The increased reactivity of esters toward aminolysis due to the presence of the hydroxyl group was attributed to hydrogen bonding of the -OH group with the approaching amino group in the intermediate reaction stage (Scheme 3).

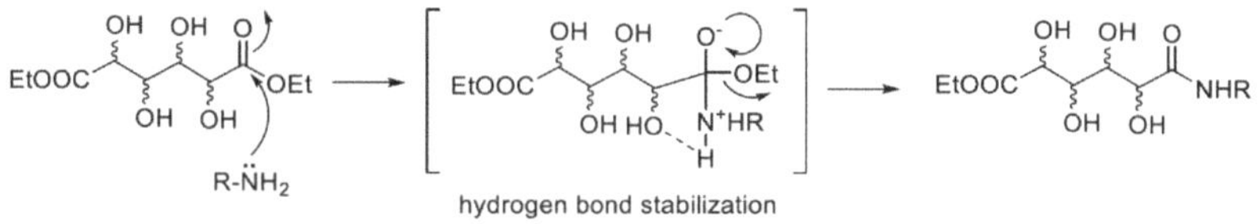

Scheme 3. Hydrogen bond stabilization of the intermediate reaction stage.

Further examination of this reaction was carried out by Hoagland et al., who investigated the reaction mechanisms of aminolysis of six-carbon galactaric acid diesters [162]. They found that the aminolysis proceeds through a two-step sequence: a fast, base-catalyzed lactonization to give the γ-lactone intermediate, followed by the slow aminolysis of the lactone (Scheme 4).

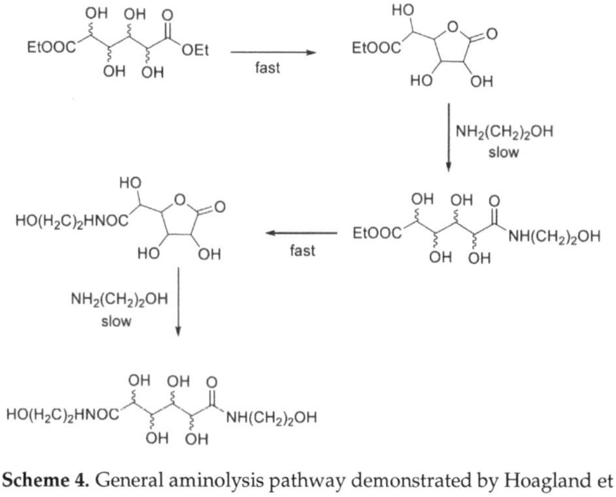

Scheme 4. General aminolysis pathway demonstrated by Hoagland et al. [162].

The lactonization of the ester appears to be much faster than the aminolysis of the ester. This study by Hoagland et al. demonstrates that the activation of the aldaric acid diesters is due to the easy formation of the highly reactive five-membered aldarolactones [162].

Polyaldaramides are typically prepared using aldaric acid monomers that are activated as esters or acyl chlorides, often with acetylated protected hydroxyl groups. Polyamides with free hydroxyl groups are then obtained via deacylation in aqueous ammonia. The synthesis of unprotected activated D-glucaric acid dimethyl or diethyl ester monomers and their polymerization were first investigated and patented by Kiely et al. [163,164]. The initial polyhydroxypolyamide (PHPA) reported by Kiely was prepared under Fisher esterification conditions in methanol. Subsequent reports on the synthesis of D-glucaric acid-based PHPAs have focused on the use of methyl D-glucaro-1,4 lactone as the starting monomer [165]. Carbohydrate-derived polyamides exhibit interesting properties, particularly in terms of crystallinity, which is typically associated with chain stereoregularity. AABB-type polyamides obtained through conventional polycondensation are stereoregular when their monomers have a 2-fold axis of symmetry. Otherwise, they may enter the polymer chain in two opposite orientations, giving rise to non-stereoregular polyamides. The polyglucaramides obtained using the above-mentioned protocol are termed *stereo-random* because D-glucaric acid does not have a symmetrical structure and therefore gives rise to randomly oriented glucaric acid units in the backbone of the macromolecule [166]. In fact, such symmetry restriction is met only by the aldaric- or alditol-based monomers having *threo-*, *manno-*, and *ido*-configuration. Nonetheless, stereoregular AABB-type polyamides derived from non-centrosymmetric D-glucaric acid have been prepared by Kiely et al. using synthetic methods that can differentiate between the reactivity toward the aliphatic diamine of the aldaric acid's two carboxyl groups [157].

While efforts have been focused on understanding the chemical behavior and finding convenient synthesis pathways for PHPAs derived from D-glucaric acid, meso-xylaric acid, and L-mannaric acid, little investment has been made in using galactaric acid derivatives as starting monomers (Scheme 5).

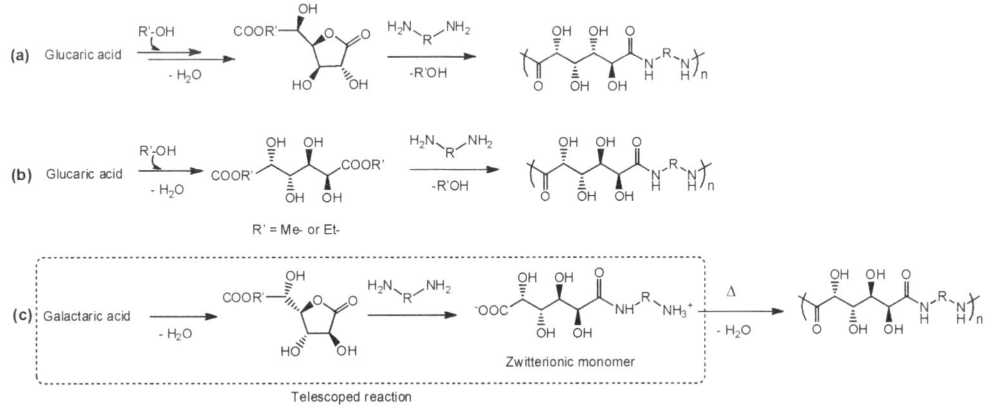

Scheme 5. Synthetic procedure for the synthesis of polyhydroxy polyamides, modified from reference [167], (**a**) synthetic path involving protection and lactone formation first, (**b**) polycondensation using diester monomers and (**c**) more recent polymerization technique using zwitterionic monomers.

Only a few examples of poly(galactaramides) being synthesized and partially characterized are present in the literature. Usually, the most convenient method reported for the synthesis of polyhydroxy polyamides from galactaric acid is using galactaric acid esters as starting monomers (Scheme 5b) [168–170]. With the aim of avoiding the use of protective groups and favoring a rigorous 1:1 diacid-diamine ratio control, Gambarotti et al. reported the synthesis of a new zwitterionic monoamide of galactaric acid to be used as a building block in the polymerization reaction (Scheme 5c). This protocol involved the conversion of galactaric acid into the corresponding γ-galactaro lactone first, followed by its reaction with diamines leading to the formation of the corresponding zwitterionic monoamides with good yields and selectivity. This new class of monomers was then employed in polymerization to obtain the corresponding poly(galactaramides), showing a high atom economy and comparable degree of polymerization (DP) usually obtained by using the protected galactaric ester form [167].

In terms of application, polyaldaramides, which appear to be biocompatible and biodegradable, offer easy access to a wide range of polymers that can be applied as biodegradable adhesives, timed release fertilizers, industrial chemicals for the textile and paper industries, water treatment chemicals, detergent components, hydrogel components, and film and fiber materials [171]. Also, an interesting study was published demonstrating that poly(glucaramides) can form a nanoparticulate system that can be used for slow and controlled release applications [172].

3. Conclusions

In conclusion, the exploration of bio-based polymers such as PLA, PHA, and polyhydroxy polyamides unveils a promising frontier in sustainable material science. These polymers offer a compelling alternative to traditional petroleum-based plastics, exhibiting comparable or even superior properties while significantly reducing environmental impacts. Their versatility and compatibility with various applications ranging from packaging to biomedical devices underscore their potential to revolutionize the polymeric material sector and industry. While challenges such as cost-effectiveness, scalability, and processability remain, ongoing research and technological advancements continue to drive innovation in this field. As we move towards a more sustainable future, the adoption of bio-based polymers stands as a crucial step in mitigating plastic pollution and fostering a greener, more resilient global economy. Through collaborative efforts between academia, industry, and policymakers, the integration of these eco-friendly materials into mainstream usage holds the promise of a brighter, more sustainable tomorrow.

Author Contributions: Conceptualization, G.I.C.R. and A.F.; methodology, G.I.C.R., F.F. and A.F.; validation, G.I.C.R. and F.F.; formal analysis, G.I.C.R. and F.F.; investigation, G.I.C.R. and F.F.; resources, G.I.C.R., F.F. and A.F.; data curation, G.I.C.R. and F.F.; writing—original draft preparation, G.I.C.R. and F.F.; writing—review and editing, G.I.C.R., F.F. and A.F.; visualization, G.I.C.R., F.F. and A.F.; supervision, G.I.C.R. and A.F.; project administration, G.I.C.R. and A.F.; funding acquisition, A.F. All authors have read and agreed to the published version of the manuscript.

Funding: This research did not receive any external funding.

Institutional Review Board Statement: Not applicable.

Data Availability Statement: Not applicable.

Acknowledgments: The authors warmly thank Lee for their assistance.

Conflicts of Interest: The authors declare no conflicts of interest.

References

1. Galbis, J.A.; García-Martín, M.G. *Monomers, Polymers and Composites from Renewable Resources*; Belgacem, M.N., Gandini, A., Eds.; Elsevier: Amsterdam, The Netherlands, 2008.
2. Nakajima, H.; Dijkstra, P.; Loos, K. The Recent Developments in Biobased Polymers toward General and Engineering Applications: Polymers That Are Upgraded from Biodegradable Polymers, Analogous to Petroleum-Derived Polymers, and Newly Developed. *Polymers* **2017**, *9*, 523. [CrossRef] [PubMed]
3. Isikgor, F.H.; Becer, C.R. Lignocellulosic Biomass: A Sustainable Platform for the Production of Bio-Based Chemicals and Polymers. *Polym. Chem.* **2015**, *6*, 4497–4559. [CrossRef]
4. Iwata, T. Biodegradable and Bio-Based Polymers: Future Prospects of Eco-Friendly Plastics. *Angew. Chem. Int. Ed.* **2015**, *54*, 3210–3215. [CrossRef] [PubMed]
5. Garrison, T.; Murawski, A.; Quirino, R. Bio-Based Polymers with Potential for Biodegradability. *Polymers* **2016**, *8*, 262. [CrossRef] [PubMed]
6. Hottle, T.A.; Bilec, M.M.; Landis, A.E. Sustainability Assessments of Bio-Based Polymers. *Polym. Degrad. Stab.* **2013**, *98*, 1898–1907. [CrossRef]
7. Luckachan, G.E.; Pillai, C.K.S. Biodegradable Polymers—A Review on Recent Trends and Emerging Perspectives. *J. Polym. Environ.* **2011**, *19*, 637–676. [CrossRef]

8. Karthika, M.; Shaji, N.; Johnson, A.; Neelakandan, M.S.; AGopakumar, D.; Thomas, S. Biodegradation of Green Polymeric Composites Materials. In *Bio Monomers for Green Polymeric Composite Materials*; Wiley: Hoboken, NJ, USA, 2019; pp. 141–159.
9. Engineer, C.; Parikh, J.; Raval, A. Review on Hydrolytic Degradation Behavior of Biodegradable Polymers from Controlled Drug Delivery Systems. *Trends Biomater. Artif. Organs* **2011**, *25*, 79–85.
10. Samir, A.; Ashour, F.H.; Hakim, A.A.A.; Bassyouni, M. Recent Advances in Biodegradable Polymers for Sustainable Applications. *Npj Mater. Degrad.* **2022**, *6*, 68. [CrossRef]
11. Vert, M.; Doi, Y.; Hellwich, K.-H.; Hess, M.; Hodge, P.; Kubisa, P.; Rinaudo, M.; Schué, F. Terminology for Biorelated Polymers and Applications (IUPAC Recommendations 2012). *Pure Appl. Chem.* **2012**, *84*, 377–410. [CrossRef]
12. Joseph, T.M.; Unni, A.B.; Joshy, K.S.; Kar Mahapatra, D.; Haponiuk, J.; Thomas, S. Emerging Bio-Based Polymers from Lab to Market: Current Strategies, Market Dynamics and Research Trends. *C* **2023**, *9*, 30. [CrossRef]
13. Muringayil Joseph, T.; Mariya, H.J.; Haponiuk, J.T.; Thomas, S.; Esmaeili, A.; Sajadi, S.M. Electromagnetic Interference Shielding Effectiveness of Natural and Chlorobutyl Rubber Blend Nanocomposite. *J. Compos. Sci.* **2022**, *6*, 240. [CrossRef]
14. PlasticsEurope. Plastics—The Fast Facts 2023 Preview. 2023. Available online: https://plasticseurope.org/knowledge-hub/plastics-the-fast-facts-2023/ (accessed on 28 February 2024).
15. Grand View Research Bioplastics Market Size, Share & Trends Analysis Report by Product (Biodegradable, Non-Biodegradable), by Application, by Region, and Segment Forecasts, 2023–2030. Available online: https://www.grandviewresearch.com/industry-analysis/bioplastics-industry#:~:text=The%20global%20bioplastics%20market%20size,18.8%25%20from%202023%20to%202030 (accessed on 28 February 2024).
16. Naser, A.Z.; Deiab, I.; Defersha, F.; Yang, S. Expanding Poly(Lactic Acid) (PLA) and Polyhydroxyalkanoates (PHAs) Applications: A Review on Modifications and Effects. *Polymers* **2021**, *13*, 4271. [CrossRef] [PubMed]
17. Chaveriat, L.; Stasik, I.; Demailly, G.; Beaupère, D. The Direct Synthesis of 6-Amino-6-Deoxyaldonic Acids as Monomers for the Preparation of Polyhydroxylated Nylon 6. *Tetrahedron Asymmetry* **2006**, *17*, 1349–1354. [CrossRef]
18. Winnacker, M.; Rieger, B. Biobased Polyamides: Recent Advances in Basic and Applied Research. *Macromol. Rapid Commun.* **2016**, *37*, 1391–1413. [CrossRef] [PubMed]
19. Paul, A.; Sreedevi, K.; Sharma, S.S.; Anjana, V.N. Polylactic Acid (PLA). In *Handbook of Biopolymers*; Springer Nature: Singapore, 2023; pp. 1195–1227.
20. Babu, R.P.; O'Connor, K.; Seeram, R. Current Progress on Bio-Based Polymers and Their Future Trends. *Prog. Biomater.* **2013**, *2*, 8. [CrossRef] [PubMed]
21. Ajioka, M.; Enomoto, K.; Suzuki, K.; Yamaguchi, A. The Basic Properties of Poly(Lactic Acid) Produced by the Direct Condensation Polymerization of Lactic Acid. *J. Environ. Polym. Degrad.* **1995**, *3*, 225–234. [CrossRef]
22. Laonuad, P.; Chaiyut, N.; Ksapabutr, B. Poly(Lactic Acid) Preparation by Polycondensation Method. *Optoelectron. Adv. Mater. Rapid Commun.* **2010**, *4*, 1200–1202.
23. Pang, X.; Zhuang, X.; Tang, Z.; Chen, X. Polylactic Acid (PLA): Research, Development and Industrialization. *Biotechnol. J.* **2010**, *5*, 1125–1136. [CrossRef] [PubMed]
24. Cabedo, L.; Luis Feijoo, J.; Pilar Villanueva, M.; Lagarón, J.M.; Giménez, E. Optimization of Biodegradable Nanocomposites Based on APLA/PCL Blends for Food Packaging Applications. *Macromol. Symp.* **2006**, *233*, 191–197. [CrossRef]
25. Bassi, M.B.; Padias, A.B.; Hall, H.K. The Hydrolytic Polymerization of Epsilon-Caprolactone by Triphenyltin Acetate. *Polym. Bull.* **1990**, *24*, 227–232. [CrossRef]
26. Bero, M.; Adamus, G.; Kasperczyk, J.; Janeczek, H. Synthesis of Block-Copolymers of Epsilon-Caprolactone and Lactide in the Presence of Lithium t-Butoxide. *Polym. Bull.* **1993**, *31*, 9–14. [CrossRef]
27. Nijenhuis, A.J.; Grijpma, D.W.; Pennings, A.J. Lewis Acidcatalyzed Polymerization of L-Lactide—Kinetics and Mechanism of the Bulk-Polymerization. *Macromolecules* **1992**, *24*, 6419–6424. [CrossRef]
28. Van de Velde, K.; Kiekens, P. Biopolymers: Overview of Several Properties and Consequences on Their Applications. *Polym. Test.* **2002**, *21*, 433–442. [CrossRef]
29. Righetti, G.I.C.; Nasti, R.; Beretta, G.; Levi, M.; Turri, S.; Suriano, R. Unveiling the Hidden Properties of Tomato Peels: Cutin Ester Derivatives as Bio-Based Plasticizers for Polylactic Acid. *Polymers* **2023**, *15*, 1848. [CrossRef] [PubMed]
30. Farah, S.; Anderson, D.G.; Langer, R. Physical and Mechanical Properties of PLA, and Their Functions in Widespread Applications—A Comprehensive Review. *Adv. Drug Deliv. Rev.* **2016**, *107*, 367–392. [CrossRef]
31. Pawlak, A.; Galeski, A. Crystallization of Polypropylene. In *Polypropylene Handbook*; Springer International Publishing: Cham, Switzerland, 2019; pp. 185–242.
32. Maddah, H.A. Polypropylene as a Promising Plastic: A Review. *Am. J. Polym. Sci.* **2016**, *16*, 1–11.
33. Kontou, E.; Farasoglou, P. Determination of the True Stress–Strain Behaviour of Polypropylene. *J. Mater. Sci.* **1998**, *33*, 147–153. [CrossRef]
34. Greene, J.P. Microstructures of Polymers. In *Automotive Plastics and Composites*; Elsevier: Amsterdam, The Netherlands, 2021; pp. 27–37.
35. Jordan, J.L.; Casem, D.T.; Bradley, J.M.; Dwivedi, A.K.; Brown, E.N.; Jordan, C.W. Mechanical Properties of Low Density Polyethylene. *J. Dyn. Behav. Mater.* **2016**, *2*, 411–420. [CrossRef]
36. Kalay, G.; Sousa, R.A.; Reis, R.L.; Cunha, A.M.; Bevis, M.J. The Enhancement of the Mechanical Properties of a High-Density Polyethylene. *J. Appl. Polym. Sci.* **1999**, *73*, 2473–2483. [CrossRef]

37. MatWeb Overview of Materials for High Density Polyethylene (HDPE), Ultra High Molecular Weight—Web Page. Available online: https://matweb.com/search/DataSheet.aspx?MatGUID=fce23f90005d4fbe8e12a1bce53ebdc8&ckck=1 (accessed on 25 March 2024).
38. Tarani, E.; Arvanitidis, I.; Christofilos, D.; Bikiaris, D.N.; Chrissafis, K.; Vourlias, G. Calculation of the Degree of Crystallinity of HDPE/GNPs Nanocomposites by Using Various Experimental Techniques: A Comparative Study. *J. Mater. Sci.* **2023**, *58*, 1621–1639. [CrossRef]
39. Pillin, I.; Montrelay, N.; Bourmaud, A.; Grohens, Y. Effect of Thermo-Mechanical Cycles on the Physico-Chemical Properties of Poly(Lactic Acid). *Polym. Degrad. Stab.* **2008**, *93*, 321–328. [CrossRef]
40. Younes, H.; Cohn, D. Phase Separation in Poly(Ethylene Glycol)/Poly(Lactic Acid) Blends. *Eur. Polym. J.* **1988**, *24*, 765–773. [CrossRef]
41. Shirai, M.A.; Grossmann, M.V.E.; Mali, S.; Yamashita, F.; Garcia, P.S.; Müller, C.M.O. Development of Biodegradable Flexible Films of Starch and Poly(Lactic Acid) Plasticized with Adipate or Citrate Esters. *Carbohydr. Polym.* **2013**, *92*, 19–22. [CrossRef] [PubMed]
42. Dominguez-Candela, I.; Ferri, J.M.; Cardona, S.C.; Lora, J.; Fombuena, V. Dual Plasticizer/Thermal Stabilizer Effect of Epoxidized Chia Seed Oil (*Salvia hispanica* L.) to Improve Ductility and Thermal Properties of Poly(Lactic Acid). *Polymers* **2021**, *13*, 1283. [CrossRef]
43. Quiles-Carrillo, L.; Duart, S.; Montanes, N.; Torres-Giner, S.; Balart, R. Enhancement of the Mechanical and Thermal Properties of Injection-Molded Polylactide Parts by the Addition of Acrylated Epoxidized Soybean Oil. *Mater. Des.* **2018**, *140*, 54–63. [CrossRef]
44. Ljungberg, N.; Wesslén, B. Preparation and Properties of Plasticized Poly(Lactic Acid) Films. *Biomacromolecules* **2005**, *6*, 1789–1796. [CrossRef] [PubMed]
45. Hu, Y.; Rogunova, M.; Topolkaraev, V.; Hiltner, A.; Baer, E. Aging of Poly(Lactide)/Poly(Ethylene Glycol) Blends. Part 1. Poly(Lactide) with Low Stereoregularity. *Polymer* **2003**, *44*, 5701–5710. [CrossRef]
46. Hu, Y.; Hu, Y.S.; Topolkaraev, V.; Hiltner, A.; Baer, E. Crystallization and Phase Separation in Blends of High Stereoregular Poly(Lactide) with Poly(Ethylene Glycol). *Polymer* **2003**, *44*, 5681–5689. [CrossRef]
47. Kulinski, Z.; Piorkowska, E.; Gadzinowska, K.; Stasiak, M. Plasticization of Poly(L-Lactide) with Poly(Propylene Glycol). *Biomacromolecules* **2006**, *7*, 2128–2135. [CrossRef]
48. Okamoto, K.; Ichikawa, T.; Yokohara, T.; Yamaguchi, M. Miscibility, Mechanical and Thermal Properties of Poly(Lactic Acid)/Polyester-Diol Blends. *Eur. Polym. J.* **2009**, *45*, 2304–2312. [CrossRef]
49. Schlemmer, D.; de Oliveira, E.R.; Araújo Sales, M.J. Polystyrene/Thermoplastic Starch Blends with Different Plasticizers. *J. Therm. Anal. Calorim.* **2007**, *87*, 635–638. [CrossRef]
50. Martin, O.; Avérous, L. Poly(Lactic Acid): Plasticization and Properties of Biodegradable Multiphase Systems. *Polymer* **2001**, *42*, 6209–6219. [CrossRef]
51. Auras, R.A.; Singh, S.P.; Singh, J.J. Evaluation of Oriented Poly(Lactide) Polymers vs. Existing PET and Oriented PS for Fresh Food Service Containers. *Packag. Technol. Sci.* **2005**, *18*, 207–216. [CrossRef]
52. Guerra, A.; Cano, P.; Rabionet, M.; Puig, T.; Ciurana, J. 3D-Printed PCL/PLA Composite Stents: Towards a New Solution to Cardiovascular Problems. *Materials* **2018**, *11*, 1679. [CrossRef] [PubMed]
53. Leroux, J.-C.; Allémann, E.; De Jaeghere, F.; Doelker, E.; Gurny, R. Biodegradable Nanoparticles—From Sustained Release Formulations to Improved Site Specific Drug Delivery. *J. Control. Release* **1996**, *39*, 339–350. [CrossRef]
54. Matsumoto, J.; Nakada, Y.; Sakurai, K.; Nakamura, T.; Takahashi, Y. Preparation of Nanoparticles Consisted of Poly(l-Lactide)–Poly(Ethylene Glycol)–Poly(l-Lactide) and Their Evaluation In Vitro. *Int. J. Pharm.* **1999**, *185*, 93–101. [CrossRef]
55. Fishbein, I.; Chorny, M.; Rabinovich, L.; Banai, S.; Gati, I.; Golomb, G. Nanoparticulate Delivery System of a Tyrphostin for the Treatment of Restenosis. *J. Control. Release* **2000**, *65*, 221–229. [CrossRef]
56. Xing, J.; Zhang, D.; Tan, T. Studies on the Oridonin-Loaded Poly(d,l-Lactic Acid) Nanoparticles in Vitro and in Vivo. *Int. J. Biol. Macromol.* **2007**, *40*, 153–158. [CrossRef]
57. Rancan, F.; Papakostas, D.; Hadam, S.; Hackbarth, S.; Delair, T.; Primard, C.; Verrier, B.; Sterry, W.; Blume-Peytavi, U.; Vogt, A. Investigation of Polylactic Acid (PLA) Nanoparticles as Drug Delivery Systems for Local Dermatotherapy. *Pharm. Res.* **2009**, *26*, 2027–2036. [CrossRef]
58. Gao, H.; Wang, Y.; Fan, Y.; Ma, J. Synthesis of a Biodegradable Tadpole-Shaped Polymer via the Coupling Reaction of Polylactide onto Mono(6-(2-Aminoethyl)Amino-6-Deoxy)-β-Cyclodextrin and Its Properties as the New Carrier of Protein Delivery System. *J. Control. Release* **2005**, *107*, 158–173. [CrossRef]
59. Athanasiou, K.; Agrawal, C.; Barber, F.; Burkhart, S. Orthopaedic Applications for PLA-PGA Biodegradable Polymers. *Arthrosc. J. Arthrosc. Relat. Surg.* **1998**, *14*, 726–737. [CrossRef] [PubMed]
60. DeStefano, V.; Khan, S.; Tabada, A. Applications of PLA in Modern Medicine. *Eng. Regen.* **2020**, *1*, 76–87. [CrossRef]
61. Capuana, E.; Lopresti, F.; Ceraulo, M.; La Carrubba, V. Poly-l-Lactic Acid (PLLA)-Based Biomaterials for Regenerative Medicine: A Review on Processing and Applications. *Polymers* **2022**, *14*, 1153. [CrossRef] [PubMed]
62. Madhavan Nampoothiri, K.; Nair, N.R.; John, R.P. An Overview of the Recent Developments in Polylactide (PLA) Research. *Bioresour. Technol.* **2010**, *101*, 8493–8501. [CrossRef] [PubMed]
63. Siracusa, V.; Rocculi, P.; Romani, S.; Rosa, M.D. Biodegradable Polymers for Food Packaging: A Review. *Trends Food Sci. Technol.* **2008**, *19*, 634–643. [CrossRef]

64. Mensitieri, G.; Di Maio, E.; Buonocore, G.G.; Nedi, I.; Oliviero, M.; Sansone, L.; Iannace, S. Processing and Shelf Life Issues of Selected Food Packaging Materials and Structures from Renewable Resources. *Trends Food Sci. Technol.* **2011**, *22*, 72–80. [CrossRef]
65. Wang, X.; Huang, L.; Li, Y.; Wang, Y.; Lu, X.; Wei, Z.; Mo, Q.; Zhang, S.; Sheng, Y.; Huang, C.; et al. Research Progress in Polylactic Acid Processing for 3D Printing. *J. Manuf. Process.* **2024**, *112*, 161–178. [CrossRef]
66. Yu, W.; Dong, L.; Lei, W.; Zhou, Y.; Pu, Y.; Zhang, X. Effects of Rice Straw Powder (RSP) Size and Pretreatment on Properties of FDM 3D-Printed RSP/Poly(Lactic Acid) Biocomposites. *Molecules* **2021**, *26*, 3234. [CrossRef] [PubMed]
67. Yang, Z.; Feng, X.; Xu, M.; Rodrigue, D. Printability and Properties of 3D-Printed Poplar Fiber/Polylactic Acid Biocomposite. *BioResources* **2021**, *16*, 2774–2788. [CrossRef]
68. Razali, M.S.; Khimeche, K.; Melouki, R.; Boudjellal, A.; Vroman, I.; Alix, S.; Ramdani, N. Preparation and Properties Enhancement of Poly(Lactic Acid)/Calcined-seashell Biocomposites for 3D Printing Applications. *J. Appl. Polym. Sci.* **2022**, *139*, 51591. [CrossRef]
69. Hu, L.; Long, H.; Yang, F.; Cai, Q.; Wu, Y.; Li, Z.; Xiao, J.; Xiao, D.; Zhang, S.; Guan, L.; et al. Novel Lignin Microspheres Reinforced Poly (Lactic Acid) Composites for Fused Deposition Modeling. *Polym. Compos.* **2022**, *43*, 6817–6828. [CrossRef]
70. Pavan, M.V.; Balamurugan, K.; Srinivasadesikan, V.; Lee, S.-L. Impact and Shear Behavior of PLA/12%Cu Reinforced Composite Filament Printed at Different FDM Conditions. *Arab. J. Sci. Eng.* **2021**, *46*, 12709–12720. [CrossRef]
71. Jing, J.; Zhang, Y.; Fang, Z.-P.; Wang, D.-Y. Core-Shell Flame Retardant/Graphene Oxide Hybrid: A Self-Assembly Strategy towards Reducing Fire Hazard and Improving Toughness of Polylactic Acid. *Compos. Sci. Technol.* **2018**, *165*, 161–167. [CrossRef]
72. Sun, Y.; Yu, B.; Liu, Y.; Bai, F.; Yan, J.; Wang, J.; Huang, F.; Gao, S. Design of 2d Charring-Foaming Agent for Highly Efficient Intumescent Flame Retardant Polylactic Acid Composites. *Compos. Commun.* **2023**, *43*, 101720. [CrossRef]
73. Shi, Y.; Wang, Z.; Liu, C.; Wang, H.; Guo, J.; Fu, L.; Feng, Y.; Wang, L.; Yang, F.; Liu, M. Engineering Titanium Carbide Ultra-Thin Nanosheets for Enhanced Fire Safety of Intumescent Flame Retardant Polylactic Acid. *Compos. B Eng.* **2022**, *236*, 109792. [CrossRef]
74. Wang, H.; Wang, Z.; Shi, Y.; Fu, L.; Liu, M.; Feng, Y.; Yu, B.; Gao, J.; Yang, F. Interface Assembly of Hypophosphite/Ultrathin MXene Nanosheets towards Fire Safe Polylactic Acid Composites. *Compos. Commun.* **2022**, *34*, 101270. [CrossRef]
75. Wang, H.; Wang, Z.; Shi, Y.; Liu, M.; Yao, A.; Feng, Y.; Fu, L.; Lv, Y.; Yang, F.; Yu, B. Supramolecular Engineered Ultrathin MXene towards Fire Safe Polylactic Acid Composites. *Compos. Commun.* **2023**, *37*, 101405. [CrossRef]
76. Yu, X.; Su, X.; Liu, Y.; Yu, D.; Ren, Y.; Liu, X. Biomass Intumescent Flame Retardant Polyacrylonitrile Composite: Flame Retardancy, Smoke Suppression and Recycling. *Compos. Part A Appl. Sci. Manuf.* **2023**, *173*, 107647. [CrossRef]
77. Sun, Y.; Yu, B.; Liu, Y.; Cheng, B.; Wang, J.; Yan, J.; Huang, F. Novel Bio-Based Nanosheets: Improving the Fire Safety, Electromagnetic Shielding and Mechanical Properties of Polylactic Acid. *Compos. Part A Appl. Sci. Manuf.* **2024**, *179*, 108044. [CrossRef]
78. Fernández-Dacosta, C.; Posada, J.A.; Kleerebezem, R.; Cuellar, M.C.; Ramirez, A. Microbial Community-Based Polyhydroxyalkanoates (PHAs) Production from Wastewater: Techno-Economic Analysis and Ex-Ante Environmental Assessment. *Bioresour. Technol.* **2015**, *185*, 368–377. [CrossRef]
79. Dwivedi, R.; Pandey, R.; Kumar, S.; Mehrotra, D. Poly Hydroxyalkanoates (PHA): Role in Bone Scaffolds. *J. Oral Biol. Craniofac. Res.* **2020**, *10*, 389–392. [CrossRef] [PubMed]
80. Kumar, P.; Jun, H.-B.; Kim, B.S. Co-Production of Polyhydroxyalkanoates and Carotenoids through Bioconversion of Glycerol by Paracoccus Sp. Strain LL1. *Int. J. Biol. Macromol.* **2018**, *107*, 2552–2558. [CrossRef] [PubMed]
81. Griebel, R.; Smith, Z.; Merrick, J.M. Metabolism of Poly-Hydroxybutyrate. I. Purification, Composition, and Properties of Native Poly-Hydroxybutyrate Granules from Bacillus Megateriu. *Biochemistry* **1968**, *7*, 3676–3681. [CrossRef] [PubMed]
82. Shahid, S.; Razzaq, S.; Farooq, R.; Nazli, Z.-H. Polyhydroxyalkanoates: Next Generation Natural Biomolecules and a Solution for the World's Future Economy. *Int. J. Biol. Macromol.* **2021**, *166*, 297–321. [CrossRef] [PubMed]
83. Koller, M. Production of Polyhydroxyalkanoate (PHA) Biopolyesters by Extremophiles? *MOJ Polym. Sci.* **2017**, *1*, 69–85. [CrossRef]
84. Sharma, V.; Sehgal, R.; Gupta, R. Polyhydroxyalkanoate (PHA): Properties and Modifications. *Polymer* **2021**, *212*, 123161. [CrossRef]
85. Li, Z.; Yang, J.; Loh, X.J. Polyhydroxyalkanoates: Opening Doors for a Sustainable Future. *NPG Asia Mater.* **2016**, *8*, e265. [CrossRef]
86. Guimarães, T.C.; Araújo, E.S.; Hernández-Macedo, M.L.; López, J.A. Polyhydroxyalkanoates: Biosynthesis from Alternative Carbon Sources and Analytic Methods: A Short Review. *J. Polym. Environ.* **2022**, *30*, 2669–2684. [CrossRef]
87. Handali, S.; Moghimipour, E.; Rezaei, M.; Ramezani, Z.; Dorkoosh, F.A. PHBV/PLGA Nanoparticles for Enhanced Delivery of 5-Fluorouracil as Promising Treatment of Colon Cancer. *Pharm. Dev. Technol.* **2020**, *25*, 206–218. [CrossRef]
88. Li, Z.; Loh, X.J. Water Soluble Polyhydroxyalkanoates: Future Materials for Therapeutic Applications. *Chem. Soc. Rev.* **2015**, *44*, 2865–2879. [CrossRef]
89. Raza, Z.A.; Riaz, S.; Banat, I.M. Polyhydroxyalkanoates: Properties and Chemical Modification Approaches for Their Functionalization. *Biotechnol. Prog.* **2018**, *34*, 29–41. [CrossRef] [PubMed]
90. Rai, R.; Keshavarz, T.; Roether, J.A.; Boccaccini, A.R.; Roy, I. Medium Chain Length Polyhydroxyalkanoates, Promising New Biomedical Materials for the Future. *Mater. Sci. Eng. R Rep.* **2011**, *72*, 29–47. [CrossRef]

91. Nomura, C.T.; Tanaka, T.; Gan, Z.; Kuwabara, K.; Abe, H.; Takase, K.; Taguchi, K.; Doi, Y. Effective Enhancement of Short-Chain-Length–Medium-Chain-Length Polyhydroxyalkanoate Copolymer Production by Coexpression of Genetically Engineered 3-Ketoacyl-Acyl-Carrier-Protein Synthase III (FabH) and Polyhydroxyalkanoate Synthesis Genes. *Biomacromolecules* **2004**, *5*, 1457–1464. [CrossRef] [PubMed]
92. Thomas, T.; Elain, A.; Bazire, A.; Bruzaud, S. Complete Genome Sequence of the Halophilic PHA-Producing *Bacterium halomonas* sp. SF2003: Insights into Its Biotechnological Potential. *World J. Microbiol. Biotechnol.* **2019**, *35*, 50. [CrossRef] [PubMed]
93. Samrot, A.V.; Avinesh, R.B.; Sukeetha, S.D.; Senthilkumar, P. Accumulation of Poly[(R)-3-Hydroxyalkanoates] in Enterobacter Cloacae SU-1 During Growth with Two Different Carbon Sources in Batch Culture. *Appl. Biochem. Biotechnol.* **2011**, *163*, 195–203. [CrossRef] [PubMed]
94. Khunthongkaew, P.; Murugan, P.; Sudesh, K.; Iewkittayakorn, J. Biosynthesis of Polyhydroxyalkanoates Using Cupriavidus Necator H16 and Its Application for Particleboard Production. *J. Polym. Res.* **2018**, *25*, 131. [CrossRef]
95. Chanasit, W.; Hodgson, B.; Sudesh, K.; Umsakul, K. Efficient Production of Polyhydroxyalkanoates (PHAs) from Pseudomonas Mendocina PSU Using a Biodiesel Liquid Waste (BLW) as the Sole Carbon Source. *Biosci. Biotechnol. Biochem.* **2016**, *80*, 1440–1450. [CrossRef] [PubMed]
96. Sagong, H.-Y.; Son, H.F.; Choi, S.Y.; Lee, S.Y.; Kim, K.-J. Structural Insights into Polyhydroxyalkanoates Biosynthesis. *Trends Biochem. Sci.* **2018**, *43*, 790–805. [CrossRef]
97. Tsuge, T.; Yano, K.; Imazu, S.; Numata, K.; Kikkawa, Y.; Abe, H.; Taguchi, S.; Doi, Y. Biosynthesis of Polyhydroxyalkanoate (PHA) Copolymer from Fructose Using Wild-Type and Laboratory-Evolved PHA Synthases. *Macromol. Biosci.* **2005**, *5*, 112–117. [CrossRef]
98. Verlinden, R.A.J.; Hill, D.J.; Kenward, M.A.; Williams, C.D.; Radecka, I. Bacterial Synthesis of Biodegradable Polyhydroxyalkanoates. *J. Appl. Microbiol.* **2007**, *102*, 1437–1449. [CrossRef]
99. Cruz, M.V.; Araújo, D.; Alves, V.D.; Freitas, F.; Reis, M.A.M. Characterization of Medium Chain Length Polyhydroxyalkanoate Produced from Olive Oil Deodorizer Distillate. *Int. J. Biol. Macromol.* **2016**, *82*, 243–248. [CrossRef] [PubMed]
100. Zou, H.; Shi, M.; Zhang, T.; Li, L.; Li, L.; Xian, M. Natural and Engineered Polyhydroxyalkanoate (PHA) Synthase: Key Enzyme in Biopolyester Production. *Appl. Microbiol. Biotechnol.* **2017**, *101*, 7417–7426. [CrossRef] [PubMed]
101. Andrade, A.P.; Witholt, B.; Chang, D.; Li, Z. Synthesis and Characterization of Novel Thermoplastic Polyester Containing Blocks of Poly[(R)-3-Hydroxyoctanoate] and Poly[(R)-3-Hydroxybutyrate]. *Macromolecules* **2003**, *36*, 9830–9835. [CrossRef]
102. Mitra, R.; Xu, T.; Xiang, H.; Han, J. Current Developments on Polyhydroxyalkanoates Synthesis by Using Halophiles as a Promising Cell Factory. *Microb. Cell Fact.* **2020**, *19*, 86. [CrossRef] [PubMed]
103. Raza, Z.A.; Khalil, S.; Abid, S. Recent Progress in Development and Chemical Modification of Poly(Hydroxybutyrate)-Based Blends for Potential Medical Applications. *Int. J. Biol. Macromol.* **2020**, *160*, 77–100. [CrossRef] [PubMed]
104. Dalton, B.; Bhagabati, P.; De Micco, J.; Padamati, R.B.; O'Connor, K. A Review on Biological Synthesis of the Biodegradable Polymers Polyhydroxyalkanoates and the Development of Multiple Applications. *Catalysts* **2022**, *12*, 319. [CrossRef]
105. Work, W.J.; Horie, K.; Hess, M.; Stepto, R.F.T. Definition of Terms Related to Polymer Blends, Composites, and Multiphase Polymeric Materials (IUPAC Recommendations 2004). *Pure Appl. Chem.* **2004**, *76*, 1985–2007. [CrossRef]
106. Utracki, L.A. Reasons for, Benefits and Problems of Blending. In *Commercial Polymer Blends*; Springer: Boston, MA, USA, 1998; pp. 85–97.
107. Wang, C.; Venditti, R.A.; Zhang, K. Tailor-Made Functional Surfaces Based on Cellulose-Derived Materials. *Appl. Microbiol. Biotechnol.* **2015**, *99*, 5791–5799. [CrossRef] [PubMed]
108. Zhang, L.; Deng, X.; Huang, Z. Miscibility, Thermal Behaviour and Morphological Structure of Poly(3-Hydroxybutyrate) and Ethyl Cellulose Binary Blends. *Polymer* **1997**, *38*, 5379–5387. [CrossRef]
109. El-Shafee, E.; Saad, G.R.; Fahmy, S.M. Miscibility, Crystallization and Phase Structure of Poly(3-Hydroxybutyrate)/Cellulose Acetate Butyrate Blends. *Eur. Polym. J.* **2001**, *37*, 2091–2104. [CrossRef]
110. Jayarathna, S.; Andersson, M.; Andersson, R. Recent Advances in Starch-Based Blends and Composites for Bioplastics Applications. *Polymers* **2022**, *14*, 4557. [CrossRef]
111. Godbole, S. Preparation and Characterization of Biodegradable Poly-3-Hydroxybutyrate–Starch Blend Films. *Bioresour. Technol.* **2003**, *86*, 33–37. [CrossRef] [PubMed]
112. Lai, S.; Sun, W.; Don, T. Preparation and Characterization of Biodegradable Polymer Blends from Poly(3-hydroxybutyrate)/Poly (Vinyl Acetate)-modified Corn Starch. *Polym. Eng. Sci.* **2015**, *55*, 1321–1329. [CrossRef]
113. El Bouhali, A.; Gnanasekar, P.; Habibi, Y. Chemical Modifications of Lignin. In *Lignin-Based Materials for Biomedical Applications*; Elsevier: Amsterdam, The Netherlands, 2021; pp. 159–194.
114. Mousavioun, P.; Halley, P.J.; Doherty, W.O.S. Thermophysical Properties and Rheology of PHB/Lignin Blends. *Ind. Crop. Prod.* **2013**, *50*, 270–275. [CrossRef]
115. Weihua, K.; He, Y.; Asakawa, N.; Inoue, Y. Effect of Lignin Particles as a Nucleating Agent on Crystallization of Poly(3-hydroxybutyrate). *J. Appl. Polym. Sci.* **2004**, *94*, 2466–2474. [CrossRef]
116. Mousavioun, P.; Doherty, W.O.S.; George, G. Thermal Stability and Miscibility of Poly(Hydroxybutyrate) and Soda Lignin Blends. *Ind. Crop. Prod.* **2010**, *32*, 656–661. [CrossRef]
117. Kurth, N.; Renard, E.; Brachet, F.; Robic, D.; Guerin, P.; Bourbouze, R. Poly(3-Hydroxyoctanoate) Containing Pendant Carboxylic Groups for the Preparation of Nanoparticles Aimed at Drug Transport and Release. *Polymer* **2002**, *43*, 1095–1101. [CrossRef]

118. Kai, D.; Loh, X.J. Polyhydroxyalkanoates: Chemical Modifications Toward Biomedical Applications. *ACS Sustain. Chem. Eng.* **2014**, *2*, 106–119. [CrossRef]
119. Arkin, A.H.; Hazer, B.; Borcakli, M. Chlorination of Poly(3-Hydroxy Alkanoates) Containing Unsaturated Side Chains. *Macromolecules* **2000**, *33*, 3219–3223. [CrossRef]
120. Timbart, L.; Renard, E.; Langlois, V.; Guerin, P. Novel Biodegradable Copolyesters Containing Blocks of Poly(3-hydroxyoctanoate) and Poly(E-caprolactone): Synthesis and Characterization. *Macromol. Biosci.* **2004**, *4*, 1014–1020. [CrossRef]
121. Bear, M.-M.; Leboucher-Durand, M.-A.; Langlois, V.; Lenz, R.W.; Goodwin, S.; Guérin, P. Bacterial Poly-3-Hydroxyalkenoates with Epoxy Groups in the Side Chains. *React. Funct. Polym.* **1997**, *34*, 65–77. [CrossRef]
122. Nguyen, S.; Marchessault, R.H. Graft Copolymers of Methyl Methacrylate and Poly([R]-3-hydroxybutyrate) Macromonomers as Candidates for Inclusion in Acrylic Bone Cement Formulations: Compression Testing. *J. Biomed. Mater. Res. B Appl. Biomater.* **2006**, *77B*, 5–12. [CrossRef] [PubMed]
123. Macit, H.; Hazer, B.; Arslan, H.; Noda, I. The Synthesis of PHA-g -(PTHF-b -PMMA) Multiblock/Graft Copolymers by Combination of Cationic and Radical Polymerization. *J. Appl. Polym. Sci.* **2009**, *111*, 2308–2317. [CrossRef]
124. Liu, J.; Zhou, Z.; Li, H.; Yang, X.; Wang, Z.; Xiao, J.; Wei, D.-X. Current Status and Challenges in the Application of Microbial PHA Particles. *Particuology* **2024**, *87*, 286–302. [CrossRef]
125. Grage, K.; Jahns, A.C.; Parlane, N.; Palanisamy, R.; Rasiah, I.A.; Atwood, J.A.; Rehm, B.H.A. Bacterial Polyhydroxyalkanoate Granules: Biogenesis, Structure, and Potential Use as Nano-/Micro-Beads in Biotechnological and Biomedical Applications. *Biomacromolecules* **2009**, *10*, 660–669. [CrossRef] [PubMed]
126. Draper, J.L.; Rehm, B.H. Engineering Bacteria to Manufacture Functionalized Polyester Beads. *Bioengineered* **2012**, *3*, 203–208. [CrossRef] [PubMed]
127. Ole Bahls, M.; Kardashliev, T.; Panke, S.; Rehm, F.B.H.; Grage, K.; Rehm, B.H.A. *Consequences of Microbial Interactions with Hydrocarbons, Oils, and Lipids: Production of Fuels and Chemicals*; Lee, S.Y., Ed.; Springer International Publishing: Cham, Switzerland, 2017; ISBN 978-3-319-50435-3.
128. Nigmatullin, R.; Thomas, P.; Lukasiewicz, B.; Puthussery, H.; Roy, I. Polyhydroxyalkanoates, a Family of Natural Polymers, and Their Applications in Drug Delivery. *J. Chem. Technol. Biotechnol.* **2015**, *90*, 1209–1221. [CrossRef]
129. Xiong, Y.-C.; Yao, Y.-C.; Zhan, X.-Y.; Chen, G.-Q. Application of Polyhydroxyalkanoates Nanoparticles as Intracellular Sustained Drug-Release Vectors. *J. Biomater. Sci. Polym. Ed.* **2010**, *21*, 127–140. [CrossRef]
130. Pavic, A.; Stojanovic, Z.; Pekmezovic, M.; Veljović, Đ.; O'Connor, K.; Malagurski, I.; Nikodinovic-Runic, J. Polyenes in Medium Chain Length Polyhydroxyalkanoate (Mcl-PHA) Biopolymer Microspheres with Reduced Toxicity and Improved Therapeutic Effect against Candida Infection in Zebrafish Model. *Pharmaceutics* **2022**, *14*, 696. [CrossRef]
131. Bokrova, J.; Marova, I.; Matouskova, P.; Pavelkova, R. Fabrication of Novel PHB-Liposome Nanoparticles and Study of Their Toxicity in Vitro. *J. Nanoparticle Res.* **2019**, *21*, 49. [CrossRef]
132. Voinova, V.; Bonartseva, G.; Bonartsev, A. Effect of Poly(3-Hydroxyalkanoates) as Natural Polymers on Mesenchymal Stem Cells. *World J. Stem Cells* **2019**, *11*, 764–786. [CrossRef] [PubMed]
133. Fan, F.; Wu, X.; Zhao, J.; Ran, G.; Shang, S.; Li, M.; Lu, X. A Specific Drug Delivery System for Targeted Accumulation and Tissue Penetration in Prostate Tumors Based on Microbially Synthesized PHBHHx Biopolyester and IRGD Peptide Fused PhaP. *ACS Appl. Bio Mater.* **2018**, *1*, 2041–2053. [CrossRef] [PubMed]
134. Gonzalez-Miro, M.; Chen, S.; Gonzaga, Z.J.; Evert, B.; Wibowo, D.; Rehm, B.H.A. Polyester as Antigen Carrier toward Particulate Vaccines. *Biomacromolecules* **2019**, *20*, 3213–3232. [CrossRef]
135. Duvernoy, O.; Malm, T.; Ramström, J.; Bowald, S. A Biodegradable Patch Used as a Pericardial Substitute after Cardiac Surgery: 6- and 24-Month Evaluation with CT. *Thorac. Cardiovasc. Surg.* **1995**, *43*, 271–274. [CrossRef] [PubMed]
136. Gregory, D.A.; Taylor, C.S.; Fricker, A.T.R.; Asare, E.; Tetali, S.S.V.; Haycock, J.W.; Roy, I. Polyhydroxyalkanoates and Their Advances for Biomedical Applications. *Trends Mol. Med.* **2022**, *28*, 331–342. [CrossRef]
137. Lee, J.W.; Parlane, N.A.; Wedlock, D.N.; Rehm, B.H.A. Bioengineering a Bacterial Pathogen to Assemble Its Own Particulate Vaccine Capable of Inducing Cellular Immunity. *Sci. Rep.* **2017**, *7*, 41607. [CrossRef] [PubMed]
138. Rubio Reyes, P.; Parlane, N.A.; Wedlock, D.N.; Rehm, B.H.A. Immunogencity of Antigens from Mycobacterium Tuberculosis Self-Assembled as Particulate Vaccines. *Int. J. Med. Microbiol.* **2016**, *306*, 624–632. [CrossRef]
139. Parlane, N.A.; Chen, S.; Jones, G.J.; Vordermeier, H.M.; Wedlock, D.N.; Rehm, B.H.A.; Buddle, B.M. Display of Antigens on Polyester Inclusions Lowers the Antigen Concentration Required for a Bovine Tuberculosis Skin Test. *Clin. Vaccine Immunol.* **2016**, *23*, 19–26. [CrossRef]
140. Rubio-Reyes, P.; Parlane, N.A.; Buddle, B.M.; Wedlock, D.N.; Rehm, B.H.A. Immunological Properties and Protective Efficacy of a Single Mycobacterial Antigen Displayed on Polyhydroxybutyrate Beads. *Microb. Biotechnol.* **2017**, *10*, 1434–1440. [CrossRef]
141. Parlane, N.A.; Grage, K.; Mifune, J.; Basaraba, R.J.; Wedlock, D.N.; Rehm, B.H.A.; Buddle, B.M. Vaccines Displaying Mycobacterial Proteins on Biopolyester Beads Stimulate Cellular Immunity and Induce Protection against Tuberculosis. *Clin. Vaccine Immunol.* **2012**, *19*, 37–44. [CrossRef]
142. Gor, D.O.; Ding, X.; Briles, D.E.; Jacobs, M.R.; Greenspan, N.S. Relationship between Surface Accessibility for PpmA, PsaA, and PspA and Antibody-Mediated Immunity to Systemic Infection by Streptococcus Pneumoniae. *Infect. Immun.* **2005**, *73*, 1304–1312. [CrossRef]

143. Qadir, M.I.; Sajjad, S. Phage Therapy against Streptococcus Pneumoniae: Modern Tool to Control Pneumonia. *Crit. Rev. Eukaryot. Gene Expr.* **2017**, *27*, 289–295. [CrossRef]
144. González-Miro, M.; Rodríguez-Noda, L.; Fariñas-Medina, M.; García-Rivera, D.; Vérez-Bencomo, V.; Rehm, B.H.A. Self-Assembled Particulate PsaA as Vaccine against Streptococcus Pneumoniae Infection. *Heliyon* **2017**, *3*, e00291. [CrossRef]
145. González-Miró, M.; Rodríguez-Noda, L.M.; Fariñas-Medina, M.; Cedré-Marrero, B.; Madariaga-Zarza, S.; Zayas-Vignier, C.; Hernández-Cedeño, M.; Kleffmann, T.; García-Rivera, D.; Vérez-Bencomo, V.; et al. Bioengineered Polyester Beads Co-Displaying Protein and Carbohydrate-Based Antigens Induce Protective Immunity against Bacterial Infection. *Sci. Rep.* **2018**, *8*, 1888. [CrossRef]
146. Ali, I.; Jamil, N. Polyhydroxyalkanoates: Current Applications in the Medical Field. *Front. Biol.* **2016**, *11*, 19–27. [CrossRef]
147. Shishatskaya, E.I.; Nikolaeva, E.D.; Vinogradova, O.N.; Volova, T.G. Experimental Wound Dressings of Degradable PHA for Skin Defect Repair. *J. Mater. Sci. Mater. Med.* **2016**, *27*, 165. [CrossRef]
148. Wang, C.; Zeng, H.-S.; Liu, K.-X.; Lin, Y.-N.; Yang, H.; Xie, X.-Y.; Wei, D.-X.; Ye, J.-W. Biosensor-Based Therapy Powered by Synthetic Biology. *Smart Mater. Med.* **2023**, *4*, 212–224. [CrossRef]
149. González, E.; Herencias, C.; Prieto, M.A. A Polyhydroxyalkanoate-based Encapsulating Strategy for 'Bioplasticizing' Microorganisms. *Microb. Biotechnol.* **2020**, *13*, 185–198. [CrossRef]
150. Saxena, A.; Tiwari, A. Polyhydroxyalkonates: Green Plastics of the Future. *Int. J. Biomed. Adv. Res.* **2011**, *2*, 356–367. [CrossRef]
151. Carraher, C.E., Jr. *Polymer Chemistry*, 6th ed.; Lagowski, J.J., Ed.; Marcel Dekker, Inc.: New York, NY, USA, 2006.
152. Varela, O.; Orgueira, H.A. Synthesis of Chiral Polyamides from Carbohydrate-Derived Monomers. *Adv. Carbohydr. Chem. Biochem.* **2000**, *55*, 137–174. [CrossRef]
153. Mungara, P.M.; Gonsalves, K.E. Synthesis of Funcionalized Targeted Polyamides. *MRS Online Proc. Libr.* **1993**, *330*, 13–18. [CrossRef]
154. Kumar, G.S.; Kalpagam, V. Polymer Reviews Biodegradable Polymers: Prospects, Problems, and Progress. *J. Macromol. Sci. Part C* **2007**, *22*, 225–260. [CrossRef]
155. Vert, M. Bioresorbable Polymers for Temporary Therapeutic Applications. *Angew. Makromol. Chem.* **1989**, *166*, 155–168. [CrossRef]
156. Sakuta, R.; Nakamura, N. Production of Hexaric Acids from Biomass. *Int. J. Mol. Sci.* **2019**, *20*, 3660. [CrossRef]
157. Galbis, J.A.; García-Martín, M.D.G.; De Paz, M.V.; Galbis, E. Synthetic Polymers from Sugar-Based Monomers. *Chem. Rev.* **2016**, *116*, 1600–1636. [CrossRef]
158. Ogata, N.; Sanui, K.; Hosoda, Y. Active Polycondensation of Diethyl2,3,4,5-Tetra h Ydr. *J. Polym. Sci.* **1976**, *14*, 783–792.
159. Wolfrom, M.L.; Toy, M.S.; Chaney, A. Condensation Polymers from Tetra-O-Acetylgalactaroyl Dichloride and Diamines. *J. Am. Chem. Soc.* **1958**, *80*, 6328–6330. [CrossRef]
160. Ogata, N.; Sanui, K.; Kayama, Y. Copolycondensation of Hydroxyl Diesters and Active Diesters with Hexamethylenediamine. *J. Polym. Sci.* **1977**, *15*, 1523–1526. [CrossRef]
161. Ogata, N.; Sanui, K.; Kato, A. Polycondensation of Diester and Diamines Having Hetero Atom Groups in Polar Solvents. *J. Polym. Sci.* **1979**, *11*, 827–833. [CrossRef]
162. Hoagland, P.D. The Formation of Intermediate Lactones during Amino-Lysis of Diethyl Galactarate. *Carbohydr. Res.* **1981**, *98*, 203–208. [CrossRef]
163. Kiely, D.E.; Chen, L.; Lin, T.H. Hydroxylated Nylons Based on Unprotected Esterified D-Glucaric Acid by Simple Condensation Reactions. *J. Am. Chem. Soc.* **1994**, *116*, 571–578. [CrossRef]
164. Kiely, D.E.; Chen, L.; Lin, T.-H. Simple Preparation of Hydroxylated Nylons—Polyamides Derived from Aldaric Acids. *ACS Symp. Ser.* **1994**, *575*, 149–158. [CrossRef]
165. Chen, L.; Kiely, D.E. Synthesis of Stereoregular Head, Tail Hydroxylated Nylons Derived from D-Glucose. *J. Org. Chem.* **1996**, *61*, 5847–5851. [CrossRef]
166. Kiely, D.E.; Kramer, K.; Zhang, J. Method for Preparing High Molecular Weight Random Polyhydroxypolyamides. U.S. Patent 6,894,135, 9 July 2005.
167. Righetti, G.I.C.; Truscello, A.; Li, J.; Sebastiano, R.; Citterio, A.; Gambarotti, C. Sustainable Synthesis of Zwitterionic Galactaric Acid Monoamides as Monomers of Hydroxylated Polyamides. *J. Carbohydr. Chem.* **2022**, *41*, 314–328. [CrossRef]
168. Kiely, D.E.; Chen, L.; Lin, T.H. Synthetic Polyhydroxypolyamides from Galactaric, Xylaric, D-Glucaric, and D-Mannaric Acids and Alkylenediamine Monomers—Some Comparisons. *J. Polym. Sci. A Polym. Chem.* **1999**, *38*, 594–603. [CrossRef]
169. Morton, D.W.; Kiely, D.E. Synthesis of Poly(Azaalkylene Aldaramide)s and Poly(Oxaalkylene Aldaramide)s Derived from D-Glucaric and D-Galactaric Acids. *J. Polym. Sci. A Polym. Chem.* **2000**, *38*, 604–613. [CrossRef]
170. Kiely, D.E.; Vishwanathan, A.; Jarman, B.P.; Manley-Harris, M. Synthesis of Poly(Galactaramides) from Alkylene- and Substituted Alkylenediammonium Galactarates. *J. Carbohydr. Chem.* **2009**, *28*, 348–368. [CrossRef]
171. Kiely, D.E.; Chen, L.; Morton, D.W. Polyaldaramide Polymers Useful for Films and Ashesives. U.S. Patent 5,434,233, 1995.
172. Johnston, E.R.; Smith, T.N.; Serban, M.A. Nanoparticulate Poly(Glucaramide)-Based Hydrogels for Controlled Release Applications. *Gels* **2017**, *3*, 17. [CrossRef]

MDPI AG
Grosspeteranlage 5
4052 Basel
Switzerland
Tel.: +41 61 683 77 34

Polymers Editorial Office
E-mail: polymers@mdpi.com
www.mdpi.com/journal/polymers

www.ingramcontent.com/pod-product-compliance
Lightning Source LLC
LaVergne TN
LVHW072333090526
838202LV00019B/2409